Oberwolfach Seminars

Volume 56

The workshops organized by the *Mathematisches Forschungsinstitut Oberwolfach* are intended to introduce students and young mathematicians to current fields of research. By means of these well-organized seminars, also scientists from other fields will be introduced to new mathematical ideas. The publication of these workshops in the series *Oberwolfach Seminars* (formerly *DMV seminar*) makes the material available to an even larger audience.

Wuchen Li • Bernhard Schmitzer •
Gabriele Steidl • François-Xavier Vialard •
Christian Wald

Variational and Information Flows in Machine Learning and Optimal Transport

Wuchen Li
Department of Mathematics
University of South Carolina
Columbia, SC, USA

Bernhard Schmitzer
Institute of Computer Science
University of Göttingen
Göttingen, Germany

Gabriele Steidl
Institute of Mathematics
Technical University of Berlin
Berlin, Germany

François-Xavier Vialard
Laboratoire d'Informatique Gaspard-Monge
Université Gustave Eiffel
Marne-la-Vallée Cedex 2, France

Christian Wald
Institute for Mathematics
Technical University of Berlin
Berlin, Germany

ISSN 1661-237X ISSN 2296-5041 (electronic)
Oberwolfach Seminars
ISBN 978-3-031-92730-0 ISBN 978-3-031-92731-7 (eBook)
https://doi.org/10.1007/978-3-031-92731-7

© The Editor(s) (if applicable) and The Author(s), under exclusive license to Springer Nature Switzerland AG 2025

This work is subject to copyright. All rights are solely and exclusively licensed by the Publisher, whether the whole or part of the material is concerned, specifically the rights of translation, reprinting, reuse of illustrations, recitation, broadcasting, reproduction on microfilms or in any other physical way, and transmission or information storage and retrieval, electronic adaptation, computer software, or by similar or dissimilar methodology now known or hereafter developed.
The use of general descriptive names, registered names, trademarks, service marks, etc. in this publication does not imply, even in the absence of a specific statement, that such names are exempt from the relevant protective laws and regulations and therefore free for general use.
The publisher, the authors and the editors are safe to assume that the advice and information in this book are believed to be true and accurate at the date of publication. Neither the publisher nor the authors or the editors give a warranty, expressed or implied, with respect to the material contained herein or for any errors or omissions that may have been made. The publisher remains neutral with regard to jurisdictional claims in published maps and institutional affiliations.

This book is published under the imprint Birkhäuser, www.birkhauser-science.com by the registered company Springer Nature Switzerland AG
The registered company address is: Gewerbestrasse 11, 6330 Cham, Switzerland

If disposing of this product, please recycle the paper.

Preface

This book is based on lectures given at the Mathematisches Forschungsinstitut Oberwolfach within the Seminar on "Computational Variational Flows in Machine Learning and Optimal Transport" in autumn 2023. As machine learning is a rapidly developing and lively field of research, we extended the contribution by recent novel ideas.

Variational and stochastic flows on measure spaces are now ubiquitous in machine learning and generative modeling. Indeed, many such models can be interpreted as flows from a latent distribution to the sample distribution and training corresponds to learning the flow vector field. Optimal transport and diffeomorphic flows provide powerful frameworks to analyze such trajectories of distributions with elegant notions from differential geometry, such as geodesics, gradient, and Hamiltonian flows. Recently, mean field control and mean field games offered a general optimal control variational view on learning problems. The book addresses the question how these tools lead us to a better understanding and further development of machine learning and generative models by **four independent overview chapters**. Our intention is to provide basic material for lectures and seminars on the topic.

Chapter "A Dynamic Perspective of Optimal Transport" by B. Schmitzer starts with an overview on optimal transport and its entropic regularization with special emphasis on their dynamic interpretations.

The unregularized part centers around the primal-dual structure of the dynamic Benamou–Brenier formula. How does the dynamic dual potential direct the movement of mass particles via its gradient in a drift equation and how is this dynamic dual potential linked to the static Kantorovich dual potentials via the Hopf–Lax formula? Interestingly, this is directly related to convex duality of suitable intermediate multi-marginal optimal transport problems.

The entropic regularization gives rise to a dynamic interpretation via Schrödinger bridges. Some intuition of the origin of this interpretation is given, which leads once more via an auxiliary multi-marginal formulation, to a system of drift-diffusion equations and gives rise to an entropic variant of the Benamou–Brenier formula.

Chapter "A Geometric Perspective on Diffeomorphic and Optimal Transport Flows and Their Applications" by F.-X. Vialard provides a geometric perspective on diffeomorphic and optimal transport flows driven by applications in shape analysis and computational anatomy, two fields where registration problems have gotten significant attention in the last years. The intention is to give a unified geometric framework that integrates diffeomorphic flows, optimal transport, and related models as well as to provide a rigorous treatment of the analysis of these flows, particularly the geodesic flows on diffeomorphism groups. Finally, gradient flows arising in machine learning are studied from the above geometric and analytic viewpoint.

Chapter "Generalized Wasserstein Dynamics in Mathematical Data Sciences" by W. Li provides an overview of recent developments in generalized optimal transport on an informal level. The note addresses the basic ideas and is suited for high year undergraduate students, graduate students, and postdocs working in different disciplines. It covers three different components of generalized optimal transport, namely, original dynamic optimal transport, Wasserstein gradient flows, and Hamiltonian flows arising from a flow optimal control formulation of generalized optimal transport problems known as mean field control problems. The note also presents some mathematics formulations and computational methods. This explains why optimal transport-related equations or algorithms are essential in mathematical data sciences. Toward this goal, we particularly prepare readers to digest some terminologies in modern data sciences, including tangent space, co-tangent space, Wasserstein metrics, Jordan–Kinderlehrer–Otto (JKO) schemes, generative models, neural ODEs, Deep JKO methods, etc.

Chapter "Flow Matching: Markov Kernels, Stochastic Processes and Transport Plans" by G. Steidl and Ch. Wald was written to understand recent generative models, namely, so-called Flow Matching from a mathematical point of view. Indeed, the velocity fields that have to be learned by a neural network are approached from three different directions: transport plans (couplings) between latent and target distributions, Markov kernels, and stochastic processes. Due to their close relation, continuous normalizing flows and score matching diffusion models are briefly recalled. From the application side it is not only interesting to generating samples but also to solve Bayesian inverse problems. This requires a generalization to Bayesian Flow Matching based on the introduction of conditional Wasserstein distances.

Columbia, SC, USA	Wuchen Li
Göttingen, Germany	Bernhard Schmitzer
Berlin, Germany	Gabriele Steidl
Marne-la-Vallée Cedex 2, France	François-Xavier Vialard
Berlin, Germany	Christian Wald

Acknowledgments First of all, we are grateful to the Mathematisches Forschungsinstitut Oberwolfach (MFO) for providing the opportunity and optimal environment to run our seminar.

Further, we like to thank all participants of the seminar for their great contributions and many lively discussions. It was a pleasure for us to have such a group of enthusiastic, creative, and well-prepared young researchers at the MFO.

G. Steidl acknowledges support from the German Research Foundation (DFG) under Germany's Excellence Strategy–The Berlin Mathematics Research Center MATH+; and Ch. Wald by the DFG within the SFB "Tomography Across the Scales."

B. Schmitzer acknowledges support by the Emmy Noether program of the DFG.

W. Li acknowledges support from AFOSR YIP award No. FA9550-23-1-0087, NSF DMS-2245097, and NSF RTG: 2038080.

The work of F.-X. Vialard is supported by the Bézout Labex (New Monge Problems), funded by ANR, reference ANR-10-LABX-58.

Contents

A Dynamic Perspective of Optimal Transport 1
Bernhard Schmitzer
1 Introduction 1
2 Optimal Transport 2
   2.1 Kantorovich Formulation 2
   2.2 Wasserstein Distance 6
   2.3 Benamou–Brenier Formula 11
   2.4 Multi-Marginal Formulation 21
3 Entropic Optimal Transport 32
   3.1 Entropic Kantorovich Problem 32
   3.2 Interlude: Diffusion and Schrödinger Bridges 36
   3.3 Entropic Multi-Marginal Formulation 42
   3.4 Entropic Benamou–Brenier Formula 46
References 50

A Geometric Perspective on Diffeomorphic and Optimal Transport Flows and Their Applications 53
François-Xavier Vialard
1 Introduction 53
2 Elementary Notions of Group Actions 55
   2.1 Group Actions 55
   2.2 Cotangent Actions 58
   2.3 About Lie Groups and Infinite Dimensions 61
   2.4 Infinitesimal Generator 61
   2.5 Momentum Maps 65
   2.6 Riemannian Metrics Induced by Group Actions 66
3 Geodesics and the Euler-Arnold-Poincaré Equation 72
   3.1 General Euler-Arnold-Poincaré Equation 73
   3.2 Fluid Dynamic Examples 76
4 Local Existence in Lagrangian Coordinates and Analytical Properties 77
   4.1 Action on Operators 78

	4.2	Application to the Incompressible Euler Equation	80
	4.3	Application to Sobolev Type Right-Invariant Riemannian Metrics	81
5	Diffeomorphisms Groups Generated by Flows of Vector Fields		84
6	Induced Metrics and the Case of Measures		89
	6.1	Induced Metrics on Groups of Points (a.k.a. Space of Landmarks)	90
	6.2	Induced Metrics on Measures	91
	6.3	The Optimal Transport Metric via the Benamou–Brenier Formulation	92
	6.4	The Wasserstein-Fisher-Rao Metric on Positive Densities	95
7	A Geometric Point of View on Gradient Flows on Measures		98
	7.1	The Importance of the Co-metric in Gradient Flows	99
	7.2	A Geometric Perspective on Different Entropy Flows, Including Stein Variational Gradient Descent	101
	7.3	Wasserstein Gradient Flow of the Coulomb Discrepancy	105
	7.4	Infinite Width/Infinite Depth Neural Networks	106
Appendix 1: Sobolev Spaces			110
Appendix 2: Reproducing Kernel Hilbert Spaces			113
Appendix 3: Bochner Integral			116
Appendix 4: ODE on Banach Spaces			118
Appendix 5: Polyak-Lojasiewicz Inequality			121
References			125

Generalized Wasserstein Dynamics in Mathematical Data Sciences 129
Wuchen Li

1	Introduction		130
2	Basics of Wasserstein Dynamics		131
	2.1	Review of Dynamics in Euclidean Space	131
	2.2	Mean-Field Information Metric Spaces with Gradient Flows and Hamiltonian Flows	134
3	Deep Neural Network Based Minimization Movement Schemes		150
	3.1	Dynamic JKO Schemes in Lagrangian Coordinates	152
	3.2	A Particle Method	155
	3.3	Nonlinear Mobility	157
	3.4	Neural ODE Empowered JKO Schemes	158
	3.5	Learning the Potential Function	159
	3.6	Deep JKO Algorithms	159
	3.7	Numerical Examples	162
4	Acceleration Optimization Methods in Probability Density Spaces		165
	4.1	Reviews	166
	4.2	Accelerated Information Gradient Flow	169
	4.3	Convergence Rate Analysis on AIG Flows	172
	4.4	Discrete-Time Algorithms for AIG Flows	174

	4.5	Numerical Experiments	177
	4.6	Bayesian Neural Network..	180
5	Conclusions and Future Directions		180
References			182

Flow Matching: Markov Kernels, Stochastic Processes and Transport Plans 185
Christian Wald and Gabriele Steidl

1	Introduction		186
2	Preliminaries and Notation		190
	2.1	Special Metric Spaces..	190
	2.2	Push-Forward Operator	192
	2.3	Disintegration and Markov Kernels	194
	2.4	Couplings and Wasserstein Distance	196
3	Absolutely Continuous Curves in $(\mathcal{P}_2(\mathbb{R}^d), W_2)$		198
4	Curves Induced by Couplings		202
5	Curves via Markov Kernels		212
	5.1	General Construction...	212
	5.2	Curves Induced by Couplings via Disintegration	216
6	Curves via Stochastic Processes...		220
	6.1	Conditional Expectation	221
	6.2	Curve via Conditional Expectation	222
7	Flow Matching		225
8	Numerical Examples for Flow Matching		229
	8.1	Flow Matching Algorithms	229
	8.2	Numerical Examples	230
9	Flow Matching in Bayesian Inverse Problems		231
	9.1	Conditional Wasserstein Distance	232
	9.2	Almost Conditional Couplings	235
	9.3	Bayesian Flow Matching..	237
	9.4	Numerical Examples of Bayesian Flow Matching.......	238
10	Continuous Normalizing Flows		240
	10.1	Curves of Probability Measures from ODEs...............	240
	10.2	Likelihood Loss for Continuous Normalizing Flow	242
	10.3	Computing Gradients with the Adjoint Method	244
	10.4	Numerical Examples of Continuous Normalizing Flows	247
11	A Glimpse at Score-Based Diffusion		248
Appendix A: Proof of Theorem 3.1			250
Appendix B: Measurability of v_t in Lemma 5.7			251
Appendix C: Remark on Normalizing Flows			252
References			253

About the Authors

Photo credit: Archives of the Mathematisches Forschungsinstitut Oberwolfach. With kind permission.

Wuchen Li received his BSc in Mathematics from the Shandong University in 2009. He obtained a PhD in Mathematics from the Georgia Institute of Technology in 2016. He was a CAM Assistant Adjunct Professor in the Department of Mathematics at the University of California, Los Angeles, from 2016 to 2020. Currently, he is an Assistant Professor at the University of South Carolina. His research interests include optimal transport, information geometry, and mean field games with applications in data science and scientific computing.

Bernhard Schmitzer completed his PhD in 2014 at the University of Heidelberg under the direction of Christoph Schnörr. Afterwards he was a postdoc at Université Paris-Dauphine and University of Münster under the mentorship of Gabriel Peyré and Benedikt Wirth. In 2019 he became a junior research group leader at TU Munich, and since 2020, he is a Professor at the University of Göttingen.

Gabriele Steidl completed her PhD at the University of Rostock under supervision of Manfred Tasche. Afterwards, she held positions as Assistant and Full Professor at the TU Darmstadt, the University of Mannheim, and the TU Kaiserslautern. Since 2020, she is Professor at the Institut of Mathematics at the TU Berlin. She worked as consultant of the Fraunhofer Institute for Industrial Mathematics and is in the Scientific Advisory Board of the Helmholtz Imaging Platform of the Helmholtz Association. She is a SIAM fellow (2022) and Editor-in-Chief of the *SIAM Journal on Imaging Sciences*.

François-Xavier Vialard completed his PhD in 2009 at ENS Paris-Saclay under the supervision of Alain Trouvé. He was a post-doc at Imperial College, London, in a project led by Darryl D. Holm before he started an Assistant Professor position at the University Paris-Dauphine in 2011. In 2018, he moved to the University Gustave Eiffel in the Paris area.

Christian Wald studied mathematics at the University of Stuttgart and Humboldt University Berlin. He completed his PhD in algebraic number theory under the supervision of Elmar Groesse-Kloenne. Following his PhD, he was a postdoctoral researcher at Charité Berlin, working on artificial intelligence in cardiac CT imaging. He is currently a postdoc at the TU of Berlin. His research focuses on curves in Wasserstein spaces, gradient flows, generative modeling, and sampling from Boltzmann densities.

A Dynamic Perspective of Optimal Transport

Bernhard Schmitzer

1 Introduction

In this chapter we explore some aspects of dynamic optimal transport, in particular for the Wasserstein-2 distance on $\mathbb{R}^d$, with and without entropic regularization. The unregularized setting is treated in Sect. 2. We put a special emphasis on the primal-dual structure of the dynamic Benamou–Brenier formula and explore how the dynamic dual potential directs the movement of mass particles via its gradient in a drift equation and how this dynamic dual potential is linked to the static Kantorovich dual potentials via the Hopf–Lax formula. We obtain this formula directly via convex duality of suitable intermediate multi-marginal problems.

In Sect. 3 we add entropic regularization. While this destroys the metric structure, it gives rise to a dynamic interpretation via the notion of Schrödinger bridges. We give some intuition for the origin of this interpretation and then study the behaviour of these bridges in more detail, once more via an auxiliary multi-marginal formulation, which leads to a system of drift-diffusion equations. This then gives rise to an entropic variant of the Benamou–Brenier formula.

For accessibility, we make an effort to use only rather basic tools from convex analysis on compact spaces and avoid more advanced tools such as measures on paths or stochastic calculus. References to more in-depth treatments are given throughout the text. While all results in this chapter are well known and covered in the literature, we hope that our exploration in this chapter via basic tools and the discussion of various

B. Schmitzer (✉)
Institute of Computer Science, University of Göttingen, Göttingen, Germany
e-mail: schmitzer@cs.uni-goettingen.de

equivalent reformulations will help to broaden the readers' understanding of dynamic optimal transport.

Remark 1.1 (Notation) Throughout this chapter, (X, d) is a compact metric space. We denote by $C(X)$ the Banach space of continuous functions on X, equipped with the sup-norm. Its topological dual can be identified with the space of Radon measures $\mathcal{M}(X)$, equipped with the total variation norm. The weak* topology on $\mathcal{M}(X)$ is the one induced by the pairing with $C(X)$. Probability measures are noted by $\mathcal{P}(X)$, non-negative measures by $\mathcal{M}_+(X)$. $\langle \cdot, \cdot \rangle$ denotes duality pairing, the implied paired spaces are always clear from context.

2 Optimal Transport

2.1 Kantorovich Formulation

Definition 2.1 (Kantorovich Formulation of Optimal Transport) Let (X, d) be a compact metric space, $c \in C(X \times X)$, and $\mu, \nu \in \mathcal{P}(X)$. Then the Kantorovich optimal transport problem between measures μ and ν for cost function c is given by

$$C(\mu, \nu) := \inf \left\{ \int_{X \times X} c(x, y) \, d\gamma(x, y) \,\bigg|\, \gamma \in \Gamma(\mu, \nu) \right\} \tag{2.1}$$

where $\Gamma(\mu, \nu)$ is the set of *transport plans* or *couplings* between μ and ν, given by

$$\Gamma(\mu, \nu) := \left\{ \gamma \in \mathcal{M}_+(X \times X) \,\bigg|\, P_1 \gamma = \mu, \, P_2 \gamma = \nu \right\} \tag{2.2}$$

and $P_i \gamma = p_{i\#} \gamma$ where $p_i(x_1, x_2) = x_i$ is the projection onto the i-th coordinate.

Proposition 2.2 *Minimizers in* (2.1) *exist.*

Proof The objective $\gamma \mapsto \langle c, \gamma \rangle$ is continuous. The set $\Gamma(\mu, \nu)$ is weak* closed and compact (by Banach–Alaoglu or Prokhorov's theorems). The same arguments can be generalized to lower-semicontinuous c and Polish X, see [51, Theorem 4.1]. □

Equation (2.1) is a prototypical linear program and convex duality is an essential tool for its analysis. We will establish convex duality results via the Fenchel–Rockafellar theorem between topologically paired spaces. Two vector spaces U and U^* with locally convex topologies are said to be topologically paired if the bounded linear functionals of each can be identified with elements of the other. The pairing between these spaces is a bilinear form $\langle \cdot, \cdot \rangle : U \times U^* \to \mathbb{R}$. Our example of interest is the space of continuous functions

$C(X)$ with the sup-norm topology paired with the Radon measures $\mathcal{M}(X)$ with the weak* topology.

Theorem 2.3 (Fenchel–Rockafellar [44]) *Let (U, U^*) and (V, V^*) be two couples of topologically paired spaces. Let $A : U \to V$ be a bounded linear operator and $A^* : V^* \to U^*$ its adjoint. Let F and G be lower-semicontinuous proper convex functions on U and V respectively, with values in $\mathbb{R} \cup \{\infty\}$. If there exists an $x \in U$ such that $F(-x) < \infty$ and G is continuous at Ax, then*

$$\sup_{x \in U} -F(-x) - G(Ax) = \min_{y \in V^*} F^*(A^*y) + G^*(y) \tag{2.3}$$

and the minimum is attained. Moreover, there exists a maximizer $x \in U$ if and only if there is some $y \in V^*$ such that $Ax \in \partial G^*(y)$ and $A^*y \in \partial F(-x)$.

Note that the two optimality conditions at the end of the statement are equivalent to $y \in \partial G(Ax)$ and $x \in -\partial F^*(A^*y)$, respectively.

We are now ready to state the duality result. A more general duality statement for non-compact X and lower-semicontinuous cost c can be found in [51, Theorem 5.10].

Proposition 2.4 (Dual Kantorovich Problem)

$$C(\mu, \nu) = \sup \left\{ \int_X \phi \, d\mu + \int_X \psi \, d\nu \,\Big|\, \phi, \psi \in C(X), \phi \oplus \psi \leq c \right\} \tag{2.4}$$

where $\phi \oplus \psi$ denotes the function $(x, y) \mapsto \phi(x) + \psi(y)$ and the inequality $\phi \oplus \psi \leq c$ is to be enforced on all of $X \times X$. Minimizers in (2.1) exist.

Proof This follows from Theorem 2.3 as follows: Let $U = C(X) \times C(X)$, $V = C(X \times X)$, $A : U \to V$, $(\phi, \psi) \mapsto \phi \oplus \psi$, $F : (\phi, \psi) \mapsto \int \phi \, d\mu + \int \psi \, d\nu$ and $G : \eta \mapsto 0$ if $\eta \leq c$ and $+\infty$ otherwise. Then (2.4) has the form $\sup_{(\phi,\psi) \in U} -F(-(\phi, \psi)) - G(A(\phi, \psi))$. F is finite everywhere and G is continuous at any function η that is strictly smaller than c, e.g. at $\eta : (x, y) \mapsto C$ for $C := \min_{(x',y') \in X \times X} c(x', y') - 1$ and $\eta = A(\phi, \phi)$ for $\phi : x \mapsto C/2$. Therefore, (2.4) is dual to $\min_{\gamma \in V^*} F^*(A^*\gamma) + G^*(\gamma)$. One finds that $A^*\gamma = (P_1\gamma, P_2\gamma)$, $F^*(\rho, \sigma) = 0$ if $\rho = \mu$ and $\sigma = \nu$ and $+\infty$ otherwise, and $G^*(\gamma) = \int c \, d\gamma$ if $\gamma \geq 0$ and $+\infty$ otherwise, which means that the dual problem is equal to (2.1). □

Proposition 2.4 does not directly imply existence of dual maximizers (ϕ, ψ). Fortunately, their existence can be established with some simple explicit arguments that will also be helpful later on. Consider the dual problem (2.4) and assume for now that $\psi \in C(X)$ is fixed. What is the best possible choice for the remaining supremum over ϕ? Formally, ignoring the constraint that ϕ must be continuous, this partial optimization problem can be solved point-wise for each value $\phi(x)$, by setting $\phi(x)$ such that the first of the constraints

$\phi(x) \leq c(x,y) - \psi(y)$ for some $y \in X$ becomes active. This motivates the following definition.

Definition 2.5 (*c*-**Transform**) For a cost function $c \in C(X \times X)$ and a potential $\psi \in C(X)$ the c-transform of ψ is defined as the function ψ^c given by

$$\psi^c : x \mapsto \inf_{y \in X} c(x,y) - \psi(y).$$

In the same vein we define the $\bar{c}$-transform as

$$\psi^{\bar{c}} : y \mapsto \inf_{x \in X} c(x,y) - \psi(x).$$

Of course, when c is symmetric, both transformations are identical. For continuous c and ψ it turns out that ψ^c is indeed continuous and therefore a maximizer for (2.4) with respect to ϕ for fixed ψ. In fact, ψ^c is even more regular, which is the key for the following existence proof.

Proposition 2.6 *Maximizing* (ϕ, ψ) *for* (2.4) *exist.*

Proof We sketch here the key steps of the proof. More details can be found in [48, Section 1.2].

As discussed above, for fixed ψ the maximizing ϕ is given by $\phi = \psi^c$. Conversely, for fixed ϕ, the maximizing ψ is given by $\psi = \phi^{\bar{c}}$. In addition, it is easy to verify that $((\psi^c)^{\bar{c}})^c = \psi^c$. Consequently, we can restrict ourselves to maximizing sequences $(\phi_n, \psi_n)_n$ where the potentials are c-transforms of each other, i.e. $\phi_n = \psi_n^c$ and $\psi_n = \phi_n^{\bar{c}}$.

Next, observe that ψ^c inherits the modulus of continuity of c: Since c is continuous on a compact domain, there is a continuous function $\omega : \mathbb{R}_+ \to \mathbb{R}_+$ with $\omega(0) = 0$ such that $|c(x,y) - c(x',y)| \leq \omega(d(x,x'))$ (and likewise for the second argument). Assume now that $\phi = \psi^c$. Then find

$$\phi(x) \leq c(x,y) - \psi(y) \leq c(x',y) - \psi(y) + \omega(d(x,x')).$$

Taking now the infimum over y on the right-hand side one obtains $\phi(x) \leq \phi(x') + \omega(d(x,x'))$. By the symmetric argument we obtain $|\phi(x) - \phi(x')| \leq \omega(d(x,x'))$. The argument for ψ is identical.

Therefore, the maximizing sequence $(\phi_n, \psi_n)_n$ is equicontinuous. The objective (2.4) is invariant under applying constant shifts of the form $(\psi, \psi) \mapsto (\phi + \lambda, \psi - \lambda)$ for $\lambda \in \mathbb{R}$. Thus we may assume that $\phi_n(x_0) = 0$ for some fixed $x_0 \in X$, which then implies that $(\phi_n, \psi_n)_n$ is also equi-bounded. Then, by the Arzelà–Ascoli theorem (e.g. [46, Thm. 11.28]) there exists a converging subsequence with limit (ϕ, ψ). The objective

$\langle\phi,\mu\rangle+\langle\psi,\nu\rangle$ is continuous, and the admissible set imposed by the constraint $\phi\oplus\psi\leq c$ is closed, hence the limit is a maximizer. □

Definition 2.7 (Contact Set) Let $(\phi,\psi)\in C(X)^2$ such that

$$\phi(x)+\psi(y)\leq\frac{|x-y|^2}{2}\quad\text{for all}\quad(x,y)\in X\times X.$$

We call the pairs $(x,y)\in X\times X$ where one has equality the *contact set* of (ϕ,ψ).

Proposition 2.8 $\gamma\in\Gamma(\mu,\nu)$ and $(\phi,\psi)\in C(X)^2$ with $\phi\oplus\psi\leq c$ are primal and dual optimal in (2.1) and (2.4) if and only if $\phi\oplus\psi=c$ γ-almost everywhere, i.e. γ is concentrated on the contact set of (ϕ,ψ).

Proof Consider the primal-dual gap between (2.1) and (2.4) for admissible candidates:

$$0\leq\int_{X\times X}c\,\mathrm{d}\gamma-\int_X\phi\,\mathrm{d}\mu-\int_X\psi\,\mathrm{d}\nu=\int_{X\times X}(c-\phi\oplus\psi)\,\mathrm{d}\gamma\tag{2.5}$$

The integrand $c-\phi\oplus\psi$ on the right-hand side is non-negative and thus the integral equals zero if and only if $\phi\oplus\psi=c$ γ-almost everywhere. □

Remark 2.9 (Primal-Dual Optimality Conditions via Fenchel–Rockafellar Duality) The primal-dual optimality condition (2.8) can also be obtained via the condition given in Theorem 2.3. In the following we reuse the conventions established in the proof of Proposition 2.4.

In this case $F^*(A^*\gamma)=F^*((\mathrm{P}_1\gamma,\mathrm{P}_2\gamma))=0$ if $(\mathrm{P}_1\gamma,\mathrm{P}_2\gamma)=(\mu,\nu)$ and $+\infty$ otherwise, i.e. $\partial F^*(A^*\gamma)=C(X)\times C(X)$ if $(\mathrm{P}_1\gamma,\mathrm{P}_2\gamma)=(\mu,\nu)$ and $\emptyset$ otherwise. Therefore the condition $(\phi,\psi)\in-\partial F^*(A^*\gamma)$ is equivalent to the constraint that γ has the correct marginals (but not necessarily that γ is non-negative).

If $\gamma\in\partial G(A(\phi,\psi))$ then one must have that

$$G(A(\phi,\psi)+\eta)\geq G(A(\phi,\psi))+\int_{X\times X}\eta\,\mathrm{d}\gamma$$

for all $\eta\in C(X\times X)$, which implies that $\gamma\geq 0$ and $A(\phi,\psi)=c$ γ-almost everywhere.

Remark 2.10 (Relaxation of Dual Function Space) The inequality (2.5) holds for any $\phi\in L^1(\mu)$, $\psi\in L^1(\nu)$ with $\phi\oplus\psi\leq c$, so the inequality (2.4) $\leq$ (2.1) still holds when the function spaces for ϕ and ψ are relaxed to $L^1(\mu)$ and $L^1(\nu)$ (but we still impose the inequality constraint for all $(x,y)\in X^2$).

2.2 Wasserstein Distance

A particularly important instance of the optimal transport problem (2.1) is when the ground transport cost function $c \in C(X \times X)$ is chosen to be a power $p \in [1, \infty)$ of a metric on the base space X. This induces the celebrated Wasserstein distances. We focus here on the choice $p = 2$ and restrict ourselves to compact metric spaces. Of course, other p and non-compact X can also be considered. A more complete treatment including some valuable historical context and bibliographical notes are given in [51, Chapter 6]. We add a factor $1/2$ in the following definition for notational convenience in the remainder of this chapter.

Definition 2.11 (Wasserstein Distance) Let (X, d) be a compact metric space. Then the Wasserstein distance between two measures $\mu, \nu \in \mathcal{P}(X)$ is given by

$$W(\mu, \nu) := \inf \left\{ \int_{X \times X} \tfrac{1}{2} d(x, y)^2 \, d\gamma(x, y) \,\Big|\, \gamma \in \Gamma(\mu, \nu) \right\}^{1/2}. \tag{2.6}$$

Theorem 2.12 W *is a metric on the set $\mathcal{P}(X)$ and metrizes the weak* topology.*

A more general statement for non-compact spaces is shown in [51, Chapter 6], where a suitable notion of weak convergence must be considered and the convergence (or boundedness) of the second moment (or more generally, the p-th moment) must be verified in addition to the value of W. On compact domains this moment condition can be ignored. As preparation for later it will be instructive to recall the proof for the triangle inequality here, based on the gluing lemma.

Lemma 2.13 (Gluing Lemma) *Let $\mu_1, \mu_2, \mu_3 \in \mathcal{P}(X)$ and let $\gamma_{12} \in \Gamma(\mu_1, \mu_2)$, $\gamma_{23} \in \Gamma(\mu_2, \mu_3)$. Then there is some $\gamma \in \mathcal{P}(X^3)$ such that $P_{12}\gamma = \gamma_{12}$ and $P_{23}\gamma = \gamma_{23}$. Here P_{12} and P_{23} denote the projection operators onto the joint $(1, 2)$-marginal or $(2, 3)$-marginal of γ.*

Proof This can be proved by an explicit construction using disintegration [3, Theorem 5.3.1], which formalizes the concept of conditional probabilities. For instance, there is a family of probability measures $(\gamma_{12,y})_{y \in X}$, unique for μ_2-almost all y, such that for a continuous test function $\psi \in C(X \times X)$ one has

$$\int_{X \times X} \psi(x, y) d\gamma_{12}(x, y) = \int_X \left[\int_X \psi(x, y) \, d\gamma_{12,y}(x) \right] d\mu_2(y).$$

We call $(\gamma_{12,y})_{y \in X}$ the disintegration of γ_{12} with respect to its second marginal (which is μ_2). If $(\boldsymbol{x}, \boldsymbol{y})$ is a pair of X-valued random variables with joint law γ_{12}, then $\gamma_{12,y}$ is the law of $\boldsymbol{x}$ when conditioned on $\boldsymbol{y} = y$.

Likewise, we note by $(\gamma_{23,y})_y$ the disintegration of γ_{23} with respect to its first marginal (which is μ_2). Let then $\gamma \in \mathcal{P}(X^3)$ be the measure that is characterized by

$$\int_{X \times X \times X} \psi(x,y,z)\, d\gamma(x,y,z) = \int_X \left[\int_X \psi(x,y,z)\, d\gamma_{12,y}(x)\, d\gamma_{23,y}(z) \right] d\mu_2(y)$$

for $\psi \in C(X^3)$. By choosing ψ that are constant with respect to the first or third argument one can then verify that γ satisfies the above requirements on marginals. □

Proof of the Triangle Inequality Let $\mu_1, \mu_2, \mu_3 \in \mathcal{P}(X)$ and let $\gamma_{12} \in \Gamma(\mu_1, \mu_2)$, $\gamma_{23} \in \Gamma(\mu_2, \mu_3)$. Let $\gamma \in \mathcal{P}(X^3)$, as provided by the gluing lemma and set $\gamma_{13} := P_{13}\gamma$ to be the joint $(1,3)$-marginal. Then one finds that $\gamma_{13} \in \Gamma(\mu_1, \mu_3)$ and therefore

$$\begin{aligned}
W(\mu_1, \mu_3) &\le \left(\int_{X \times X} \tfrac{1}{2} d(x,z)^2 d\gamma_{13}(x,z) \right)^{1/2} \\
&= \left(\int_{X \times X \times X} \tfrac{1}{2} d(x,z)^2 d\gamma(x,y,z) \right)^{1/2} \\
&\le \left(\int_{X \times X \times X} \tfrac{1}{2} [d(x,y) + d(y,z)]^2 d\gamma(x,y,z) \right)^{1/2} \\
&\le \left(\int_{X \times X \times X} \tfrac{1}{2} d(x,y)^2 d\gamma(x,y,z) \right)^{1/2} + \left(\int_{X \times X \times X} \tfrac{1}{2} d(y,z)^2 d\gamma(x,y,z) \right)^{1/2} \\
&= \left(\int_{X \times X \times X} \tfrac{1}{2} d(x,y)^2 d\gamma_{12}(x,y) \right)^{1/2} + \left(\int_{X \times X \times X} \tfrac{1}{2} d(y,z)^2 d\gamma_{23}(y,z) \right)^{1/2}.
\end{aligned}$$

Here we used the triangle inequality on d in the second inequality and the Minkowski inequality in $L^2(X^3, \gamma)$ in the third inequality. The claim then follows by taking the infimum over γ_{12} and γ_{23} over $\Gamma(\mu_1, \mu_2)$ and $\Gamma(\mu_2, \mu_3)$, respectively. □

Theorem 2.14 (Brenier) *Let X be a compact, convex subset of $\mathbb{R}^d$, let $\mu, \nu \in \mathcal{P}(X)$ and assume $\mu \ll \mathcal{L}$. Then the optimal transport plan for the Wasserstein distance from μ to ν is unique and supported on the graph of an optimal transport* map $T : X \to X$. *The map T is given by*

$$T = \mathrm{id} - \nabla \phi = \nabla \left(\tfrac{1}{2} |\cdot|^2 - \phi \right) \tag{2.7}$$

where ϕ is a maximizer of the dual problem of (2.6), i.e. for (2.4) with $c = \tfrac{1}{2} d^2$. In particular, all such maximizers are differentiable for μ-almost every $x \in X$ and these gradients agree μ-almost everywhere for all maximizers. The function $\tfrac{1}{2} |\cdot|^2 - \phi$ is convex.

Remark 2.15 There are many versions of this theorem with varying levels of generality for the domain X, the measure μ, and the cost function, see for instance [12,27,42,48,50, 51]. The above simple version is a special case of [48, Theorem 1.17].

Definition 2.16 (Geodesic Space) We say a metric space (X, d) is geodesic, if for any two $x, y \in X$ there exists a curve $Z : [0, 1] \to X$ with $Z(0) = x$ and $Z(1) = y$, such that $d(Z(s), Z(t)) = |s - t| \cdot d(x, y)$ for all $s, t \in [0, 1]$. For given x, y we call the corresponding curve Z a *(constant speed) geodesic* from x to y.

Example 2.17 Let X be a convex subset of $\mathbb{R}^d$ with $d(x, y) = |x-y|$. For given $x, y \in X$, the unique constant speed geodesic between them is given by $Z(x, y, \cdot) : [0, 1] \ni t \mapsto (1 - t) \cdot x + t \cdot y$.

Theorem 2.18 (Wasserstein Space Inherits Geodesic Property) *Assume that (X, d) is a geodesic space and that there is a measurable selection of constant speed geodesics, i.e. there is a measurable map $Z : [0, 1] \times X \times X \to X$, such that for fixed $(x, y) \in X \times X$, the map $t \mapsto Z(t, x, y)$ is a constant speed geodesic from x to y. Then $(\mathcal{P}(X), W)$ is a geodesic space. For $\mu, \nu \in \mathcal{P}(X)$ a constant speed geodesic is given by*

$$[0, 1] \ni t \mapsto \rho_t := Z(t, \cdot, \cdot)_\# \gamma$$

where γ is an optimal transport plan for $W(\mu, \nu)$ in (2.6).

Proof The proof uses similar ideas as that for the triangle inequality of W in Theorem 2.12. We need to show that $W(\rho_s, \rho_t) = |s - t| \cdot W(\mu, \nu)$ for $s, t \in [0, 1]$. W.l.o.g. assume $s < t$. For this we will first construct admissible transport plans $\gamma_{s,t} \in \Gamma(\rho_s, \rho_t)$ as follows. ρ_s is constructed by observing that mass of γ at (x, y) should at time s be at position $Z(s, x, y)$. At time t it should have moved to $Z(t, x, y)$. From this we intuit that

$$\gamma_{s,t} := \hat{Z}(s, t, \cdot, \cdot)_\# \gamma$$

for

$$\hat{Z} : [0, 1] \times [0, 1] \times X \times X \to X \times X, \quad (s, t, x, y) \mapsto (Z(s, x, y), Z(t, x, y))$$

is a reasonable transport plan. Indeed, it is easy to verify that $\gamma_{s,t} \in \Gamma(\rho_s, \rho_t)$. For its induced cost we obtain

$$W(\rho_s, \rho_t)^2 \leq \int_{X \times X} \tfrac{1}{2} d(v, w)^2 \, d\gamma_{s,t}(v, w)$$

A Dynamic Perspective of Optimal Transport

$$= \int_{X \times X} \tfrac{1}{2} d\big(Z(s, x, y), Z(t, x, y)\big)^2 \mathrm{d}\gamma(x, y)$$
$$= \int_{X \times X} \tfrac{1}{2} \cdot |s - t|^2 \cdot d(x, y)^2 \mathrm{d}\gamma(x, y)$$
$$= |s - t|^2 \cdot \mathrm{W}(\mu, \nu)^2 \tag{2.8}$$

where we used the constant speed geodesic property of $Z(\cdot, x, y)$ in the second equality. We still need to show that this inequality is actually an equality. For this we use (2.8) on the time pairs $(0, s)$, (s, t), and $(t, 1)$ to obtain

$$\mathrm{W}(\mu, \rho_s) + \mathrm{W}(\rho_s, \rho_t) + \mathrm{W}(\rho_t, \nu) \leq \mathrm{W}(\mu, \nu) \cdot (|0 - s| + |s - t| + |t - 1|) = \mathrm{W}(\mu, \nu).$$

The triangle inequality on $(\mathcal{P}(X), \mathrm{W})$ yields the reverse inequality and therefore (2.8) must actually be an equality. □

Example 2.19 Returning to Example 2.17, we find that the map Z is clearly measurable as it is continuous. Therefore, in a Wasserstein geodesic, mass particles move with constant speed on a straight line from their initial to their target location.

Example 2.20 (Geodesics in the Setting of Brenier's Theorem) Consider the setting of Brenier's theorem (Theorem 2.14). If an optimal plan $\gamma \in \Gamma(\mu, \nu)$ for $\mathrm{W}(\mu, \nu)$ is induced by some map $T = \mathrm{id} - \nabla\phi$, then the induced geodesic in the sense of Theorem 2.18 is given by

$$\rho_t := T_{t\#}\mu \quad \text{with} \quad T_t = (1 - t) \cdot \mathrm{id} + t \cdot T = \mathrm{id} - t \cdot \nabla\phi. \tag{2.9}$$

μ-almost everywhere, mass particles initially located at x, will move to $T(x)$ along the straight line between the points with velocity $T(x) - x = -\nabla\phi(x)$. $-\nabla\phi(x)$ should therefore be interpreted as the Lagrangian velocity field of the moving particles and this Lagrangian velocity is constant in time for particles in a Wasserstein geodesic.

We have just seen that once an optimal transport plan γ is known, a whole geodesic with respect to W can be constructed and we know the position of any mass particle at all times. In fact, more knowledge about these intermediate positions can be extracted. We will now show that mass particles do not collide at intermediate times.

Proposition 2.21 *Consider the setting of Example 2.19. Let $(\phi, \psi) \in C(X)^2$ such that $\phi \oplus \psi \leq c$ on X^2 and let $t \in (0, 1)$. The map $(x, y) \mapsto Z(t, x, y) = (1 - t) \cdot x + t \cdot y$ is injective on the contact set of (ϕ, ψ).*

Proof Assume that there exists $z \in Z$ such that $z = Z(t, x, y) = Z(t, x', y')$ where (x, y) and (x', y') are two points in the contact set of (ϕ, ψ). One obtains

$$|x - y|^2 = \frac{|x-z|^2}{t} + \frac{|z-y|^2}{1-t}$$

and likewise for (x', y') (cf. Lemma 2.45). An explicit geometric calculation then yields that

$$\tfrac{1}{2}|x - y|^2 + \tfrac{1}{2}|x' - y'|^2 \geq \tfrac{1}{2}|x - y'|^2 + \tfrac{1}{2}|x' - y|^2$$

with equality if and only if $(x, y) = (x', y')$. One then obtains the chain of inequalities

$$[\phi(x) + \psi(y)] + [\phi(x') + \psi(y')] = \tfrac{1}{2}|x - y|^2 + \tfrac{1}{2}|x' - y'|^2$$
$$\geq \tfrac{1}{2}|x - y'|^2 + \tfrac{1}{2}|x' - y|^2 \geq [\phi(x) + \psi(y')] + [\phi(x') + \psi(y)]$$

where the first inequality must now be an equality and therefore $(x, y) = (x', y')$. □

The following statement is then an immediate consequence of the primal-dual optimality condition Propositions 2.8 and 2.21.

Corollary 2.22 (Mass Particles Do Not Collide at Intermediate Times) *Let $t \in (0, 1)$. For an optimal transport plan γ the map $Z(t, \cdot, \cdot)$ is injective γ-almost everywhere.*

Remark 2.23 When restricting to the setting of Brenier's theorem, this implies that the map $T_t := \mathrm{id} - t \cdot \nabla \phi$ as introduced in Example 2.20 is invertible for $t \in (0, 1)$. This can also be deduced from the fact that $\tfrac{1}{2}|\cdot|^2 - \phi$ is convex by Brenier's theorem and thus for $t \in (0, 1)$ the map $\tfrac{1}{2}|\cdot|^2 - t \cdot \phi$ is strictly convex, i.e. its gradient (which exists μ-almost everywhere) is invertible.

Another corollary of Proposition 2.21 is that for $0 < t < s < 1$ the intermediate transport plans $\gamma_{s,t} \in \Gamma(\rho_s, \rho_t)$ constructed in the proof of Theorem 2.18 are deterministic in the setting of Example 2.19, i.e. they are supported on the graph of a map, obtained by composing the inverse of $Z(s, \cdot, \cdot)$ with $Z(t, \cdot, \cdot)$.

A deeper exploration of these ideas is given in [51, Chapter 8] on the *Monge–Mather shortening principle*, which includes local spatial Lipschitz regularity of the optimal transport from intermediate times.

2.3 Benamou–Brenier Formula

2.3.1 Definition and Basic Properties

The Kantorovich formulation of the Wasserstein distance, (2.6) can be seen as a Lagrangian description: particles are identified by their initial and final location; which imply their location at intermediate times (and thus the whole Wasserstein geodesic) via Theorem 2.18. One can also adopt a Eulerian point of view: What is the distribution of mass at some intermediate time $t \in (0, 1)$? And what is the momentary velocity of a mass particle that passes through position $z \in X$ at time $t \in (0, 1)$? As it turns out, it is also possible to find Wasserstein geodesics in the Eulerian picture via the celebrated Benamou–Brenier formula [9].

Remark 2.24 (Intuitive Motivation) Consider the setting of Example 2.20. Recall that the intermediate distribution of particles at some time $t \in (0, 1)$ is given by $\rho_t = T_{t\#}\mu$ and since the Lagrangian velocity field $-\nabla \phi$ is constant in time, the intermediate Eulerian velocity field is given by $v_t := -\nabla \phi \circ T_t^{-1}$ (recall that by Remark 2.23 T_t is indeed invertible).

Let now $\psi \in C^1([0, 1] \times X)$. Then a simple computation (similar to the proof of Proposition 2.34) yields that

$$\int_0^1 \int_X (\partial_t \psi) \, d\rho_t dt + \int_0^1 \int_X \nabla \psi \cdot v_t d\rho_t dt = \int_X \psi(1, \cdot) \, dv - \int_X \psi(0, \cdot) \, d\mu.$$

By formally using integration by parts on the derivatives acting on ψ we say that the pair of curves $(\rho_t, v_t)_t$ solves the continuity equation

$$\partial_t \rho_t + \nabla \cdot (v_t \cdot \rho_t) = 0$$

in a weak sense with temporal boundary conditions $\rho_0 = \mu$, $\rho_1 = \nu$.

So the Wasserstein geodesic with the corresponding velocity field constructed by the Kantorovich optimal transport plan (or the Brenier map) is a solution $(\rho_t, v_t)_t$ to this equation. In this section we will show that among all such solutions it is the one which minimizes the action functional

$$(\rho_t, v_t)_t \mapsto \frac{1}{2} \int_0^1 \int_X |v_t(x)|^2 d\rho_t \, dt. \tag{2.10}$$

In the following we will formulate this minimization problem rigorously by using two transformations. First, we switch from a velocity field v_t to a momentum measure $\omega_t := v_t \cdot \rho_t$. One then recovers v_t as density $\frac{d\omega_t}{d\rho_t}$. This change of variables turns the continuity equation into a linear PDE constraint and the action functional into a convex function. Second, instead of considering curves $(\rho_t, \omega_t)_t$ in $\mathcal{M}(X)^{1+d}$, we only consider two measures $(\rho, \omega) \in \mathcal{M}([0, 1] \times X)^{1+d}$ over space and time. Intuitively, (ρ_t, ω_t) can be recovered by considering the disintegration of (ρ, ω) with respect to the time-axis at t. The validity of this change of variables and the disintegration is established in Proposition 2.30.

We will now introduce rigorous definitions for the concepts introduced in Remark 2.24, starting with the continuity equation for measures on space and time.

Definition 2.25 (Distributional Continuity Equation) Let $\mu, \nu \in \mathcal{P}(X)$. A pair $(\rho, \omega) \in \mathcal{M}([0, 1] \times X) \times \mathcal{M}([0, 1] \times X)^d$ is said to solve the distributional continuity equation with temporal boundary conditions μ and ν if

$$\int_{[0,1] \times X} (\partial_t \psi) \, d\rho + \int_{[0,1] \times X} \nabla \psi \cdot d\omega = \int_X \psi(1, \cdot) \, d\nu - \int_X \psi(0, \cdot) \, d\mu \qquad (2.11)$$

for all $\psi \in C^1([0, 1] \times X)$. We denote the set of solutions by $\mathcal{CE}(\mu, \nu)$.

Next, we prepare a rigorous definition of the action functional (2.10).

Remark 2.26 (1-Homogeneous Functions, Integral Functionals, and Conjugation) A function $f : \mathbb{R}^n \to (-\infty, \infty]$ is 1-homogeneous if

$$f(\lambda \cdot s) = \lambda \cdot f(s)$$

for all $\lambda \geq 0$, $s \in \mathbb{R}^n$ (with the convention $0 \cdot \infty = 0$ for $f(s) = \infty$ and $\lambda = 0$).

In the following, assume that f only takes values in $[0, \infty]$ and that f is convex and lower-semicontinuous. Let $\mu \in \mathcal{M}(X)^n$ and let $\sigma \in \mathcal{M}_+(X)$ such that $\mu \ll \sigma$. Then one can consider the following integral functional:

$$F(\mu, \sigma) := \int_X f\left(\frac{d\mu}{d\sigma}\right) d\sigma$$

Due to the 1-homogeneity of f the integral does not actually depend on the choice of σ as long as $\mu \ll \sigma$, i.e. F can be interpreted as being solely a function of μ. Under the above assumptions on f, F is convex and weak* lower-semicontinuous [2, Proposition 2.37 and Theorem 2.38]. Note that one can also consider spatially varying f that also explicitly depend on the position x in the integral.

As f is 1-homogeneous, its Fenchel–Legendre conjugate f^* is the indicator function of a closed, convex set $C \subset \mathbb{R}^n$, i.e. $f^* = \iota_C$ and in fact $C = \partial f(0)$. The conjugate of F is then the functional

$$F^* : C(X) \to \mathbb{R} \cup \{\infty\}, \qquad \eta \mapsto \begin{cases} 0 & \text{if } \eta(x) \in \partial f(0) \text{ for all } x \in X, \\ +\infty & \text{else.} \end{cases}$$

A Dynamic Perspective of Optimal Transport

Definition 2.27 (Action Functional) Let

$$\Phi : \mathbb{R} \times \mathbb{R}^d \to \mathbb{R} \cup \{\infty\}, \qquad (r, w) \mapsto \begin{cases} \frac{|w|^2}{2r} & \text{if } r > 0, \\ 0 & \text{if } (r, w) = (0, 0), \\ +\infty & \text{else.} \end{cases}$$

Φ is convex and lower semi-continuous (jointly in (r, w)), since its sub-level sets are closed and convex. Moreover, Φ is 1-homogeneous. The action is then given by

$$\mathcal{A}(\rho, \omega) = \int_{[0,1] \times X} \Phi\left(\frac{\mathrm{d}(\rho, \omega)}{\mathrm{d}\sigma}\right) \mathrm{d}\sigma(t, x)$$

where σ is any measure in $\mathcal{M}_+([0, 1] \times X)$ such that $(\rho, \omega) \ll \sigma$. Due to Remark 2.26 the definition does not depend on the choice of σ.

Remark 2.28 If $\omega \ll \rho$ so that in particular $\omega = v \cdot \rho$ for some velocity field $v \in L^1([0, 1] \times X, \mathbb{R}^d)$, then one can choose $\sigma = \rho$ for the evaluation of $\mathcal{A}(\rho, \omega)$ and obtains

$$\mathcal{A}(\rho, \omega) = \frac{1}{2} \int_{[0,1] \times X} |v(t, x)|^2 \mathrm{d}\rho(t, x)$$

in line with Remark 2.24.

Now we have gathered the ingredients for the rigorous definition of the Benamou–Brenier formula.

Definition 2.29 (Benamou–Brenier Formula)

$$W_{\mathrm{BB}}(\mu, \nu)^2 := \inf \left\{ \mathcal{A}(\rho, \omega) \middle| (\rho, \omega) \in \mathcal{CE}(\mu, \nu) \right\} \tag{2.12}$$

In (2.12) ρ and ω were only defined as measures on $[0, 1] \times X$, which does not necessarily imply a well-defined notion of mass distribution at intermediate times $t \in (0, 1)$. Fortunately, for pairs $(\rho, \omega) \in \mathcal{CE}(\mu, \nu)$ with $\mathcal{A}(\rho, \omega) < \infty$, one can show that both measures can be decomposed into 'time-slices' such that for Lebesgue-almost every time t the spatial arrangement of mass is fixed. Moreover, ω is actually dominated by ρ and thus it can be written as $\omega = v \cdot \rho$ for a suitable velocity field v. This justifies both transformations sketched in Remark 2.24.

Proposition 2.30 (Time-Disintegration of ρ and ω, and Velocity Field) *Let $(\rho, \omega) \in \mathcal{CE}(\mu, \nu)$ such that $\mathcal{A}(\rho, \omega) < \infty$. Then the following holds:*

(i) $\rho \in \mathcal{M}_+([0, 1] \times X)$, $\|\rho\|_{\mathcal{M}([0,1]\times X)} = 1$.
(ii) $T_\#\rho = \mathcal{L}\llcorner[0, 1]$ where $T : [0, 1] \times X \to [0, 1]$, $(t, x) \mapsto t$ and $\mathcal{L}\llcorner[0, 1]$ denotes the Lebesgue measure on $[0, 1]$. There is a Lebesgue-a.e. unique measurable family of measures $(\rho_t)_{t\in[0,1]}$ with $\rho_t \in \mathcal{P}(X)$ such that for any $\psi \in C([0, 1] \times X)$ one has

$$\int_{[0,1]\times X} \psi(t, x) \, d\rho(t, x) = \int_{[0,1]} \left[\int_X \psi(t, x) \, d\rho_t(x) \right] dt. \qquad (2.13)$$

(iii) $\omega \ll \rho$. *This implies that ω can also be decomposed into time-slices and that it can be written as $\omega = v \cdot \rho$ for some $v \in L^1(\rho, \mathbb{R}^d)$. $v(t, \cdot)$ is then the Eulerian velocity field of particles at time $t \in [0, 1]$.*

Proof First claim: If $\rho \notin \mathcal{M}_+([0, 1] \times X)$ then $\frac{d\rho}{d\sigma} < 0$ for a set that is not σ-negligible, for every $\sigma \in \mathcal{M}_+([0, 1] \times X)$ with $\rho \ll \sigma$ and thus $\mathcal{A}(\rho, \omega) = \infty$. Using the test function $\psi(t, x) = t$ in the definition of $\mathcal{CE}(\mu, \nu)$ we get

$$\int_{[0,1]\times X} d\rho(t, x) = \int_X d\nu(x) = 1.$$

This is the total variation norm of ρ since ρ is non-negative.

Second claim: Let $f \in C([0, 1])$ and let

$$F(t) := \int_0^t f(s) \, ds \text{ for } t \in [0, 1].$$

Also let $T : [0, 1] \times X \to [0, 1]$, $(t, x) \mapsto t$ be the projection to the time coordinate. By construction $\partial_t F = f$ and $F \circ T \in C^1([0, 1] \times X)$. So from the continuity equation we know that

$$\int_{[0,1]} f \, d(T_\#\rho) = \int_{[0,1]\times X} (\partial_t F) \circ T \, d\rho$$

$$= \int_X (F \circ T)(1, \cdot) \, d\nu - \int_X (F \circ T)(0, \cdot) \, d\mu = F(1) - F(0).$$

So the integral against $T_\#\rho$ coincides with the Lebesgue measure for all continuous test functions, hence $T_\#\rho = \mathcal{L}\llcorner[0, 1]$.

Third claim: Let σ be some reference measure with $(\rho, \omega) \ll \sigma$. If $\omega \not\ll \rho$ then there must be a set $A \subset [0, 1] \times X$ with $\sigma(A) > 0$ where $\frac{d\rho}{d\sigma} = 0$ but $\frac{d\omega}{d\sigma} \neq 0$ and consequently $\Phi(\frac{d\rho}{d\sigma}, \frac{d\omega}{d\sigma}) = \infty$ and thus $\mathcal{A}(\rho, \omega) = \infty$. $\square$

With these basic regularity properties it is then easy to establish existence of minimizers.

Proposition 2.31 *Minimizers of the Benamou–Brenier formulation* (2.12) *exist.*

Proof If $W_{\mathrm{BB}}(\mu, \nu) = \infty$, one merely needs to show that $\mathcal{CE}(\mu, \nu)$ is non-empty. This will be established in the proof of Proposition 2.34. So assume from now on that $W_{\mathrm{BB}}(\mu, \nu) < \infty$.

Let $(\rho_n, \omega_n)_n$ be a minimizing sequence, which implies $(\rho_n, \omega_n) \in \mathcal{CE}(\mu, \nu)$ for all n and that $\mathcal{A}(\rho_n, \omega_n) \leq C$ for some $C < \infty$. By Proposition 2.30 $\rho_n \geq 0$, $\|\rho_n\|_{\mathcal{M}([0,1] \times X)} = 1$ and $\omega_n \ll \rho_n$. Therefore we can pick $\sigma = \rho_n$ as reference measure in the definition of the action $\mathcal{A}(\rho_n, \omega_n)$ and obtain

$$\mathcal{A}(\rho_n, \omega_n) = \int_{[0,1] \times X} \Phi\left(\frac{\mathrm{d}\rho_n}{\mathrm{d}\rho_n}, \frac{\mathrm{d}\omega_n}{\mathrm{d}\rho_n}\right) \mathrm{d}\rho_n = \frac{1}{2} \int_{[0,1] \times X} \left|\frac{\mathrm{d}\omega_n}{\mathrm{d}\rho_n}\right|^2 \mathrm{d}\rho_n.$$

With this we can bound the total variation norm of ω_n:

$$\|\omega_n\|_{\mathcal{M}([0,1] \times X)^d} = \int_{[0,1] \times X} \left|\frac{\mathrm{d}\omega_n}{\mathrm{d}\rho_n}\right| \mathrm{d}\rho_n \leq \left(\int_{[0,1] \times X} \left|\frac{\mathrm{d}\omega_n}{\mathrm{d}\rho_n}\right|^2 \mathrm{d}\rho_n\right)^{1/2}$$
$$= (2\,\mathcal{A}(\omega_n, \rho_n))^{1/2} \leq (2C)^{1/2}$$

where we used Jensen's inequality. Therefore, the sequence is uniformly bounded in norm and thus, by the Banach–Alaoglu theorem it must have a convergent subsequence with limit (ρ, ω). Since $\mathcal{CE}(\mu, \nu)$ is weak$*$-closed, one has $(\rho, \omega) \in \mathcal{CE}(\mu, \nu)$. Moreover, since Φ is convex, lower semi-continuous and 1-homogeneous, the functional $\mathcal{A}$ is lower semi-continuous (Remark 2.26). Therefore, (ρ, ω) is a minimizer of W_{BB}. □

Remark 2.32 (Absolutely Continuous Curves in Wasserstein Space) It turns out that one can establish much more regularity than in Proposition 2.30. When $\mathcal{A}(\rho, \omega) < \infty$ for $(\rho, \omega) \in \mathcal{CE}(\mu, \nu)$, it is possible to show that the family $(\rho_t)_t$ can be chosen such that the path $[0, 1] \ni t \mapsto \rho_t$ is continuous in $(\mathcal{P}(X), \mathrm{W})$, in fact *absolutely continuous*. It then becomes meaningful to consider continuity equations without prescribed temporal boundary conditions. (This is achieved by enforcing (2.11) only for C^1 test functions ψ that are compactly supported in $(0, 1) \times X$.) Conversely, one finds that if a curve $t \mapsto \rho_t$ is absolutely continuous, then there is some ω such that (ρ, ω) solve the continuity equation and $\mathcal{A}(\rho, \omega) < \infty$. We refer to [48, Section 5.3] for a more detailed discussion of this topic.

Remark 2.33 (Applications and Generalizations) The functional $\mathcal{A}$ can be used for much more than for merely finding geodesics in Wasserstein space. It can be used to model problems where the curve ρ_t interacts with other things at intermediate times, such as measurement constraints. An example for dynamic medical imaging where $\mathcal{A}$ acts as temporal regularizer is developed in [41, 49]. It is also a natural starting point for generalizing the Wasserstein distance, for instance to *unbalanced transport* between measures of unequal mass [8, 21, 22, 32, 37] or to vector-valued measures [14, 19].

2.3.2 Equivalence to Kantorovich Formulation

Proposition 2.34 *The Benamou–Brenier formula provides a lower bound for the Kantorovich formulation, i.e.* (2.12) $\leq$ (2.6).

For the proof we will use the following small Lemma, which ca be found, for instance in [13, Lemma 3.15]. It is a corollary of the disintegration theorem and Jensen's inequality.

Lemma 2.35 *Let A, B be measurable spaces, $T : A \to B$ measurable, $\mu \in \mathcal{M}(A)^n$, $\nu \in \mathcal{M}_+(A)$, $\mu \ll \nu$ (which implies $T_\#\mu \ll T_\#\nu$), and $f : \mathbb{R}^n \to \mathbb{R} \cup \{\infty\}$ convex. Then*

$$\int_B f\left(\frac{dT_\#\mu}{dT_\#\nu}\right) dT_\#\nu \leq \int_A f\left(\frac{d\mu}{d\nu}\right) d\nu.$$

Proof of Proposition 2.34 The strategy is to construct admissible candidates for (2.12) from admissible candidates for (2.6) that have a potentially lower objective value. Let $\gamma \in \Gamma(\mu, \nu)$ and let $Z : X \times X \times [0, 1]$, $Z(x, y, t) = (1 - t) \cdot x + t \cdot y$ be the map that parametrizes all pairwise constant speed geodesics. Let $v : X \times X \to \mathbb{R}^d$ be given by $v(x, y) = y - x$. Then for $t \in [0, 1]$ introduce the measures $\rho_t := Z(\cdot, \cdot, t)_\#\gamma$ and $\omega_t := Z(\cdot, \cdot, t)_\#(v \cdot \gamma)$. By Theorem 2.18, the map $t \mapsto \rho_t$ describes a Wasserstein geodesic from μ to ν (if γ is minimal) and ω_t is the Eulerian momentum field, which is constructed from the Lagrangian momentum field $v \cdot \gamma$, by putting the mass to the appropriate intermediate location at intermediate time t in the analogous way as for ρ_t. Observe that $\partial_t Z(x, y, t) = v(x, y)$.

From the family $(\rho_t)_t$ we can then construct a measure ρ in $\mathcal{P}([0, 1] \times X)$ via the characterization

$$\int_{[0,1]\times X} \psi\, d\rho := \int_0^1 \left[\int_X \psi(t, \cdot) d\rho_t\right] dt$$

for all $\psi \in C([0, 1] \times X)$. This is essentially the reversal of the decomposition (2.13). Analogously, we construct ω from the family $(\omega_t)_t$.

We find that the pair (ρ, ω) is in $\mathcal{CE}(\mu, \nu)$, since for $\psi \in C^1([0,1] \times X)$ one has

$$\int_{[0,1] \times X} \partial_t \psi \, d\rho + \int_{[0,1] \times X} \nabla \psi \cdot d\omega$$

$$= \int_0^1 \left[\int_X \partial_t \psi(t, \cdot) d\rho_t + \int_X \nabla \psi(t, \cdot) \cdot d\omega_t \right] dt$$

$$= \int_0^1 \left[\int_{X \times X} [\partial_t \psi(t, Z(x, y, t)) + \nabla \psi(t, Z(x, y, t)) \cdot v(x, y)] d\gamma(x, y) \right] dt$$

$$= \int_0^1 \left[\int_{X \times X} \left[\frac{d}{dt} \psi(t, Z(x, y, t)) \right] d\gamma(x, y) \right] dt$$

$$= \int_{X \times X} [\psi(1, Z(x, y, 1)) - \psi(0, Z(x, y, 0))] d\gamma(x, y)$$

$$= \int_X \psi(1, \cdot) d\nu - \int_X \psi(0, \cdot) d\mu.$$

And finally, as claimed, we find that $\mathcal{A}(\rho, \omega) \leq \frac{1}{2} \int |x - y|^2 d\gamma(x, y)$. For this we use that $v\gamma \ll \gamma$ implies that $\omega \ll \rho$ and therefore we can use ρ as reference measure σ in the definition of $\mathcal{A}$. We also use Lemma 2.35 on the map $Z(\cdot, \cdot, t)$. One finds:

$$\mathcal{A}(\rho, \omega) = \int_{[0,1] \times X} \Phi\left(\frac{d\rho}{d\rho}, \frac{d\omega}{d\rho}\right) d\rho = \int_0^1 \left[\int_X \Phi\left(\frac{d\rho_t}{d\rho_t}, \frac{d\omega_t}{d\rho_t}\right) d\rho_t \right] dt$$

$$\leq \int_0^1 \left[\int_{X \times X} \Phi(1, v) d\gamma \right] dt = \frac{1}{2} \int |x - y|^2 d\gamma(x, y).$$

□

Remark 2.36 (Reverse Inequality) The proof for the reverse inequality (and thus the equivalence of the definitions (2.6) and (2.12) for the Wasserstein distance) is more involved. The basic idea is to consider an admissible pair $(\rho, \omega) \in \mathcal{CE}(\mu, \nu)$ with $\mathcal{A}(\rho, \omega) < \infty$ and to construct a corresponding transport plan $\gamma \in \Gamma(\mu, \nu)$ with a potentially lower objective value.

Intuitively, this plan γ is constructed by tracing the trajectories of mass particles as described by the Eulerian velocity field $v = \frac{d\omega}{d\rho}$ to obtain once more a Lagrangian description. Formally this is done by solving the initial value problems

$$\partial_t T(t, x) = v(t, T(t, x)), \qquad T(0, x) = x \qquad (2.14)$$

for $t \in [0, 1]$ and μ-almost all $x \in X$. However, the fact that $v \in L^1(\rho; \mathbb{R}^d)$ provides far too little regularity for this ODE to be well-posed and problems may arise from the fact that μ contains atoms, and that their mass may need to be split for the optimal transport. This can be remedied by smoothing arguments, as outlined in [50, Theorem 8.1] or [48, Section 5.3.3]. We skip these arguments here and assume for simplicity that (2.14) has a unique solution and that $\mu \ll \mathcal{L}$. Using a calculation analogous to that in the proof of Proposition 2.34 for the continuity equation, one can then show that $\rho_t = T(t, \cdot)_\# \mu$ for almost all t, and in fact, ρ_t can be chosen such that this holds for all t, see Remark 2.32. Therefore in particular $\nu = T(1, \cdot)_\# \mu$ and thus the plan $\gamma := (\mathrm{id}, T(1, \cdot))_\# \mu$ is admissible, i.e. $\gamma \in \Gamma(\mu, \nu)$. One can then show that the Kantorovich objective for γ is bounded from above by the Benamou–Brenier objective along the following lines:

$$\int_{X \times X} \tfrac{1}{2}|x-y|^2 \mathrm{d}\gamma(x,y) = \int_X \tfrac{1}{2}|T(1,x) - T(0,x)|^2 \mathrm{d}\mu(x)$$

$$\leq \int_X \int_0^1 \tfrac{1}{2}|\partial_t T(t,x)|^2 \mathrm{d}\mu(x)$$

$$= \int_X \int_0^1 \tfrac{1}{2}|v(t,x)|^2 \mathrm{d}T(t,\cdot)_\# \mu(x)$$

2.3.3 Duality

Now we consider a dual formulation of (2.12). Applying Theorem 2.3 to (2.12) yields the following problem. We will subsequently analyze its structure and interpretation in more depth.

Proposition 2.37 (Dual Benamou–Brenier Formula)

$$W_{BB}(\mu, \nu) = \sup \left\{ \int_X \Psi(1, \cdot) \, d\nu - \int_X \Psi(0, \cdot) \, d\mu \, \middle| \, \Psi \in \mathrm{C}^1([0,1] \times X), \right. \\ \left. \partial_t \Psi + \tfrac{1}{2}|\nabla \Psi|^2 \leq 0 \right\} \quad (2.15)$$

Proof We will proceed analogously to the proof of Proposition 2.4 and write the supremum in (2.15) in the form of (2.3). For this choose $U = \mathrm{C}^1([0,1] \times X)$ and $V = \mathrm{C}([0,1] \times X)^{1+d}$, $A : \Psi \mapsto (\partial_t \Psi, \nabla \Psi)$, $F(\Psi) = \int_X \Psi(1, \cdot) d\nu - \int_X \Psi(0, \cdot) d\mu$ and

$$G : \mathrm{C}([0,1] \times X) \times \mathrm{C}([0,1] \times X)^d \ni (\xi, \zeta) \mapsto \begin{cases} 0 & \text{if } \xi + \tfrac{1}{2}|\zeta|^2 \leq 0, \\ +\infty & \text{else.} \end{cases}$$

F is finite for all $\Psi \in U$. Choosing $\Psi(t, x) := -t$ yields a candidate such that G is continuous at $A\Psi$.

The dual problem is then an optimization over $V^* = \mathcal{M}([0, 1] \times X)^{1+d}$ and we denote the candidates as pairs (ρ, ω) as above. Observing that

$$\Phi^*(a, b) = \begin{cases} 0 & \text{if } a + |b|^2/2 \leq 0, \\ +\infty & \text{else.} \end{cases}$$

one has that $G^*(\rho, \omega) = \mathcal{A}(\rho, \omega)$ (see Remark 2.26). Since $F(\Psi)$ is linear, it can be written as $\langle l, \Psi \rangle_{C^{1*}, C^1}$ for some suitable $l \in C^{1*}$ characterized by

$$\langle l, \Psi \rangle_{C^{1*}, C^1} = \int_X \Psi(1, \cdot) \mathrm{d}\nu - \int_X \Psi(0, \cdot) \mathrm{d}\mu$$

and F^* is then consequently the indicator function of that single element set $\{l\}$, i.e. $F^*(l') = 0$ if $l' = l$ and $+\infty$ otherwise. From the definition of A we find that

$$\langle A^*(\rho, \omega), \Psi \rangle_{C^{1*}, C^1} = \langle (\rho, \omega), A\Psi \rangle_{\mathcal{M}, C} = \int_{[0,1] \times X} \partial_t \Psi \mathrm{d}\rho + \int_{[0,1] \times X} \nabla \Psi \cdot \mathrm{d}\omega$$

and thus $F^*(A^*(\rho, \omega)) = 0$ if and only if $(\rho, \omega) \in \mathcal{CE}(\mu, \nu)$ (cf. (2.11)). □

Proposition 2.38 (Primal-Dual Optimality Conditions for the Benamou–Brenier Formulation) *A pair $(\rho, \omega) \in \mathcal{M}([0, 1] \times X)^{1+d}$ and $\Psi \in C^1([0, 1] \times X)$ is primal-dual optimal for (2.12) and (2.15) if and only if*

$$\partial_t \rho + \nabla \cdot \omega = 0 \quad \text{with temporal boundary conditions } \rho(0, \cdot) = \mu, \ \rho(1, \cdot) = \nu$$

in the distributional sense of (2.11),

$$\rho \geq 0, \qquad \omega = \nabla \Psi \cdot \rho,$$

and

$$\partial_t \Psi + \tfrac{1}{2} |\nabla \Psi|^2 \leq 0 \text{ with equality } \rho\text{-almost everywhere.}$$

Proof The primal-dual optimality conditions for (2.12) ad (2.15) implied by Theorem 2.3 are (using the conventions of the proof of Proposition 2.37) $\Psi \in -\partial F^*(A^*(\rho, \omega))$ and $(\rho, \omega) \in \partial G(A\Psi)$. The former implies $(\rho, \omega) \in \mathcal{CE}(\mu, \nu)$ (argument similar as in Remark 2.9).

The latter implies first that $\partial G(A\Psi) \neq \emptyset$, i.e. $G(A\Psi) < \infty$, which means $G(A\Psi) = 0$, which implies that Ψ satisfies the inequality constraint $\partial_t \Psi + \frac{1}{2}|\nabla \Psi|^2 \leq 0$. Then using the equivalent condition (cf. Fenchel–Young inequality)

$$G^*(\rho, \omega) = \langle (\rho, \omega), A\Psi \rangle - G(A\Psi)$$

we conclude that $G^*(\rho, \omega) < \infty$, i.e. $\rho \geq 0$, $\omega = v \cdot \rho$ for some density velocity field v (as in the proof of Proposition 2.30) and therefore the above equality becomes

$$G^*(\rho, \omega) = \int_{[0,1] \times X} \Phi(1, v) d\rho = \int_{[0,1] \times X} [1 \cdot \partial_t \Psi + v \cdot \nabla \Psi] d\rho$$

which implies that $(\partial_t \Psi, \nabla \Psi) \in \partial \Phi(1, v)$ ρ-almost everywhere, which finally means that ρ-almost everywhere $\partial_t \Psi + \frac{1}{2}|\nabla \Psi|^2 \leq 0$ is actually an equality and $v = \nabla \Psi$. □

Remark 2.39 (Gradient Structure of Optimal Velocity Fields) Intuitively it makes sense that the optimal velocity field $v = \nabla \Psi$ in Proposition 2.38 is a gradient: if it were not so, one could (at least formally) subtract the divergence-free (with respect to ρ) component of v (akin to the Helmholtz decomposition) to reduce the objective $\frac{1}{2} \int |v|^2 d\rho$ while preserving the continuity equation. This will result in a gradient vector field. In fact, one may interpret Brenier's theorem as a non-linear version of the Helmholtz decomposition [12, Section 1.3], and this intuition is the mechanism behind the celebrated Angenent–Haker–Tannenbaum scheme [4].

Remark 2.40 ((Non-)existence of Dual Maximizers) At this point we should start to suspect that continuously differentiable dual maximizers of (2.15) are unlikely to exist in general: We have seen that the initial velocity field of the particles (which equals their Lagrangian velocity field), given by Brenier's theorem (Theorem 2.14, see also Example 2.20) will in general be defined only μ-almost everywhere (and only if μ has no atoms), and we could not establish sufficient regularity for v in (2.14) to integrate the flow. We will revisit the question of dual regularity in Sect. 2.4.3.

Remark 2.41 (Equivalence with Kantorovich Dual Problem) Given the equivalence between the primal Kantorovich and Benamou–Brenier formulations (Sect. 2.3.2) one might wonder about the equivalence between the dual problems (2.4) (for $c = \frac{1}{2}d^2$) and (2.15). By comparing the respective objective functions it seems tempting to seek for an equivalence between (ϕ, ψ) in (2.4) and $(-\Psi(0, \cdot), \Psi(1, \cdot))$ in (2.15). This intuition is furthered by the observation that the initial velocity field is given by $-\nabla \phi$, consistent with the optimality condition that $v = \nabla \Psi$ (Proposition 2.38).

However, at this point it seems unclear how the respective constraints $\phi \oplus \psi \leq c$ and $\partial_t \Psi + \frac{1}{2}|\nabla \psi|^2 \leq 0$ can be in correspondence with each other. This connection can be made via the celebrated Hopf–Lax formula (see for instance [25]). We will naturally stumble upon this formula by means of convex analysis via an auxiliary multi-marginal formulation in Sect. 2.4.2 and show its connection to the Benamou–Brenier formulation in Sect. 2.4.3.

Remark 2.42 (Connection to Burger's Equation and Geodesic Equation for W) In Example 2.19 we established that mass particles move with constant speed along straight lines in a Wasserstein geodesic. In the special case of Brenier's theorem (Example 2.20) this constant velocity field was shown to be $v_0 = -\nabla \phi$ (in a Lagrangian frame, where we identify particles by their initial position). At a formal level, the Eulerian velocity field should then be a velocity field that *propagates itself*, i.e. one should have

$$\partial_t v + (\nabla v) \cdot v = 0 \qquad (2.16)$$

with initial condition $v(0, \cdot) = v_0$. This is known as (inviscid) Burger's equation, which is known for its tendency to develop shocks [25, Section 3.4.1].

From Brenier's theorem we know that $\frac{1}{2}|\cdot|^2 - \phi$ is convex, which implies that $\frac{1}{2}|\cdot|^2 - t \cdot \phi$ is strictly convex for $t \in [0, 1)$ and therefore that $\text{id} + t \cdot v$ is invertible for $t \in [0, 1)$ (Remark 2.23). Hence, no shocks develop at intermediate times along a Wasserstein geodesic. However, shocks may develop at $t = 1$, or when one tries to extrapolate a Wasserstein geodesic beyond the interval $[0, 1]$. This is consistent with Proposition 2.21.

From Proposition 2.38 we learn that the Eulerian velocity field $v_t = \nabla \Psi(t, \cdot)$ is formally given as a gradient at all intermediate times where Ψ satisfies

$$\partial_t \Psi + \tfrac{1}{2}|\nabla \Psi|^2 = 0 \quad \rho\text{-almost everywhere.} \qquad (2.17)$$

Formally applying a spatial gradient to (2.17) one indeed obtains (2.16). Thus, (2.17) can be seen as reformulation of (2.16) to explicitly incorporate the gradient property of v. If for a given μ and some $\Psi(0, \cdot) \in C^1(X)$ there would be a classical solution of (2.17) on some non-empty time interval $[0, t]$, the corresponding velocity field would induce a path of measures via the continuity equation, which would then indeed be a Wasserstein geodesic. Therefore, (2.17) has been dubbed the *geodesic equation* for W [39].

2.4 Multi-Marginal Formulation

2.4.1 Primal Problem

Multi-marginal optimal transport is a versatile tool with many applications, such as modeling in economics, fluid dynamics, and density function theory, see [48, Section 1.7.4] for some references. In this section we merely use it in the context of dynamic optimal transport, similar to [11]. Our main motivation is to serve as an intermediate step between the Kantorovich and the Benamou–Brenier formulations of optimal transport, to provide a better understanding of the dual formulation of the latter, and to prepare similar constructions for the entropic setting in Sect. 3. From a modeling perspective this reformulation is also interesting since it allows to couple the movement of mass to other observations and constraints at intermediate times (Remark 2.46).

Definition 2.43 (Multi-Marginal Cost Function) Let X be a compact, convex subset of $\mathbb{R}^d$, and let $c(x, y) = \frac{1}{2}|x - y|^2$. For a natural number $N \geq 2$ and time points $t_0 = 0 < t_1 < t_2 < \ldots < t_N = 1$, the multi-marginal cost function is given by

$$c_{\mathrm{MM}} : X^{N+1} \mapsto \mathbb{R}, \qquad (x_0, \ldots, x_N) \mapsto \sum_{i=1}^{N} \frac{1}{2(t_i - t_{i-1})} |x_i - x_{i-1}|^2. \qquad (2.18)$$

We will occasionally denote tuples $(x_0, \ldots, x_N) \in X^{N+1}$ more compactly as $\vec{x}$.

We now show that the Kantorovich formulation of W, (2.6), can be rewritten as a multi-marginal transport problem with cost c_{MM}, where only the first and last marginal are prescribed.

Proposition 2.44 (Multi-Marginal Formulation of W) *Problem* (2.6) *is equivalent to the problem*

$$\inf \left\{ \int_{X^{N+1}} c_{\mathrm{MM}}(\vec{x}) \, d\gamma_{\mathrm{MM}}(\vec{x}) \,\middle|\, \gamma_{\mathrm{MM}} \in \mathcal{P}(X^{N+1}),\, P_0 \gamma_{\mathrm{MM}} = \mu,\, P_N \gamma_{\mathrm{MM}} = \nu \right\} \qquad (2.19)$$

in the following sense: First, their minimal values are equal. Moreover, if γ_{MM} is a minimizer of (2.19), *then $P_{0,N} \gamma_{\mathrm{MM}}$ is a minimizer of* (2.6). *Here $P_{0,N}$ denotes the push-forward by the map $X^{N+1} \to X^2$, $(x_0, \ldots, x_N) \mapsto (x_0, x_N)$, i.e. $P_{0,N} \gamma_{\mathrm{MM}}$ extracts the joint $(0, N)$-th marginal of γ_{MM}. Conversely, if γ is a minimizer of* (2.6) *then $F_{N\#}\gamma$ is a minimizer of* (2.19) *where*

$$F_N : X^2 \to X^{N+1}, \qquad (x_0, x_N) \mapsto (x_0, x_1, \ldots, x_{N-1}, x_N)$$

with $x_i := (1 - t_i) \cdot x_0 + t_i \cdot x_N$ for $i \in \{1, \ldots, N-1\}$.

The proof hinges on the following lemma, which can be proved by a simple explicit calculation.

Lemma 2.45 *For $(x_0, x_1) \in X^2$ one has*

$$c(x_0, x_1) = \min \left\{ c_{\mathrm{MM}}(x_0, \ldots, x_N) \,\middle|\, x_1, \ldots, x_{N-1} \in X \right\}$$

and the unique minimizer on the r.h.s. is given by $x_i = (1 - t_i) \cdot x_0 + t_i \cdot x_N$ for $i \in \{1, \ldots, N-1\}$.

Proof of Proposition 2.44 Let γ_{MM} be admissible in (2.19). Then $\gamma := P_{0,N} \gamma_{MM}$ clearly lies in $\Gamma(\mu, \nu)$ and is therefore admissible in (2.6) and by Lemma 2.45 one has

$$\int_{X^2} c \, d\gamma = \int_{X^{N+1}} c \circ p_{0,N} \, d\gamma_{MM} \leq \int_{X^{N+1}} c_{MM} \, d\gamma_{MM}, \tag{2.20}$$

where $p_{0,N} : (x_0, \ldots, x_N) \mapsto (x_0, x_N)$. Thus (2.6) $\leq$ (2.19). Conversely, let $\gamma \in \Gamma(\mu, \nu)$ be an admissible candidate for (2.6) and set $\gamma_{MM} := F_{N\#}\gamma$, which is then again admissible for (2.19) with cost

$$\int_{X^{N+1}} c_{MM} \, d\gamma_{MM} = \int_{X^2} c_{MM} \circ F_M \, d\gamma = \int_{X^2} c \, d\gamma \tag{2.21}$$

where the second equality is again a consequence of Lemma 2.45. Therefore (2.19) $\leq$ (2.6). Consequently, both infima are the same, and minimizers for one can be constructed from the other as in the proof. In particular existence of minimizers for (2.19) follows from existence in (2.6) (which is in turn a consequence of Proposition 2.4). □

Remark 2.46 (Interpretation of (2.6) **and** (2.19)**)** In the Kantorovich formulation of W, (2.6), the plan γ encodes the distribution of initial and final locations of mass particles. Via Theorem 2.18 it is implied that in a dynamic interpretation these particles travel on a constant speed straight line from their initial to the final location during transport. This perspective is further developed through the equivalence with the Benamou–Brenier formulation (Sect. 2.3.2).

In contrast, a plan γ_{MM} in (2.19) can be seen as a more detailed description of the itinerary of the mass particles, where a sequence of locations $(x_0, \ldots, x_N)$ at times $(t_0 = 0, \ldots, t_N = 1)$ is specified, including intermediate times $(t_i)_{i=1}^{N-1}$. Of course, in hindsight, this description is redundant, as shown by Proposition 2.44. However, first, it provides a more gradual path from the static Kantorovich formulation to the dynamic Benamou–Brenier formula (see the remainder of this section, in particular Sect. 2.4.3), and second, it allows to model more general dynamic processes, where the particles interact with the environment at intermediate times, e.g. through volume constraints or measurement information, see for instance [11,30,34]. Note that all the references actually use entropic optimal transport. However, in [11] it appears merely as a numerical tool, whereas in the other two references entropy is actually part of the data model. We will study this in more detail in Sect. 3.2.

The dynamic description implied by (2.19) is Lagrangian, since the measure γ_{MM} keeps track of all individual particles and their trajectories. The following equivalent reformulation (stated without proof) has a more Eulerian flavour, as it keeps track of the intermediate mass distributions and their next steps. It is also more tractable numerically, see Remark 2.48.

Proposition 2.47 *Problems (2.6) and (2.19) are equivalent to the following problem*

$$\inf\left\{\sum_{i=0}^{N-1}\int_{X^2}\frac{|x-y|^2}{2(t_{i+1}-t_i)}\,d\gamma_i(x,y)\,\middle|\,(\rho_0,\ldots,\rho_N)\in\mathcal{P}(X),\,\rho_0=\mu,\,\rho_N=\nu,\right.$$

$$\left.\gamma_i\in\Gamma(\rho_i,\rho_{i+1})\,\text{for}\,i\in\{0,\ldots,N-1\}\right\} \quad (2.22)$$

in the sense that the minimal values are identical and minimizers can be constructed from each for the other in a similar spirit as in Theorem 2.18 and Proposition 2.44.

For minimizers of (2.22), the transport plans γ_i are optimal for the Wasserstein optimal transport problem $W(\rho_i,\rho_{i+1})$, *(2.6), between their marginals.*

Remark 2.48 (Markov Property) The equivalence between (2.19) and (2.22) holds since the cost c_{MM} is in fact only a sum of functions involving adjacent time steps. Thus, not the full joint distribution γ_{MM} is relevant, but merely the collection of pairwise distributions of all successive time step pairs, which is captured in (2.22). We may therefore assume, for instance, that γ_{MM} is Markov in the following sense: Let γ_{MM} be the joint law of a tuple $(x_1,\ldots,x_N)$ of X-valued random variables. Then x_i and x_k are independent when conditioned on some x_j for a triple $i<j<k$.

At first glance, in (2.19) there might also exist minimizers that are not Markov. However, Corollary 2.22 implies that this is not the case, since all particles at some intermediate position $z\in X$ at some intermediate time $t_j\in(0,1)$ must have the same initial and final positions. Therefore the positions x_i, x_k at earlier or later times become concentrated on single points when conditioned on $x_j=z$ at time t_j and are thus independent.

In the language of [30] this Markov property means that c_{MM} corresponds to a chain graph structure and it is one of the avenues to make high-dimensional multi-marginal transport problems numerically tractable [1, 7, 10, 11, 30]. This Markov property becomes also relevant in the presence of entropic regularization (Sect. 3.1) where it is again essential for numerical tractability (see some of the references above) and also crucial for the dynamic interpretation and modeling (Remark 3.23).

Remark 2.49 (Limit $N\to\infty$) Of course, formulations (2.19) and (2.22) are not fully dynamic in the sense that they are still time-discrete and we implicitly rely on the fact that the γ_i are pairwise optimal plans and then interpolate the gaps with Theorem 2.18. It appears natural to consider the limit $N\to\infty$. In (2.19) in this limit the tuples $(x_0,\ldots,x_N)$ will have to be replaced by a suitable class of paths $x:[0,1]\to X$ where $x(t)$ gives the position of a mass particle at arbitrary times $t \in [0, 1]$, and γ_{MM} will be replaced by a measure over such paths.

The proper regularity class of paths is determined by considering where the corresponding limit of the cost function c_{MM} is well-behaved, i.e. we need that for every admissible path x the supremum of

$$\sum_{i=1}^{N} (t_i - t_{i-1}) \cdot \frac{|x(t_i) - x(t_{i+1})|^2}{(t_i - t_{i-1})^2}$$

over N and intermediate time points $(t_i)_i$ is bounded. This can be seen as finite-difference approximation of the integral $\int_0^1 |\partial_t x(t)|^2 dt$ and therefore naturally leads to the space $H^1([0, 1], X)$ when X is a subset of $\mathbb{R}^d$, or to the class of *absolutely continuous curves*, denoted by $AC([0, 1], X)$, if X is a more general metric space. So the natural limit for γ_{MM} will be a measure on paths over X, concentrated on $AC([0, 1], X)$ and the mass distribution at time t can be extracted by the push-forward under the evaluation map

$$\mathrm{ev}_t : AC([0, 1], X) \to X, \qquad x \mapsto x(t).$$

In (2.22) the natural limit for all intermediate $(\rho_i)_i$ will be a curve $[0, 1] \to \mathcal{P}(X)$ and the intermediate γ_i will play the role of the momentum measure ω, as in the Benamou–Brenier formula (2.12).

Recall that in Remark 2.32 we mentioned the notion of absolutely continuous paths of measures in $(\mathcal{P}(X), W)$, which we now denote as $AC([0, 1], \mathcal{P}(X))$. It was shown in [38] that the two notions are essentially equivalent: For a measure γ_{MM} on absolutely continuous paths the curve $t \mapsto \rho_t := \mathrm{ev}_{t\#}\gamma_{MM}$ of time marginals lies in $AC([0, 1], \mathcal{P}(X))$. Conversely, if $t \mapsto \rho_t$ lies in $AC([0, 1], \mathcal{P}(X))$, then there is a γ_{MM} concentrated on $AC([0, 1], X)$ such that $\rho_t = \mathrm{ev}_{t\#}\gamma_{MM}$.

Measures on paths as a Lagrangian description of dynamic optimal transport are a fundamental and popular tool, see for instance [15, 35].

We do not consider this limit here to avoid the related technicalities. By choosing N and the time positions of the intermediate marginals flexibly, and via the equivalences of (2.6), (2.19), and (2.22) we can obtain all results of interest in this chapter.

2.4.2 Dual Problem

Now we consider a dual perspective on the multi-marginal problems considered in the previous section. This will eventually help us to better understand the dual formulation of the Benamou–Brenier formulation.

Proposition 2.50 (Dual Multi-Marginal Problem) *A dual problem for* (2.19) *is given by*

$$\sup \left\{ \int_X \phi \, d\mu + \int_X \psi \, dv \, \middle| \, \phi, \psi \in C(X), \right.$$

$$\left. \phi(x_0) + \psi(x_N) \leq c_{MM}(x_0, \ldots, x_N) \, \forall \, (x_0, \ldots, x_N) \in X^{N+1} \right\} \quad (2.23)$$

and (2.23) *is equivalent to* (2.4) *(for cost* $c = \frac{1}{2}d^2$*) in the sense that the optimal values and the set of maximizers are identical.*

Proof The form of (2.23) can be obtained via Fenchel–Rockafellar duality from (2.19) in almost the exact same way as (2.4) was obtained from (2.1).

The form of (2.23) and equivalence with (2.4) can be obtained from Lemma 2.45 by observing that the constraint

$$\phi(x_0) + \psi(x_N) \leq c_{MM}(x_0, \ldots, x_N) \, \forall \, (x_0, \ldots, x_N) \in X^{N+1}$$

is equivalent to

$$\phi(x_0) + \psi(x_N) \leq \inf_{(x_1, \ldots, x_{N-1}) \in X^{N-1}} c_{MM}(x_0, \ldots, x_N) \, \forall \, (x_0, x_N) \in X^2.$$

$\square$

To gain some intuition for the next steps, consider the above problem for $N = 2$. Then the constraint can be written as

$$\psi(x_2) \leq \frac{|x_0 - x_1|^2}{2t_1} + \frac{|x_1 - x_2|^2}{2(1 - t_1)} - \phi(x_0).$$

We can take the infimum over x_0 to obtain

$$\psi(x_2) \leq \frac{|x_1 - x_2|^2}{2(1 - t_1)} + \psi_1(x_1) \quad \text{with} \quad \psi_1(x_1) := \inf_{x_0 \in X} \frac{|x_0 - x_1|^2}{2t_1} - \phi(x_0).$$

We see that by introducing suitable auxiliary dual potentials for intermediate times, we can localize the constraints in time. This motivates the following definition.

A Dynamic Perspective of Optimal Transport

Proposition 2.51 (Dynamic Duals) *Let (ϕ, ψ) be a pair of dual maximizers of (2.4) that satisfy $\phi = \psi^c$, $\psi = \phi^c$. We then introduce dynamic dual functions $\Phi, \Psi : [0,1] \times X \to \mathbb{R}$ as follows:*

$$\Phi(t,x) := \inf_{y \in X} \frac{|x-y|^2}{2 \cdot (1-t)} - \psi(y) \quad \text{for } t \in [0,1), \quad \Phi(1,x) := -\psi(x), \quad (2.24a)$$

$$\Psi(t,x) := \inf_{y \in X} \frac{|x-y|^2}{2 \cdot t} - \phi(y) \quad \text{for } t \in (0,1], \quad \Psi(0,x) := -\phi(x). \quad (2.24b)$$

Then Φ and Ψ are Lipschitz continuous with respect to both arguments and they satisfy the recursive definitions

$$\Phi(s,x) = \inf_{y \in X} \frac{|x-y|^2}{2 \cdot (t-s)} + \Phi(t,y), \quad \Psi(t,x) = \inf_{y \in X} \frac{|x-y|^2}{2 \cdot (t-s)} + \Psi(s,y) \quad (2.25)$$

for $s, t \in [0,1]$ with $s < t$.

Expressions (2.24) and (2.25) are special cases of the *Hopf–Lax formula* (see for instance [25]) which provide suitable generalized solutions to evolution equations as they appear in the constraint of the dual Benamou–Brenier formula, (2.15). In Sect. 2.4.3 we will make this relation more explicit.

Proof The recursive definition (2.25) can be verified by using the original definitions for (2.24) for Φ and Ψ on the right-hand sides and then invoking Lemma 2.45.

The proof of Lipschitz continuity with respect to x is closely related to the arguments how dual potentials inherit regularity from the cost function via the c-transform, as used in the proof of Proposition 2.6. First, note that since X is bounded, c is Lipschitz continuous on $X \times X$ and therefore $\phi = \psi^c$ and $\psi = \phi^c$ are both Lipschitz continuous. In the following we denote a suitable Lipschitz constant by L. One can then extend ϕ and ψ to $\mathbb{R}^d$ via $\psi(x) := \sup_{y \in X} -L|x-y| + \psi(y)$ for all $x \in \mathbb{R}^d \setminus X$ (and the same formula for ϕ). These global extensions will still have Lipschitz constant L and satisfy $\psi(x) \leq \psi(p_X(x))$ where p_X denotes the projection onto the compact and convex set X. Therefore, when $\tilde{c} : \mathbb{R}^d \times \mathbb{R}^d \to \mathbb{R}$, $(x,y) \mapsto \tilde{c}(x,y)$ is an increasing function of $|x-y|$, one finds for $x \in X$ that

$$\inf_{y \in X} \tilde{c}(x,y) - \psi(y) = \inf_{y \in \mathbb{R}^d} \tilde{c}(x,y) - \psi(y).$$

As a consequence, Φ and Ψ inherit the Lipschitz constant L from ϕ and ψ with respect to the spatial argument. Indeed one has for $t \in [0,1)$ and $x, x' \in X$, $y \in \mathbb{R}^d$,

$$\Phi(t,x) \leq \frac{|x-y|^2}{2 \cdot (1-t)} - \psi(y) \leq \frac{|x' - (y + x' - x)|^2}{2 \cdot (1-t)} - \psi(y + x' - x) + L|x' - x|.$$

Taking now the infimum over $y \in \mathbb{R}^d$ on the right-hand side yields $\Phi(t, x) \leq \Phi(t, x') + L\,|x - x'|$ and thus by the symmetric argument on obtains the Lipschitz bound with respect to x. Lipschitz continuity of $\Phi(1, \cdot) = -\psi$ follows directly from the definition. Arguments for Ψ are identical.

For the Lipschitz bound with respect to t observe first that (2.25) implies $\Phi(s, x) \leq \Phi(t, x)$ for $0 \leq s < t \leq 1$. Using the spatial Lipschitz regularity one then finds

$$\Phi(s, x) = \inf_{y \in X} \frac{|x-y|^2}{2 \cdot (t-s)} + \Phi(t, y) \geq \inf_{y \in X} \frac{|x-y|^2}{2 \cdot (t-s)} + \Phi(t, x) - L|x-y|$$

$$\geq \Phi(t, x) - \frac{L^2(t-s)}{2}.$$

□

The following proposition provides additional structure on the dynamic duals Φ and Ψ.

Proposition 2.52 *Consider the setting of Proposition 2.51 and let $t \in (0, 1)$. For any $z \in X$ one has*

$$\Phi(t, z) + \Psi(t, z) \geq 0$$

with equality if and only if z can be written as $z = (1 - t) \cdot x + t \cdot y$ for some (x, y) in the contact set of (ϕ, ψ) (Definition 2.7). In this case the pair (x, y) is unique (in the contact set) and one has

$$\Phi(t, z) = \frac{|z-x|^2}{2 \cdot (1-t)} - \psi(x), \qquad \Psi(t, z) = \frac{|z-y|^2}{2 \cdot t} - \phi(y),$$

i.e. x and y are minimizers in (2.24) for $\Phi(t, z)$ and $\Psi(t, z)$ and in fact are the unique minimizers.

Proof Using the definitions (2.24) we find

$$\Phi(t, z) + \Psi(t, z) = \inf_{(x,y) \in X^2} \frac{|x-z|^2}{2 \cdot (1-t)} + \frac{|z-y|^2}{2 \cdot t} - \phi(x) - \psi(y) \tag{2.26}$$

$$\geq \inf_{(x,y) \in X^2} \frac{|x-y|^2}{2} - \phi(x) - \psi(y) \geq 0 \tag{2.27}$$

where the first inequality is a consequence of Lemma 2.45.

Now let (x, y) be in the contact set and set $z = (1 - t) \cdot x + t \cdot y$. Then we observe that the candidate (x, y) yields the value zero in both infima above and thus indeed attains both infima. Conversely, if $\Phi(t, z) + \Psi(t, z) = 0$, then both inequalities must be equalities and the infimal values must both equal zero. Let (x, y) be a minimizer of the first infimum (existence implied by compactness and continuity). For the infimum to be zero, one must have $z = (1-t) \cdot x + t \cdot y$ and (x, y) must be in the contact set. Uniqueness of this (x, y) follows from Proposition 2.21. □

In (2.26) we see that the dynamic duals take on a particularly simple form on suitable straight lines that correspond to the flow of the optimal transport velocity field (cf. Remark 2.42). This is related to the *method of characteristics*, which plays an important role for the Hopf–Lax formula.

We can now rewrite (2.23) with auxiliary potentials and constraints that are local in time. The resulting problem starts to resemble the dual Benamou–Brenier formula (2.15).

Proposition 2.53 *Problem (2.23) has the same optimal value as*

$$\sup \left\{ \int_X \psi_N \, dv - \int_X \psi_0 \, d\mu \, \middle| \, (\psi_0, \ldots, \psi_N) \in C(X)^{N+1}, \right.$$
$$\left. \psi_{i+1}(x) - \psi_i(y) \leq \frac{|x-y|^2}{2(t_{i+1} - t_i)} \, \forall x, y \in X, \, i \in \{0, \ldots, N-1\} \right\}. \quad (2.28)$$

Given maximizers (ϕ, ψ) for (2.23) that satisfy $\phi = \psi^c$, $\psi = \phi^c$ (such maximizers exist, as shown in Proposition 2.6), a maximizer for (2.28) is given constructing first the function Ψ from (ϕ, ψ) via (2.24) and then setting $\psi_i = \Psi(t_i, \cdot)$ for $i \in \{0, \ldots, N\}$. For a maximizer $(\psi_0, \ldots, \psi_N)$ in (2.28), a maximizer for (2.23) is given by setting $(\phi, \psi) := (-\psi_0, \psi_N)$.

Proof Consider the claimed construction of a maximizer for (2.28) form one of (2.23). By (2.25) the tuple $(\psi_0, \ldots, \psi_N)$ indeed satisfies the inequality constraints in (2.28). And as by construction $\psi_0 = \Psi(0, \cdot) = -\phi$, they also yield the same objective in (2.28) as (ϕ, ψ) yield in (2.23). So (2.28) $\geq$ (2.23).

Now consider the converse construction. Summing the constraints in (2.28) over i we obtain that

$$\psi_N(x_N) - \psi_0(x_0) \leq \sum_{i=0}^{N-1} \frac{|x_i - x_{i+1}|^2}{2(t_{i+1} - t_i)}$$

for all $(x_0, \ldots, x_N) \in X^{N+1}$. Taking now the infimum over $(x_1, \ldots, x_{N-1})$ and using Lemma 2.45 we obtain that $(\phi, \psi) := (-\psi_0, \psi_N)$ is admissible in (2.23) and again it has the same objective value. This yields the reverse inequality and therefore equality with the confirmation the two above constructions convert maximizers of one problem into maximizers of the other. □

Corollary 2.54 *Given maximizers* $(\psi_0, \ldots, \psi_N)$ *of* (2.28), *the pair* $(-\psi_i, \psi_{i+1})$ *for* $i \in \{0, \ldots, N-1\}$, *is a maximizer for the dual problem of one time-step of* (2.22), *given by*

$$\inf\left\{\int_{X^2} \frac{|x-y|^2}{2(t_{i+1}-t_i)}\, d\gamma_i(x,y)\,\bigg|\,\gamma_i \in \Gamma(\rho_i, \rho_{i+1})\right\}$$

for fixed (ρ_i, ρ_{i+1}) *which are taken from a primal minimizer in* (2.22), *i.e. they maximize*

$$\sup\left\{\int_X \psi_{i+1}\, d\rho_{i+1} - \int_X \psi_i\, d\rho_i\,\bigg|\,\psi_i, \psi_{i+1} \in C(X),\right.$$

$$\left.\psi_{i+1}(y) - \psi_i(x) \leq \frac{|x-y|^2}{2(t_{i+1}-t_i)}\,\forall x, y \in X\right\}. \quad (2.29)$$

Proof For simplicity we only sketch the proof for minimizers in (2.22) and maximizers in (2.28) that were constructed from minimizers in (2.6) and maximizers in (2.23), but the proof can easily be extended to general optimizers. First, the pair $(-\psi_i, \psi_{i+1})$ satisfies the constraint in (2.29), as it is also part of (2.28). By virtue of Proposition 2.52, constructing ψ_i and ψ_{i+1} via Ψ yields that the contact set of $(-\psi_i, \psi_{i+1})$ with respect to the cost function $\frac{|x-y|^2}{2(t_{i+1}-t_i)}$ is given by the image of the contact set of (ϕ, ψ) under the map $\hat{Z}(t_i, t_{i+1}, \cdot, \cdot)$ (cf. Theorem 2.18). On the primal side, the optimal plan γ_i constructed from γ via the push-forward with the map $\hat{Z}(t_i, t_{i+1}, x, y)$ is concentrated on this contact set, hence their respective optimality is provided by Proposition 2.8. □

2.4.3 Connection to Dual Benamou–Brenier Formula

Assume now for simplicity that optimal $(\rho_i)_i$ in Proposition 2.47 are absolutely continuous with respect to the Lebesgue measure, so that we may apply Brenier's theorem. Then by Corollary 2.54 the Lagrangian velocity field for each ρ_i in the time interval $[t_i, t_{i+1}]$ is given by $\nabla \psi_i = \nabla \Psi(t_i, \cdot)$ (here the factor $1/(t_{i+1} - t_i)$ in the cost function cancels with the fact that the time interval is shorter), which is consistent with Proposition 2.38. This suggests that the connection of the static Kantorovich duals of (2.4) and the dynamic potential of the Benamou–Brenier formulation (2.15) may be obtained via the time-interpolation of Proposition 2.51.

Clearly, yet another equivalent dual formulation of (2.23) and (2.28) is given by

$$\sup\left\{\int_X \Psi(1,\cdot)\, dv - \int_X \Psi(0,\cdot)\, d\mu\,\bigg|\,\Psi \in C([0,1] \times X),\right.$$

$$\left.\Psi(s,x) - \Psi(t,y) \leq \frac{|x-y|^2}{2(t-s)}\,\forall x, y \in X, s, t \in [0, s] \text{ with } s < t\right\}. \quad (2.30)$$

We now show that this is a natural relaxation of the dual Benamou–Brenier formulation. First, candidates admissible in (2.15) are admissible in (2.30).

Lemma 2.55 *If $\Psi \in C^1([0,1] \times X)$ is admissible for the dual Benamou–Brenier formula (2.15), in particular when $\partial_t \psi + \frac{1}{2}|\nabla \psi|^2 \leq 0$, then $\Psi(t,x) \leq \frac{1}{2 \cdot (t-s)}|x-y|^2 + \Psi(s,y)$ for all $s < t$, x, y.*

Proof Let $z : [s,t] \to X$, $z(r) = \frac{r-s}{t-s}x + \frac{t-r}{t-s}y$, i.e. the constant speed straight line interpolation from y to x, parametrized in the time interval $[s,t]$ with $\partial_r z(r) = (x-y)/(t-s)$. Then

$$\psi(t,x) = \int_s^t \frac{d}{dr}\psi(r, z(r))\, dr + \psi(s,y)$$

$$= \int_s^t \left[\partial_t \psi(r, z(r)) + \nabla \psi(r, z(r)) \cdot \frac{x-y}{t-s}\right] dr + \psi(s,y)$$

$$\leq \int_s^t \left[-\tfrac{1}{2}|\nabla \psi(r, z(r))|^2 + \nabla \psi(r, z(r)) \cdot \frac{x-y}{t-s}\right] dr + \psi(s,y)$$

$$\leq \int_s^t \frac{|x-y|^2}{2(t-s)^2} dr + \psi(s,y) = \frac{|x-y|^2}{2(t-s)} + \psi(s,y)$$

where we used the feasibility condition in the third line and the inequality $-\frac{1}{2}|v|^2 + v \cdot w \leq \frac{1}{2}|w|^2$ for arbitrary vectors $w, v \in \mathbb{R}^d$ in the last line. □

Conversely, we conclude by showing that if an admissible candidate of (2.30) is differentiable at a given point, it will satisfy the constraint of (2.15) in that point. Note that we may restrict ourselves to consider maximizers in (2.30) that are generated via Proposition 2.51, which are Lipschitz continuous. Hence, they are differentiable almost everywhere on $[0,1] \times X$.

Lemma 2.56 *If $\psi : [0,1] \times X \to \mathbb{R}$ is differentiable in $(s,x) \in (0,1) \times X$ and satisfies $\psi(t,y) - \psi(s,x) \leq \frac{|x-y|^2}{2(t-s)}$ for all $t \in (s,1]$ and $y \in X$, then $\partial_t \psi(s,x) + \frac{1}{2}|\nabla \psi(s,x)|^2 \leq 0$.*

Proof We set $t = s + \tau$, $\tau > 0$ and $y = x + \delta$ and approximate the finite difference in the inequality $\psi(t,y) - \psi(s,x) \leq \frac{|x-y|^2}{2(t-s)}$ by a linear expansion to obtain

$$\partial_t \psi(s,x) \cdot \tau + \nabla \psi(s,x) \cdot \delta \leq \frac{|\delta|^2}{2\tau} + o(\tau) + o(|\delta|).$$

We now choose $\delta = \tau \cdot \nabla \psi(s,x)$ and divide by τ to obtain

$$\partial_t \psi(s,x) + |\nabla \psi(s,x)|^2 \leq \frac{|\nabla \psi(s,x)|^2}{2} + o(1)$$

where $o(1)$ is to be understood with respect to the limit $\tau \to 0$ and the claim follows by passing to this limit. $\square$

3 Entropic Optimal Transport

3.1 Entropic Kantorovich Problem

In these notes we will work with the following notion of entropy.

Definition 3.1 (Entropy) For measures $\mu, \lambda \in \mathcal{M}(Z)$ on a compact metric space Z we set the relative entropy of μ with respect to λ as

$$H(\gamma|\lambda) := \begin{cases} \int_X \varphi\left(\frac{d\gamma}{d\lambda}\right) d\lambda & \text{if } \gamma, \lambda \geq 0, \gamma \ll \lambda, \\ +\infty & \text{else,} \end{cases} \quad \text{with } \varphi: \mathbb{R}_+ \to \mathbb{R}_+, \quad s \mapsto s\log(s)-s. \tag{3.1}$$

There are slight variations in the literature about the choice of the integrand φ in (3.1). The common choice $\varphi(s) = s\log(s)$ corresponds to the negative Shannon entropy, another common choice is $\varphi(s) = s\log(s) - s + 1$, which yields the Kullback–Leibler divergence. From the latter we drop the $+1$ to remove a number of constant terms in various dual problems (but which can easily be added back), but keep the $-s$ as it will yield more convenient primal-dual relations. As long as λ has finite mass (as assumed above), $H(\cdot|\lambda)$ is bounded form below.

Definition 3.2 (Entropic Kantorovich Problem) Let (X, d) be a compact metric space, $c \in C(X \times X)$, and $\mu, \nu \in \mathcal{P}(X)$. Let $\lambda_1, \lambda_2 \in \mathcal{M}_+(X)$ such that $H(\mu|\lambda_1) < \infty$, $H(\nu|\lambda_2) < \infty$ and set $\lambda := \lambda_1 \otimes \lambda_2$. Let $\varepsilon > 0$. Then the entropic optimal transport problem is given by

$$C_\varepsilon(\mu, \nu) := \inf\left\{\int_{X \times X} c(x, y)\, d\gamma(x, y) + \varepsilon\, H(\gamma|\lambda)\,\Big|\, \gamma \in \Gamma(\mu, \nu)\right\}. \tag{3.2}$$

One striking advantage of the regularized problem (3.2) compared to (2.1) is that the former can be solved by the celebrated Sinkhorn algorithm. We refer to [43] for an introduction, some historical context, and an overview on variants and modifications. Other advantages and consequences of regularization are briefly discussed in Remark 3.11.

Remark 3.3 (Choice of Reference Measure) There are various common choices for the reference measure λ in (3.2). The two most common choices for λ are $(\lambda_1, \lambda_2) = (\mu, \nu)$ and $\lambda_1 = \lambda_2 = \mathcal{L} \llcorner X$ in which case $X \subset \mathbb{R}^d$ and the marginals must satisfy $\mu, \nu \ll \mathcal{L}$. For finite X, the counting measure is sometimes used for simplicity.

Proposition 3.4 *Equation* (3.2) *has a unique minimizer.*

Proof By assumption $\gamma = \mu \otimes \nu$ satisfies $H(\gamma|\lambda) < \infty$ and $\int c \, d\gamma < \infty$ and the objective is bounded from below, thus the infimum is finite. The objective is weak* lower-semicontinuous and the admissible set is weak* compact, thus a minimizer exists. It is unique by strict convexity of $H(\cdot|\lambda)$. □

Existence of minimizers in more general settings can be found, for instance, in [24, 47], see also [36]. To proceed, we need a better characterization of the minimizers. This can be obtained via duality. We will split this into several steps. First establishing the general form of the dual problem, then existence of dual maximizers (under the assumption $\lambda = \mu \otimes \nu$) and the primal-dual relation, and finally a relaxation of the dual problem.

Proposition 3.5 (Duality for Entropic Transport, Part I)

$$C_\varepsilon(\mu, \nu) = \sup \left\{ \int_X \phi \, d\mu + \int_X \psi \, d\nu - \varepsilon \int_{X \times X} \exp([\phi \oplus \psi - c]/\varepsilon) \, d\lambda \bigg| \phi, \psi \in C(X) \right\} \quad (3.3)$$

Proof The proof is analogous to that of Proposition 2.4. We choose U, V, F and A in the same way but now set

$$G : C(X \times X) \to \mathbb{R}, \qquad \eta \mapsto \varepsilon \int_{X \times X} \exp([\eta - c]/\varepsilon) \, d\lambda$$

Now both F and G are finite and continuous everywhere, thus Theorem 2.3 can be applied. Observe that for φ in (3.1) we have $\varphi^*(s) = \exp(s)$ and therefore we obtain that $G^*(\gamma) = H(\gamma|\lambda)$ [45]. □

Proposition 3.6 (Duality for Entropic Transport, Part II) *If* $(\lambda_1, \lambda_2) = (\mu, \nu)$ *then maximizers in* (3.3) *exist. The unique minimizer of* (3.2) *takes the form* $\gamma = \exp([\phi \oplus \psi - c]/\varepsilon) \cdot \mu \otimes \nu$ *where* (ϕ, ψ) *are (arbitrary) maximizers of* (3.3).

The proof is somewhat analogous to that of Proposition 2.6 and involves an entropic version of the c-transform (Definition 2.5), obtained by a formal pointwise maximization of (3.3) with respect to ϕ for fixed ψ (and vice versa). This yields the following definition.

Definition 3.7 (Entropic c-Transform) For a cost function $c \in C(X \times X)$, a positive parameter $\varepsilon > 0$ and a potential $\psi \in C(X)$ the entropic c-transform of ψ is the function $\psi^{c,\varepsilon}$ given by

$$\psi^{c,\varepsilon} : x \mapsto -\varepsilon \log \left(\int_X \exp\left(\frac{\psi - c(x, \cdot)}{\varepsilon}\right) d\nu \right)$$

and again in analogy the entropic $\bar{c}$-transform as

$$\psi^{\bar{c},\varepsilon} : y \mapsto -\varepsilon \log \left(\int_X \exp\left(\tfrac{\psi - c(\cdot,y)}{\varepsilon} \right) d\mu \right)$$

Proof of Proposition 3.6 Similar as in the proof of Proposition 2.6, for given $\psi \in C(X)$, the function $\psi^{c,\varepsilon}$ inherits the modulus of continuity of c (in particular it is in $C(X)$ and therefore admissible) and the choice $\phi = \psi^{c,\varepsilon}$ maximizes (3.3) for fixed ψ. We may therefore restrict ourselves to maximizing sequences of the form $(\phi_n = \psi_n^{c,\varepsilon}, (\psi_n^{c,\varepsilon})^{\bar{c},\varepsilon})$, which are equicontinuous and by the same argument about constant shifts also equibounded. Once more we can extract a cluster point via the Arzelà–Ascoli theorem, which must be maximizer by continuity of the objective (3.3) (which is now unconstrained).

Let now (ϕ, ψ) be a maximizer of (3.3) and let γ be the unique minimizer in (3.2) (Proposition 3.4). By duality the primal-dual gap must vanish and since $H(\gamma|\mu \otimes \nu) < \infty$, γ must be of the form $\gamma = u \cdot \mu \otimes \nu$ with this we find

$$0 = \int c \, d\gamma + \varepsilon H(\gamma|\mu \otimes \nu) - \int \phi \, d\mu - \psi \, d\nu + \varepsilon \int \exp([\phi \oplus \psi - c]/\varepsilon) d\mu \otimes \nu$$

$$= \int (\varepsilon \varphi(u) + \varepsilon \exp([\phi \oplus \psi - c]/\varepsilon) - [\phi \oplus \psi - c] \cdot u) \, d\mu \otimes \nu$$

Recalling that $\varphi^* = \exp$ and using the Fenchel–Young inequality [6] one finds that the integrand is non-negative and zero if and only if $u = \exp([\phi \oplus \psi - c]/\varepsilon)$ $\mu \otimes \nu$-almost everywhere. □

Proposition 3.8 (Duality for Entropic Transport, Part III) *The admissible spaces in (3.3) can be relaxed to $(\phi, \psi) \in L^1(\mu, [-\infty, \infty)) \times L^1(\nu, [-\infty, \infty))$ without increasing the value of the supremum and maximizers exist in this space. The unique minimizer of (3.2) takes the form $\gamma = \exp([\phi \oplus \psi - c]/\varepsilon) \cdot \lambda$ where (ϕ, ψ) are (arbitrary) maximizers of (3.3) in the relaxed space and we use the convention $\exp(-\infty) = 0$.*

Proof Note that the dual objective (3.3) is well-defined with values in $[-\infty, \infty)$ on the relaxed space, since the first two terms are finite and the third term is bounded from above (as exp is bounded from below) and can possibly take the value $-\infty$. The primal dual gap between (3.2) and (3.3) is given by (see also proof of Proposition 3.6)

$$0 \leq \int c \, d\gamma + \varepsilon H(\gamma|\lambda) - \int \phi \, d\mu - \psi \, d\nu + \varepsilon \int \exp([\phi \oplus \psi - c]/\varepsilon) d\lambda.$$

Again, by the Fenchel–Young inequality this inequality also holds on the relaxed space, so the supremum is not increased by the relaxation of the admissible space. By assumptions in Definition 3.2 we have $H(\mu|\lambda_1) < \infty$, $H(\nu|\lambda_2) < \infty$ and therefore

$$H(\gamma|\lambda) = H(\gamma|\mu \otimes \nu) - H(\mu|\lambda_1) - H(\nu|\lambda_2).$$

This means that we can reduce the general primal problem (3.2) to the special case of Proposition 3.6 where we assumed $\lambda = \mu \otimes \nu$. In particular both problems have the same minimizer γ. Denote $\mu_\lambda = \frac{d\mu}{d\lambda_1}$ and $\nu_\lambda = \frac{d\nu}{d\lambda_2}$ (these densities must exist due to the finite entropy assumption). By Proposition 3.6 we find that $\gamma = \exp([\phi \oplus \psi - c]/\varepsilon) \cdot \mu \otimes \nu$, which can be written as $\gamma = \exp([\hat\phi \oplus \hat\psi - c]/\varepsilon) \cdot \lambda$ for $\hat\phi = \phi + \varepsilon \log(\mu_\lambda)$ and $\hat\psi = \psi + \varepsilon \log(\nu_\lambda)$ with the convention $\log(0) = -\infty$. Evaluating the primal-dual gap for γ and $(\hat\phi, \hat\psi)$ yields that the latter are dual maximizers. □

Remark 3.9 (Absorption of Cost Function into Reference Measure) It is easy to verify that in (3.2) one has

$$\int_{X \times X} c(x,y) \, d\gamma(x,y) + \varepsilon H(\gamma|\lambda) = \varepsilon H(\gamma|\exp(-c/\varepsilon) \cdot \lambda),$$

i.e. we can interpret (3.2) as the problem of finding the generalized projection of the measure $\exp(-c/\varepsilon) \cdot \lambda$ onto the set $\Gamma(\mu, \nu)$ with respect to the divergence function H.

For X being a subset of $\mathbb{R}^d$, $c(x,y) := \frac{1}{2}|x-y|^2$, and $\lambda = \mathcal{L} \otimes \mathcal{L}$ (restricted to $X \times X$) one has that $\exp(-c/\varepsilon) \cdot \lambda$ is (proportional to) a Gaussian kernel with isotropic variance ε along each direction. Re-scaling such a kernel by a factor C corresponds to subtracting a constant $\varepsilon \log(C)$ from the cost function c and therefore will not change the minimizer of the corresponding entropic transport problem (3.2) but merely shift the objective value by $-\varepsilon \log(C)$.

Remark 3.10 (Convergence to Unregularized Optimal Transport as $\varepsilon \to 0$) Clearly (3.2) can be seen as a regularized variant of (2.1) and thus the question of convergence of the former to the latter as $\varepsilon \to 0$ is of interest, in particular convergence of minimizers, minimal value, and likewise for the corresponding dual problems. A good starting point for the literature on this topic are [17, 35] and the references therein.

A simple explicit way of showing Γ-convergence of (3.2) to (2.1) is the *block approximation* introduced in [16] which can easily be adapted to the above setting.

Remark 3.11 (Entropic Bias, Sinkhorn Divergence, and Loss of Metric Structure) When comparing arbitrary probability measures in $\mathcal{P}(X)$ via (3.2) the most natural choice as reference measure appears to be $\lambda = \mu \otimes \nu$ as in Proposition 3.6, since no fixed choice for λ_1, λ_2 will satisfy the assumptions of Definition 3.2 for all μ, ν.

One drawback of this adaptive choice $\lambda = \mu \otimes \nu$ is that the function C_ε becomes non-convex, due to the added dependency of λ on the arguments (it remains convex separately in each of the two arguments, however).

An advantage is that the optimal dual potentials become more regular (since they can be chosen to be entropic c-transforms of each other, see proof of Proposition 3.6) and thus allow for more robust statistical estimation of $C_\varepsilon(\mu, \nu)$ when only empirical approximations $\hat{\mu}$ and $\hat{\nu}$ of the two measures are available, see for instance [28, 40], and ultimately also estimation of $C(\mu, \nu)$ [23].

An issue of (3.2) is the loss of the metric structure as in (2.6). Both for fixed λ and for $\lambda = \mu \otimes \nu$ one will not have in general that $\mu \in \mathrm{argmin}_{\nu \in \mathcal{P}(X)} C_\varepsilon(\mu, \nu)$. This bias can be removed by using the *Sinkhorn divergence* [26] instead, which is given as

$$S_\varepsilon(\mu, \nu) := C_\varepsilon(\mu, \nu) - \tfrac{1}{2} C_\varepsilon(\mu, \mu) - \tfrac{1}{2} C_\varepsilon(\nu, \nu)$$

where in each of the three terms on the right-hand side λ is chosen as the product of the input measures. Under suitable assumptions on c one can show that $S_\varepsilon(\mu, \nu) \geq 0$ with equality if and only if $\mu = \nu$ and that $\lim_n S_\varepsilon(\mu_n, \mu) = 0$ is equivalent to $(\mu_n)_n$ converging weak* to μ (i.e. S_ε 'metrizes' weak* convergence). However, S_ε is also no longer jointly convex and it does not satisfy the triangle inequality [33, Section 7]. A recipe to construct a metric on $\mathcal{P}(X)$ from S_ε is given in [33].

Despite this loss of metric structure, (3.2) with cost $c = \tfrac{1}{2} d^2$ on $X \subset \mathbb{R}^d$ is associated with an exciting dynamic perspective related to drift-diffusion equations, which we will explore in the following sections.

3.2 Interlude: Diffusion and Schrödinger Bridges

3.2.1 Gaussian Kernels and the Diffusion Equation

In this section we collect some basic definitions and properties of Gaussian kernels and the diffusion equation, which will be fundamental later on.

Definition 3.12 (Gaussian Kernels) For a constant $\varepsilon > 0$ we introduce the Gaussian kernel with isotropic variance ε along each direction, given by

$$K_\varepsilon : \mathbb{R}^d \times \mathbb{R}^d \to \mathbb{R}, \quad (x, y) \mapsto \mathcal{N}_\varepsilon \cdot \exp\left(-\frac{|x-y|^2}{2\varepsilon}\right) \quad \text{with } \mathcal{N}_\varepsilon := (2\pi\varepsilon)^{-d/2} \tag{3.4}$$

where the normalization constant $\mathcal{N}_\varepsilon$ is chosen such that

$$\int_{\mathbb{R}^d} K_\varepsilon(x, y) \mathrm{d}y = 1,$$

i.e. $K_\varepsilon(x, \cdot)$ can be interpreted as probability density of a normal random variable with mean x and isotropic variance ε.

Lemma 3.13 (Convolution of Gaussian Kernels) *For two constants $\varepsilon_1, \varepsilon_2 > 0$ one has*

$$\int_{\mathbb{R}^d} K_{\varepsilon_1}(x, y) \, K_{\varepsilon_2}(y, z) \, dy = K_{\varepsilon_1 + \varepsilon_2}(x, z).$$

This can be verified by a simple (but somewhat tedious) explicit computation. The identity can also be interpreted from the perspective of adding independent normal random variables: Let x_1 and x_2 be independent normal $\mathbb{R}^d$-valued random variables with isotropic covariance matrices $\varepsilon_1 \cdot \mathrm{id}$ and $\varepsilon_2 \cdot \mathrm{id}$, respectively. Then $x_1 + x_2$ is a normal $\mathbb{R}^d$-valued random variable with isotropic covariance matrix $(\varepsilon_1 + \varepsilon_2) \cdot \mathrm{id}$.

Lemma 3.14 (Diffusion Equation or Heat Equation) *For $u_0 \in L^1(\mathbb{R}^d)$, $t > 0$ and $x \in \mathbb{R}^d$ let*

$$u(t, x) := \int_{\mathbb{R}} u_0(y) \, K_{\varepsilon t}(x, y) \, dy.$$

Then u is infinitely often continuously differentiable with respect to both variables, with derivatives bounded on $(\delta, \infty] \times \mathbb{R}^d$ for any $\delta > 0$ and it solves the diffusion or heat equation

$$\partial_t u(t, x) = \frac{\varepsilon}{2} \Delta u(t, x)$$

for all $(t, x) \in \mathbb{R}_{++} \times \mathbb{R}^d$ with the boundary condition at $t = 0$ in the sense that $\|u(0, \cdot) - u(t, \cdot)\|_{L^1(\mathbb{R}^d)} \to 0$. For times $0 < s < t$ one finds with Lemma 3.13 that

$$u(t, x) = \int_{\mathbb{R}} u(s, y) \, K_{\varepsilon(t-s)}(x, y) \, dy.$$

For an introductory treatment of this equation see [25].

Remark 3.15 (Stochastic Interpretation) Let the stochastic process $(x_t)_{t \geq 0}$ describe the trajectory of a particle moving in $\mathbb{R}^d$, subjected to Brownian motion with strength ε. That is, x_t is a solution to the stochastic partial differential equation

$$dx_t = \sqrt{\varepsilon} \, dW_t$$

where W_t denotes a standard Wiener process. Let the probability density for the law of x_0 be given by u_0. Then $u(t, \cdot)$ will be the distribution of x_t at time $t > 0$, since the density of the distribution of x_t, conditioned on $x_0 = x_0$ is given by $K_{\varepsilon t}(x_0, \cdot)$. $u_0(x_0) \cdot K_{\varepsilon t}(x_0, x_t)$ will be the joint density of first observing x_0 in x_0 and then x_t at x_t, and more generally $u_0(x_0) \prod_{i=1}^{N} K_{\varepsilon(t_i - t_{i-1})}(x_{i-1}, x_i)$ will be the joint density of observing x_{t_i} in x_i for an increasing tuple $t_0 = 0 < t_1 < \ldots < t_N$. Note that this joint distribution is Markov in the sense that x_{t_i} and x_{t_k} are independent when conditioned on x_{t_j} for some triple $i < j < k$.

The following definition is the equivalent of Definition 2.43 for the entropic setting.

Definition 3.16 (Multi-Marginal Gaussian Kernel) For a natural number $N \geq 2$ and time points $t_0 = 0 < t_1 < t_2 < \ldots < t_N = 1$, the multi-marginal Gaussian kernel is given by

$$K_{\mathrm{MM}} : (\mathbb{R}^d)^{N+1} \mapsto \mathbb{R},$$

$$(x_0, \ldots, x_N) \mapsto \prod_{i=0}^{N-1} K_{\varepsilon(t_{i+1}-t_i)}(x_i, x_{i+1}) = \prod_{i=0}^{N-1} \mathcal{N}_{\varepsilon(t_{i+1}-t_i)} \exp\left(-\frac{|x_{i+1}-x_i|^2}{2\varepsilon(t_{i+1}-t_i)}\right)$$
(3.5)

with normalization factors $\mathcal{N}_{\varepsilon(t_{i+1}-t_i)}$ as defined in (3.4).

And the following is the corresponding equivalent of Lemma 2.45, which is an immediate consequence of Lemma 3.13.

Lemma 3.17 *For K_{MM} as in Definition 3.16 one has*

$$K_\varepsilon(x_0, x_N) = \int_{(\mathbb{R}^d)^{N-1}} K_{\mathrm{MM}}(x_0, \ldots, x_N) dx_1 \ldots dx_{N-1}.$$

3.2.2 Schrödinger Bridges

In this section we briefly sketch a thought experiment proposed by Erwin Schrödinger in 1931 which provides an insightful interpretation of the entropic optimal transport problem (3.2) and helps to discern its dynamic structure. A translation of the original paper with historical context is given in [20], see also [36] and [18] and references therein for more literature on this problem.

Consider once more a stochastic process $(x_t)_{t \geq 0}$, describing a particle subjected to Brownian motion of strength ε, as in Remark 3.15. Assume its initial position x_0 was fixed to x_0. Then for any $t > 0$, the law of x_t will have the marginal probability density $K_{\varepsilon t}(x_0, \cdot)$. Assume now that we 'observe' the particle at $t = 1$ in position x_1. Based on this knowledge, what do we know about the likely positions of the particle at intermediate times $t \in (0, 1)$? Hypothetically, we could consider a setup where we initialize the experiment many times and observe the position of the particle at time $t = 1$ and discard all instances where the particle is not within an infinitesimally small environment of x_1. A second observer will determine the position of the particle at a fixed agreed-upon intermediate time $t \in (0, 1)$ and only keep those measurements for which we later reported that the particle ended up in the aforementioned small environment. What will be the distribution of the non-discarded observed intermediate positions? What is the law of x_t conditioned on $x_0 = x_0$ and $x_1 = x_1$? Such a process is called a *Brownian bridge* [5].

Using that the density for $x_1 = x_1$ is given by $K_\varepsilon(x_0, x_1)$ and the joint density for $(x_t = x_t, x_1 = x_1)$ is given by $K_{\varepsilon t}(x_0, x_t) \cdot K_{\varepsilon(1-t)}(x_t, x_1)$ (Remark 3.15) a tedious but simple computation yields that the conditional density for $x_t = x_t$ conditioned on $x_0 = x_0$, $x_1 = x_1$ is given by

$$
\begin{aligned}
u(x_t | x_0 = x_0, x_1 = x_1) &= \frac{K_{\varepsilon t}(x_0, x_t) \cdot K_{\varepsilon(1-t)}(x_t, x_1)}{K_\varepsilon(x_0, x_1)} \\
&= \frac{\mathcal{N}_{\varepsilon t} \cdot \mathcal{N}_{\varepsilon(1-t)}}{\mathcal{N}_\varepsilon} \exp\left(-\frac{|x_0 - x_t|^2}{2\varepsilon t} - \frac{|x_t - x_1|^2}{2\varepsilon(1-t)} + \frac{|x_0 - x_1|^2}{2\varepsilon} \right) \\
&= \mathcal{N}_{\varepsilon t(1-t)} \exp\left(-\frac{|x_t - [(1-t)x_0 + tx_1]|^2}{2\varepsilon t(1-t)} \right) \qquad (3.6) \\
&= K_{\varepsilon t(1-t)}((1-t)x_0 + tx_1, x_t) \qquad (3.7)
\end{aligned}
$$

So we find that the conditional x_t is also a normal random variable with mean $(1-t) \cdot x_0 + t \cdot x_1$ and covariance $\varepsilon t (1-t) \cdot \mathrm{id}$.

Schrödinger considered a generalization of this problem: Assume now that the initial and final positions are no longer fixed, but instead the law of x_0 is prescribed to be given by some measure $\mu \in \mathcal{P}(\mathbb{R}^d)$ and the law of x_1 is prescribed to be given by some $\nu \in \mathcal{P}(\mathbb{R}^d)$. What will be the law of the intermediate time position x_t? This conditional process has been dubbed the *Schrödinger bridge* between μ and ν [18] and it is closely related to the entropic optimal transport problem (3.2). We will examine this question in the following by intuitive considerations on discrete spaces. A more thorough but still rather accessible approach can be found in [31, Section 6].

In this thought experiment we face the difficulty of how to 'condition' x_1 on a particular law ν instead of a fixed position. This can be resolved by the following intuitive arguments. We introduce $M \in \mathbb{N}$ independent and identically distributed copies of the process x_t, denoted as $(x_t^i)_{t \geq 0, i \in \{1, \ldots, M\}}$. We then conduct the experiment many times where we observe the initial and final positions of all particles at times $t = 0$ and $t = 1$ and the second observer records the positions at the fixed intermediate time $t \in (0, 1)$. Afterwards we consider for each run of the experiment the empirical measures

$$
\mu^M := \frac{1}{M} \sum_{i=1}^{M} \delta_{x_0^i} \qquad \text{and} \qquad \nu^M := \frac{1}{M} \sum_{i=1}^{M} \delta_{x_1^i}
$$

and only keep those realizations where μ^M and ν^M are 'close' in some suitable sense to the prescribed μ and ν. For instance, we could partition $\mathbb{R}^d$ into cells and compare the masses within the cells. What will be the distribution of empirical measures at intermediate times,

$$\rho_t^M := \tfrac{1}{M} \sum_{i=1}^{M} \delta_{x_t^i},$$

that the second observer reports?

Concretely, let now $X := \{x_1, \ldots, x_L\}$ be a finite space (e.g. a finite subset of $\mathbb{R}^d$) and let $\mu := \sum_{l=1}^{L} \mu_l \cdot \delta_{x_l}$ be a probability measure on X with mass weights given by coefficients μ_l, i.e. by a slight abuse of notation we identify μ with a vector in $\mathbb{R}^L$. Assume that the probability for observing $x_0 = x_l$ equals μ_l (e.g. by 'truncating' the above random variable x_0 from $\mathbb{R}^d$ to X). Then we are interested in the law of the empirical random measure

$$\boldsymbol{\mu}^M := \tfrac{1}{M} \sum_{i=1}^{M} \delta_{x_0^i} = \sum_{l=1}^{L} \tfrac{m_l}{M} \delta_{x_l} = \sum_{l=1}^{L} \boldsymbol{\mu}_l^M \delta_{x_l}.$$

Here $(\boldsymbol{m}_l)_{l=1}^{L}$ are random variables defined by counting how many of the x_0^i are equal to each x_l and $\boldsymbol{\mu}_l^M$ are the random empirical mass weights. Clearly the $(\boldsymbol{m}_l)_{l=1}^{L}$ follow a multinomial distribution with probability

$$\mathbb{P}(\boldsymbol{m}_1 = m_1, \ldots, \boldsymbol{m}_L = m_L) = \frac{M!}{\prod_{l=1}^{L} m_l!} \prod_{l=1}^{L} \mu_l^{m_l}.$$

Consider now the regime of very large M. Using Stirling's approximation $\log(n!) \approx n \log(n) - n$ one finds

$$-\log(\mathbb{P}(\boldsymbol{m}_1 = m_1, \ldots, \boldsymbol{m}_L = m_L)) \approx M \sum_{l=1}^{L} (m_l/M) \log((m_l/M)/\mu_l)$$

$$= M \cdot \mathrm{KL}((m_l/M)_l | \mu).$$

This simple computation underlines the intimate connection of entropy to random sampling. In conclusion, for very large M, the empirical discrete measure $\boldsymbol{\mu}^M$ will be very close to μ with high probability.

Now we expand this to the product space. Let $K = \sum_{i,j=1}^{L} K_{i,j} \delta_{(x_i,x_j)} \in \mathcal{P}(X \times X)$ be the discrete joint law of (x_0, x_1) with mass coefficients $K_{i,j}$, i.e. similar to above we identify K with a matrix in $\mathbb{R}^{L \times L}$. Again, let $\boldsymbol{m}_{j,l}$ be the random variable that counts how often we observe $(x_0^i, x_1^i) = (x_j, x_l)$ for $j, l \in \{1, \ldots, L\}$. Let $\boldsymbol{k}_{j,l}^M := \boldsymbol{m}_{j,l}/M$ be the discrete empirical weights and we identify the weight matrix $\boldsymbol{k}^M$ with the corresponding empirical measure on $\mathcal{P}(X \times X)$. Then for large M again

$$-\log(\mathbb{P}(\boldsymbol{m}_{j,l} = m_{j,l} \text{ for } j, l \in \{1, \ldots, L\})) \approx M \cdot \mathrm{KL}((m_{j,l}/M)_{j,l} | K).$$

Now we conduct this experiment many times. With high probability in most instances $\boldsymbol{k}^M$ will be close to K. However, we now choose to only retain instances where the marginals of $\boldsymbol{k}^M$ are equal to (or very close to) the prescribed measures μ and ν, i.e. we condition on $\boldsymbol{k}^M \in \Gamma(\mu, \nu)$. Then, by the above intuitive arguments, with high probability $\boldsymbol{k}^M$ will be close to the minimizer

$$\operatorname{argmin}\left\{\mathrm{KL}(\gamma|K) \middle| \gamma \in \Gamma(\mu, \nu)\right\}. \tag{3.8}$$

Once this minimizer γ has been determined, which describes the joint law of $(\boldsymbol{x}_0, \boldsymbol{x}_1)$ 'conditioned' on the marginals μ and ν in the above sense, then the law of $\boldsymbol{x}_t$ can be constructed via (3.7). The density of $\boldsymbol{x}_t$ will be given by

$$\sum_{j,l=1}^{L} u(x_t | \boldsymbol{x}_0 = x_j, \boldsymbol{x}_1 = x_l) \, \gamma(\{(x_j, x_l)\}) \tag{3.9}$$

and hence we have established the link between Schrödinger bridges and the entropic optimal transport problem (3.2) (see also Remark 3.9). We will examine this further in Sect. 3.3.

Remark 3.18 (Lazy Gas Experiment) In the above thought experiment we have intuitively conditioned a collection of many particles undergoing Brownian motion on a specific marginal distribution ν at time $t = 1$. The expectation for the random measure $\frac{1}{M} \sum_{i=1}^{M} \delta_{x_1^i}$ is given by the evolution of the initial distribution μ under the diffusion equation (Remark 3.15). For large M, deviations from this prediction become increasingly unlikely. When we decrease the amplitude of the Brownian motion by sending the parameter ε to 0, then the expected distribution at $t = 1$ will converge to μ. For small $\varepsilon > 0$, particles will barely move at all. If we still insist on conditioning on some 'exotic' law ν, then the mass particles will try to reach this configuration by moving as little as possible (or at least barely more). This means that the particles will move approximately along an unregularized Wasserstein geodesic. This thought experiment has been dubbed the *lazy gas experiment* in [51, Chapter 16]. The limit $\varepsilon \to 0$ is discussed in [36, Section 6.2], see also [35]. In [51, Chapter 16] the limit case $\varepsilon = 0$ is considered directly and it is discussed how it can reveal information on the curvature of the base space X (in cases where X is a Riemannian manifold).

3.3 Entropic Multi-Marginal Formulation

Remark 3.19 (Lebesgue Densities and Measures) Throughout this section we will only consider measures that are absolutely continuous with respect to the Lebesgue measure $\mathcal{L}$ on $\mathbb{R}^d$, or its restriction to X which we denote by $\mathcal{L} \llcorner X$, or with respect to the product of the Lebesgue measure on spaces like $(\mathbb{R}^d)^{N+1}$ et cetera. We denote their Lebesgue densities by the same symbol as the measure itself. Conversely, we will also use the symbols K_ε and K_{MM} introduced as Gaussian densities in Sect. 3.2.1 to refer to the corresponding measures. We will view K_ε as a measure restricted to $X \times X$ and K_{MM} as a measure restricted $X \times (\mathbb{R}^d)^{N-1} \times X$ such that both are finite.

Based on the discussion in Sect. 3.2.2 we now introduce the following definition.

Definition 3.20 (Static Schrödinger Bridge Problem)

$$\min \left\{ \varepsilon\, H(\gamma|K_\varepsilon) \Big| \gamma \in \Gamma(\mu, \nu) \right\} \tag{3.10}$$

Taking into account Proposition 3.8 and Remark 3.9 we obtain that (3.10) has a unique minimizer of the form $\gamma = \exp(\phi \oplus \psi/\varepsilon) \cdot K_\varepsilon$ with (ϕ, ψ) being maximizers of the following dual problem

$$\max \left\{ \int_X \phi \mathrm{d}\mu + \int_X \psi \mathrm{d}\nu - \varepsilon \int_{X \times X} \exp\left(\frac{\phi \oplus \psi}{\varepsilon}\right) \mathrm{d}K_\varepsilon \Big| \phi \in \mathrm{L}^1(\mu), \psi \times \mathrm{L}^1(\nu) \right\}. \tag{3.11}$$

In analogy to (2.19), inspired by Sect. 3.2.2, to study the distribution of particles at intermediate times, we introduce the following multi-marginal problems.

Proposition 3.21 (Multi-Marginal Schrödinger Bridge Problem) *Problems* (3.10) *and* (3.11) *are equivalent to problems*

$$\min \left\{ \varepsilon\, H(\gamma_{\mathrm{MM}}|K_{\mathrm{MM}}) \Big| \gamma_{\mathrm{MM}} \in \mathcal{P}(X \times (\mathbb{R}^d)^{N-1} \times X), P_0 \gamma_{\mathrm{MM}} = \mu, P_N \gamma_{\mathrm{MM}} = \nu \right\} \tag{3.12}$$

and

$$\max \left\{ \int_X \phi \mathrm{d}\mu + \int_X \psi \mathrm{d}\nu - \varepsilon \int_{X \times (\mathbb{R}^d)^{N-1} \times X} \exp\left(\frac{\phi(x_0) + \psi(x_N)}{\varepsilon}\right) \mathrm{d}K_{\mathrm{MM}}(x_0, \ldots, x_N) \Big| \right.$$
$$\left. \phi \in \mathrm{L}^1(\mu), \psi \in \mathrm{L}^1(\nu) \right\} \tag{3.13}$$

respectively in the following sense: their optimal values are identical, (3.11) and (3.13) have the same maximizers, and minimizers for (3.12) are given by $\gamma_{MM} = \exp(\phi \oplus \psi/\varepsilon) \cdot K_{MM}$ where (ϕ, ψ) are maximal in (3.11) or (3.13), in analogy to the minimizers for (3.10).

Proof Equivalence between (3.11) and (3.13) is immediate from Lemma 3.17. Equation (3.13) $\leq$ (3.12) follows by showing that the corresponding primal-dual gap is non-negative using the Fenchel–Young inequality, in the exact way as done, for instance in the proof of Proposition 3.8. Plugging $\gamma_{MM} = \exp(\phi \oplus \psi/\varepsilon) \cdot K_{MM}$ into (3.12) we find that the primal-dual gap vanishes, establishing minimality of γ_{MM}. Uniqueness follows from strict convexity of $H(\cdot|K_{MM})$.

Instead of showing minimality of γ_{MM} via duality, we can also use an explicit primal argument to obtain the inequality (3.12) $\geq$ (3.10) as follows: Let γ_{MM} be admissible in (3.12) with finite objective, i.e. it will have the form $\gamma_{MM} = u_{MM} \cdot K_{MM}$ for some (relative) density u_{MM}. Then $\gamma := P_{0,N}\gamma_{MM}$ will be admissible in (3.10) and it will have the form $\gamma = u \cdot K_\varepsilon$ with

$$u(x_0, x_N) = \int_{(\mathbb{R}^d)^{N-1}} u_{MM}(x_0, \ldots, x_N) \frac{K_{MM}(x_0, \ldots, x_N)}{K_\varepsilon(x_0, x_N)} dx_1 \ldots dx_{N-1}.$$

Using that $\frac{K_{MM}(x_0,\ldots,x_N)}{K_\varepsilon(x_0,x_N)}$ is a probability density on $(\mathbb{R}^d)^{N-1}$ and that φ (the integrand in (3.1)) is convex one obtains that $H(\gamma|K_\varepsilon) \leq H(\gamma_{MM}|K_{MM})$ through Jensen's inequality. $\square$

Given a minimizer γ_{MM} of (3.12) one can evaluate the marginal at intermediate times to obtain a notion of how the measure μ is gradually transformed into ν. Choosing for simplicity $N = 2$ and $t_0 = 0, t_1 = t, t_2 = 1$ one obtains for (the Lebesgue density of) the marginal at time t, which we denote by $\rho(t, \cdot)$ that

$$\rho(t, y) = \int_{X \times X} \gamma_{MM}(x, y, z) \, dx \, dz$$

$$= \int_{X \times X} \exp(\phi(x)/\varepsilon) \exp(\psi(z)/\varepsilon) K_{MM}(x, y, z) \, dx \, dz$$

$$= \int_{X \times X} \exp(\phi(x)/\varepsilon) \exp(\psi(z)/\varepsilon) K_{\varepsilon t}(x, y) \, K_{\varepsilon(1-t)}(y, z) \, dx \, dz \tag{3.14}$$

$$= \int_{X \times X} \frac{K_{\varepsilon t}(x, y) \, K_{\varepsilon(1-t)}(y, z)}{K_\varepsilon(x, z)} \, d\gamma(x, z)$$

$$= \int_{X \times X} K_{\varepsilon t(1-t)}(t \cdot x + (1-t) \cdot z, y) \, d\gamma(x, z) \tag{3.15}$$

for $y \in \mathbb{R}^d$ and γ being a corresponding minimizer of (3.10). Note that (3.15) is the same formula as (3.9). For $N > 2$ and any given intermediate marginal $i \in \{1, \ldots, N-1\}$ one obtains a similar formula by integrating over all other intermediate axes $j \neq i$.

Remark 3.22 (Applications of (3.12) and Limit $N \to \infty$) In analogy to Remark 2.46, formulation (3.12) provides a way to interpret how the measure μ is gradually transformed into the measure ν by looking at marginals of γ_{MM} at intermediate times as in (3.14). These are the time-marginals of the Schrödinger bridge between μ and ν (cf. Sect. 3.2.2). As before, formulation (3.12) has the advantage over (3.10) that one can include interactions of the measure with its environment at intermediate times, such as in [30, 34].

Analogous to Remark 2.49 one may also consider again the limit $N \to \infty$ such that γ_{MM} becomes a measure on paths. K_{MM} will then essentially turn into the Wiener measure (but without fixing the initial point). We refer to [36] and references therein for more information on this formulation.

Remark 3.23 (Markov Property) From (3.14) we can observe that the minimizing γ_{MM} is Markov in the same sense as in the unregularized setting in Remark 2.48. Therefore, it is again possible to restrict to such Markov densities in (3.12). This property remains preserved, when the objective is augmented by additional terms that interact with the intermediate densities and plays a key role for numerical tractability [7, 11, 30, 34].

Equation (3.14) motivates the following definitions, which are the entropic equivalent of the Hopf–Lax formulas (2.24).

Definition 3.24 Let (ϕ, ψ) be maximizers for (3.11) or equivalently (3.13). Then we introduce the auxiliary functions $U, V, \Phi, \Psi : [0, 1] \times \mathbb{R}^d \to [-\infty, \infty]$ as follows

$$U(0, \cdot) := \exp(\phi/\varepsilon), \quad U(t, x) := \int_X U(0, y) \, K_{\varepsilon t}(x, y) dy \quad \text{for } t \in (0, 1], \tag{3.16}$$

$$V(1, \cdot) := \exp(\psi/\varepsilon), \quad V(t, x) := \int_X V(1, y) \, K_{\varepsilon(1-t)}(x, y) dy \quad \text{for } t \in [0, 1) \tag{3.17}$$

with the convention $U(0, x) = 0$ if $\phi(0, x) = -\infty$ (and likewise for $V(1, x)$ and $\psi(1, x)$), and

$$\Phi(t, x) := \varepsilon \log(U(t, x)), \qquad \Psi(t, x) := \varepsilon \log(V(t, x)) \tag{3.18}$$

with the convention $\Phi(0, x) = -\infty$ if $\phi(0, x) = -\infty$ for $x \in X$ or $x \in \mathbb{R}^d \setminus X$ (and analogously for $\Psi(1, \cdot)$).

Proposition 3.25 *Consider the functions U, V, and Ψ as introduced in Definition 3.24. Let $\rho : [0, 1] \times \mathbb{R}^d \to \mathbb{R}$ be given by (3.14) for $t \in (0, 1)$ and by the densities of μ and ν for $t \in \{0, 1\}$ (which we extend by 0 beyond X). Then on $[0, 1] \times \mathbb{R}^d$ one has*

$$\rho = U \cdot V, \tag{3.19a}$$

and on $(0, 1) \times \mathbb{R}^d$ one has

$$\partial_t U = \tfrac{\varepsilon}{2} \Delta U, \tag{3.19b}$$

$$\partial_t V = -\tfrac{\varepsilon}{2} \Delta V, \tag{3.19c}$$

$$\partial_t \rho = -\nabla[\nabla \Psi \cdot \rho] + \tfrac{\varepsilon}{2} \Delta \rho, \tag{3.19d}$$

$$\partial_t \Psi = -\tfrac{1}{2} |\nabla \Psi|^2 - \tfrac{\varepsilon}{2} \Delta \Psi. \tag{3.19e}$$

Proof For $t \in (0, 1)$ (3.19a) follows by plugging the definitions of U and V into (3.14). For $t \in \{0, 1\}$ it follows by plugging the definitions of U and V into the condition that minimizers γ of (3.10) lie in $\Gamma(\mu, \nu)$.

Equations (3.19b) and (3.19c) follow from Lemma 3.14, taking into account that V is convolved 'backwards' in time. Equations (3.19d) and (3.19e) follow from (3.19b) and (3.19c) and exploiting that U and V are infinitely often differentiable and strictly positive on $(0, 1) \times \mathbb{R}^d$, and using the definition of Ψ. □

We observe that the density ρ is on the one hand given as product of two solutions to the diffusion equation, one forward and one backward in time. And it is given as solution to a Fokker–Planck equation with a drift potential that satisfies a suitable adjoint equation. These two equations are remarkably similar to the system of equations appearing in Proposition 2.38 up to the additional ε-terms.

The next remark (in the spirit of [10, Lemma 3.4]) shows that the Markov property also allows to decompose the objective of (3.12) into a sum of step-wise objectives, similar to (2.22) for the unregularized case.

Remark 3.26 (Entropy for Markov Plans) Assume that a density γ_{MM} admissible in (3.12) has the form

$$\gamma_{\mathrm{MM}}(x_0, \ldots, x_N) = \mu(x_0) \prod_{i=0}^{N-1} \gamma_{i,i+1}(x_i, x_{i+1})$$

for suitable pairwise densities $\gamma_{i,i+1}$. Then

$$H(\gamma_{MM}|K_{MM}) = H(\mu|\mathcal{L} \llcorner X)$$
$$+ \sum_{i=0}^{N-1} \int_{X \times X} \log \left(\frac{\gamma_{i,i+1}(x_i, x_{i+1})}{K_{\varepsilon(t_{i+1}-t_i)}(x_i, x_{i+1})} \right) \gamma_{i,i+1}(x_i, x_{i+1}) \, dx_i \, dx_{i+1}. \quad (3.20)$$

This is easy to verify by explicit computations. When γ_{MM} is the minimizer, doing a calculation as for (3.14) but for the joint distribution of two intermediate times, one finds that one can choose (ignoring potential regularity issues at $t \in \{0, 1\}$)

$$\gamma_{i,i+1}(x_i, x_{i+1}) = \frac{V(t_{i+1}, x_{i+1})}{V(t_i, x_i)} \cdot K_{\varepsilon(t_{i+1}-t_i)}(x_i, x_{i+1}).$$

With this (3.20) reduces to

$$H(\gamma_{MM}|K_{MM}) = H(\mu|\mathcal{L} \llcorner X) + \sum_{i=0}^{N-1} \int_{X \times X} \log \left(\frac{V(t_{i+1}, x_{i+1})}{V(t_i, x_i)} \right) \gamma_{i,i+1}(x_i, x_{i+1}) \, dx_i \, dx_{i+1}.$$

$$= H(\mu|\mathcal{L} \llcorner X) + \sum_{i=0}^{N-1} \left[\int_X \log(V(t_{i+1}, x_{i+1})) \rho(t_{i+1}, x_{i+1}) \, dx_{i+1} \right.$$
$$\left. - \int_X \log(V(t_i, x_i)) \rho(t_i, x_i) \, dx_i \right]$$

$$= H(\mu|\mathcal{L} \llcorner X) + \frac{1}{\varepsilon} \sum_{i=0}^{N-1} \left[\int_X \Psi(t_{i+1}, x_{i+1}) \rho(t_{i+1}, x_{i+1}) \, dx_{i+1} \right.$$
$$\left. - \int_X \Psi(t_i, x_i) \rho(t_i, x_i) \, dx_i \right]$$

$$= H(\mu|\mathcal{L} \llcorner X) + \frac{1}{\varepsilon} \int_X \Psi(1, \cdot) \, d\nu - \frac{1}{\varepsilon} \int_X \Psi(0, \cdot) \, d\mu. \quad (3.21)$$

Using the relation $\phi(x) = -\Psi(0, \cdot) + \varepsilon \log(\mu(x))$, and $\int_{X \times X} \exp\left(\frac{\phi \oplus \psi}{\varepsilon}\right) dK_\varepsilon = 1$, this then equals (3.11), thus showing that indeed in this case the primal-dual gap vanishes.

3.4 Entropic Benamou–Brenier Formula

3.4.1 Primal and Dual Formulation

In Sect. 3.3 we have obtained a Lagrangian dynamic formulation of entropic optimal transport, that provided a way to interpret how the measure μ is transformed into ν. In Sect. 2.3 we studied a Eulerian dynamic formulation for *unregularized* optimal transport. The existence of such a formulation might not be so surprising, since we saw in the unregularized setting that the moving particles do not collide at intermediate times (Corollary 2.22). However, according to (3.15) in the entropic setting, particles extensively move through each other at intermediate times. It may therefore be surprising that a Eulerian formulation even exists in the regularized setting. Such a formulation was introduced in [18] and also studied in [29]. In this section we describe this formulation and sketch the equivalence in Sect. 3.4.2. Our discussion will remain purely formal.

Inspired by (3.19) we now introduce a diffusive version of the distributional continuity equation (Definition 2.25). Formally, it corresponds to the PDE

$$\partial_t \rho + \nabla \cdot \omega = \tfrac{\varepsilon}{2} \Delta \rho \qquad (3.22)$$

with temporal boundary conditions $\rho(0, \cdot) = \mu$ and $\rho(1, \cdot) = \nu$.

Similar to Definition 2.25 we introduce a formal weak formulation. Let $\mu, \nu \in \mathcal{P}(X)$, $\varepsilon > 0$. A pair $(\rho, \omega) \in \mathcal{M}([0,1] \times \mathbb{R}^d) \times \mathcal{M}([0,1] \times \mathbb{R}^d)^d$ is said to solve the distributional continuity equation with diffusion and temporal boundary conditions μ and ν if

$$\int_{[0,1] \times \mathbb{R}^d} (\partial_t \psi) \, d\rho + \int_{[0,1] \times \mathbb{R}^d} \nabla \psi \cdot d\omega + \tfrac{\varepsilon}{2} \int_{[0,1] \times \mathbb{R}^d} \Delta \psi \, d\rho = \int_X \psi(1, \cdot) \, d\nu - \int_X \psi(0, \cdot) \, d\mu \qquad (3.23)$$

for all suitable test functions, e.g. for all $\psi \in C_b^2([0,1] \times X)$ (here the subscript indicates that ψ and its derivatives are bounded). We denote the set of solutions by $\mathcal{CE}_\varepsilon(\mu, \nu)$.

The entropic primal Benamou–Brenier formula is then obtained by simply replacing $\mathcal{CE}(\mu, \nu)$ with $\mathcal{CE}_\varepsilon(\mu, \nu)$ as admissible set in (2.12):

$$C_{\varepsilon, \mathrm{BB}}(\mu, \nu) := \inf \{ \mathcal{A}(\rho, \omega) \mid (\rho, \omega) \in \mathcal{CE}_\varepsilon(\mu, \nu) \} \qquad (3.24)$$

Similar to Proposition 2.37 we can formally obtain a corresponding dual formulation:

$$C_{\varepsilon, \mathrm{BB}}(\mu, \nu) = \sup \left\{ \int_X \Psi(1, \cdot) \, d\nu - \int_X \Psi(0, \cdot) \, d\mu \,\middle|\, \Psi \in C_b^2([0,1] \times \mathbb{R}^d), \right.$$
$$\left. \partial_t \Psi + \frac{1}{2} |\nabla \Psi|^2 + \tfrac{\varepsilon}{2} \Delta \Psi \leq 0 \right\} \qquad (3.25)$$

Sketch of Proof This works in analogy to Proposition 2.37 where one now chooses $U = C_b^2([0,1] \times X)$ and $A : \Psi \mapsto (\partial_t \Psi + \tfrac{\varepsilon}{2} \Delta \Psi, \nabla \Psi)$. □

In analogy with Proposition 2.38 one obtains the following formal primal-dual optimality conditions. We are primarily interested in the formal interaction between primal mass movement and the dynamic dual potential.

Remark 3.27 (Primal-Dual Optimality Conditions for the Entropic Benamou–Brenier Formulation) A pair $(\rho, \omega) \in \mathcal{M}([0, 1] \times \mathbb{R}^d)^{1+d}$ and $\Psi \in C_b^2([0, 1] \times \mathbb{R}^d)$ is primal-dual optimal for (3.24) and (3.25) if and only if

$$\partial_t \rho + \nabla \cdot \omega + \tfrac{\varepsilon}{2}\Delta \rho = 0 \quad \text{with temporal boundary conditions } \rho(0, \cdot) = \mu,\ \rho(1, \cdot) = \nu$$

in the distributional sense of (3.23), and further

$$\rho \geq 0, \qquad \omega = \nabla \Psi \cdot \rho,$$

and

$$\partial_t \Psi + \tfrac{1}{2}|\nabla \Psi|^2 + \tfrac{\varepsilon}{2}\Delta \Psi \leq 0 \text{ with equality } \rho\text{-almost everywhere.}$$

3.4.2 Equivalence with Entropic Kantorovich Formulation

In this section we now sketch the formal equivalence between formulations (3.24) and (3.25) with their Lagrangian counter parts (3.10) and (3.11). This equivalence was shown in [18] with tools from stochastic analysis. Again, for the sake of brevity the arguments will be merely formal as there are some issues related to the regularity of dual candidates for (3.24).

Remark 3.28 Formally one has $(3.24) \leq (3.10) - \varepsilon H(\mu)$.

Sketch of Proof Let (ϕ, ψ) be solutions to (3.3) or (3.13), let ρ be as in (3.14), and let U, V, Φ, and Ψ constructed as in Definition 3.24. Let $\omega := \nabla \Psi \cdot \rho$ be a vector-valued density. Then by (3.19d), the pair (ρ, ω) is a strong solution of (3.22) on $(0, 1) \times \mathbb{R}^d$ and therefore by partial integration as distributional solution to (3.23) relaxed to the whole $\mathbb{R}^d$. Therefore, the pair (ρ, ω) is admissible in (3.24). By plugging (ρ, ω) into the objective of (3.24) one obtains

$$(3.24) \leq \mathcal{A}(\rho, \omega)$$
$$= \int_{[0,1]\times\mathbb{R}^d} \Phi(1, \nabla\Psi)\, \rho(t, x)\, \mathrm{d}x\mathrm{d}t$$
$$= \int_{[0,1]\times\mathbb{R}^d} \tfrac{1}{2}|\nabla\Psi|^2\, \rho(t, x)\, \mathrm{d}x\mathrm{d}t$$
$$= \int_0^1 \left[\int_{\mathbb{R}^d} \partial_t(\Psi \cdot \rho)\, \mathrm{d}x\right] \mathrm{d}t$$

$$= \int_X \Psi(1,\cdot)\rho(1,\cdot)\,dx - \int_X \Psi(0,\cdot)\rho(0,\cdot)\,dx$$

$$= \int_X \psi \cdot v\,dx + \int_X \phi \cdot \mu\,dx - \varepsilon \int_X \log(\mu) \cdot \mu\,dx$$

$$= \int_X \psi \cdot v\,dx + \int_X \phi \cdot \mu\,dx - \varepsilon \int_{X\times X} \exp\left(\frac{\phi \oplus \psi}{\varepsilon}\right) dK_\varepsilon - \varepsilon H(\mu)$$

$$= (3.11) - \varepsilon H(\mu) = (3.10) - \varepsilon H(\mu).$$

In this chain of equalities we used Lemma 3.29 below, the relation $\phi(x) = -\Psi(0,\cdot) + \varepsilon \log(\mu(x))$, and that $\int_{X\times X} \exp\left(\frac{\phi \oplus \psi}{\varepsilon}\right) dK_\varepsilon = 1$. □

Lemma 3.29 *In the setting of Definition 3.24 one has for any $t \in (0,1)$ that*

$$\int_{\mathbb{R}^d} \partial_t (\Psi \cdot \rho)\,dx = \int_{\mathbb{R}^d} \tfrac{1}{2} |\nabla \Psi|^2 \cdot \rho\,dx. \tag{3.26}$$

Proof The proof is a simple explicit computation, using that on $(0,1) \times \mathbb{R}^d$ Ψ and ρ are differentiable densities that solve (3.19) in a strong sense. One has

$$\int_{\mathbb{R}^d} \partial_t(\Psi \cdot \rho)\,dx = \int_{\mathbb{R}^d} [\Psi \cdot (\partial_t \rho) + (\partial_t \Psi) \cdot \rho]\,dx$$

$$= \int_{\mathbb{R}^d} [\Psi \cdot (\tfrac{\varepsilon}{2}\Delta\rho - \nabla[\nabla\Psi \cdot \rho]) - (\tfrac{1}{2}|\nabla\Psi|^2 + \tfrac{\varepsilon}{2}\Delta\Psi) \cdot \rho]\,dx$$

$$= \int_{\mathbb{R}^d} [\tfrac{\varepsilon}{2}(\Delta\Psi) \cdot \rho + |\nabla\Psi|^2 \cdot \rho]) - (\tfrac{1}{2}|\nabla\Psi|^2 + \tfrac{\varepsilon}{2}\Delta\Psi) \cdot \rho]\,dx$$

$$= \int_{\mathbb{R}^d} \tfrac{1}{2}|\nabla\Psi|^2 \cdot \rho\,dx$$

where we used integration by parts in the third equality. □

Note that the expression $\int_{\mathbb{R}^d} \partial_t(\Psi \cdot \rho)\,dx$ can also be related to the step-wise entropy decomposition in (3.21) where a finite-difference version of this temporal derivative appears.

Remark 3.30 Formally one has $(3.25) \geq (3.11) - \varepsilon H(\mu)$.

Sketch of Proof Consider the same setting as in the sketch for Remark 3.28. Observe that by (3.19e) Ψ is formally admissible in (3.25) (but it will not be differentiable or even continuous at $t \in \{0, 1\}$). Recall that $\phi(x) = -\Psi(0, \cdot) + \varepsilon \log(\mu(x))$ and $\int_{X \times X} \exp\left(\frac{\phi \oplus \psi}{\varepsilon}\right) dK_\varepsilon = 1$. Then

$$(3.11) - \varepsilon H(\mu) = \int_X \Psi(1, \cdot) \, dv - \int_X \Psi(0, \cdot) \, d\mu \leq (3.25).$$

□

References

1. J.M. Altschuler, E. Boix-Adsera, Polynomial-time algorithms for multimarginal optimal transport problems with structure. Math. Program. **199**, 1107–1178 (2023)
2. L. Ambrosio, N. Fusco, D. Pallara, *Functions of Bounded Variation and Free Discontinuity Problems*, Oxford Mathematical Monographs (Oxford University Press, Oxford, 2000)
3. L. Ambrosio, N. Gigli, G. Savaré, *Gradient Flows in Metric Spaces and in the Space of Probability Measures*, Lectures in Mathematics (Birkhäuser Boston, Boston, 2005)
4. S. Angenent, S. Haker, A. Tannenbaum, Minimizing flows for the Monge-Kantorovich problem. SIAM J. Math. Anal. **35**(1), 61–97 (2003)
5. F. Baudoin, *Diffusion Processes and Stochastic Calculus*, EMS Textbooks in Mathematics (European Mathematical Society, Zurich, 2014)
6. H.H. Bauschke, P.L. Combettes, *Convex Analysis and Monotone Operator Theory in Hilbert Spaces*. CMS Books in Mathematics,1st edn. (Springer, 2011)
7. F. Beier, J. von Lindheim, S. Neumayer, G. Steidl, Unbalanced multi-marginal optimal transport. J. Math. Imaging Vis. **65**, 394–413 (2022)
8. J.-D. Benamou, Numerical resolution of an "unbalanced" mass transport problem. ESAIM Math. Model. Numer. Anal. **37**(5), 851–868 (2003)
9. J.-D. Benamou, Y. Brenier, A computational fluid mechanics solution to the Monge-Kantorovich mass transfer problem. Numer. Math. **84**(3), 375–393 (2000)
10. J.-D. Benamou, G. Carlier, S. Di Marino, L. Nenna, An entropy minimization approach to second-order variational mean-field games, in *Mathematical Models and Methods in Applied Sciences* (2019), pp. 1–31
11. J.-D. Benamou, G. Carlier, L. Nenna, Generalized incompressible flows, multi-marginal transport and Sinkhorn algorithm. Numer. Math. **142**(1), 33–54 (2019)
12. Y. Brenier, Polar factorization and monotone rearrangement of vector-valued functions. Comm. Pure Appl. Math. **44**(4), 375–417 (1991)
13. T. Cai, J. Cheng, B. Schmitzer, M. Thorpe, The linearized Hellinger-Kantorovich distance. SIAM J. Imaging Sci. **15**(1), 45–83 (2022)
14. E.A. Carlen, J. Maas, Gradient flow and entropy inequalities for quantum Markov semigroups with detailed balance. J. Funct. Anal. **273**(5), 1810–1869 (2017)
15. G. Carlier, C. Jimenez, F. Santambrogio, Optimal transportation with traffic congestion and Wardrop equilibria. SIAM J. Control Optim. **47**(3), 1330–1350 (2008)
16. G. Carlier, V. Duval, G. Peyré, B. Schmitzer, Convergence of entropic schemes for optimal transport and gradient flows. SIAM J. Math. Anal. **49**(2), 1385–1418 (2017)
17. G. Carlier, P. Pegon, L. Tamanini, Convergence rate of general entropic optimal transport costs. Calc. Var. Partial Differential Equations **62**, 116 (2023)
18. Y. Chen, T.T. Georgiou, M. Pavon, On the relation between optimal transport and Schrödinger bridges: a stochastic control viewpoint. J. Optim. Theory Appl. **169**, 671–691 (2016)

19. Y. Chen, T.T. Georgiou, A. Tannenbaum, Matrix optimal mass transport: A quantum mechanical approach. IEEE Trans. Autom. Control **63**(8), 2612–2619 (2018)
20. R. Chetrite, P. Muratore-Ginanneschi, K.E. Schwieger, Schrödinger's 1931 paper "on the reversal of the laws of nature" ["Über die umkehrung der Naturgesetze", Sitzungsberichte der preussischen Akademie der Wissenschaften, physikalisch-mathematische Klasse, 8 n9 144–153]. Eur. Phys. J. H **46**, 28 (2021)
21. L. Chizat, G. Peyré, B. Schmitzer, F.X. Vialard, An interpolating distance between optimal transport and Fisher-Rao metrics. Found. Comput. Math. **18**(1), 1–44 (2018)
22. L. Chizat, G. Peyré, B. Schmitzer, F.X. Vialard, Unbalanced optimal transport: dynamic and Kantorovich formulations. J. Funct. Anal. **274**(11), 3090–3123 (2018)
23. L. Chizat, P. Roussillon, F. Léger, F.-X. Vialard, G. Peyré. Faster Wasserstein distance estimation with the Sinkhorn divergence, in *Advances in Neural Information Processing Systems 33 (NeurIPS 2020)* (2020)
24. I. Csiszar, I-divergence geometry of probability distributions and minimization problems. Ann. Probab. **3**(1), 146–158 (1975)
25. L.C. Evans, *Partial Differential Equations*, Graduate Studies in Mathematics, vol. 19 (American Mathematical Society, Providence, 1999)
26. J. Feydy, T. Séjourné, F.X. Vialard, S. Amari, A. Trouvé, G. Peyré, Interpolating between optimal transport and MMD using Sinkhorn divergences, in *Proceedings of the 22nd International Conference on Artificial Intelligence and Statistics (AISTATS)* (2019)
27. W. Gangbo, R.J. McCann, The geometry of optimal transportation. Acta Math. **177**(2), 113–161 (1996)
28. A. Genevay, L. Chizat, F. Bach, M. Cuturi, G. Peyré, Sample complexity of Sinkhorn divergences, in *Proceedings of Machine Learning Research*, ed. by K. Chaudhuri, M. Sugiyama, volume 89 of Proceedings of Machine Learning Research (2019), pp. 1574–1583
29. I. Gentil, C. Léonard, L. Ripani, About the analogy between optimal transport and minimal entropy. Ann. Fac. Sci. Toulouse Math. **26**(3), 569–601 (2017)
30. I. Haasler, A. Ringh, Y. Chen, J. Karlsson, Multimarginal optimal transport with a tree-structured cost and the Schrödinger bridge problem. SIAM J. Control Optim. **59**(4), 2428–2453 (2021)
31. P. Koltai, J. von Lindheim, S. Neumayer, G. Steidl, Transfer operators from optimal transport plans for coherent set detection. Phys. D **426**, 132980 (2021)
32. S. Kondratyev, L. Monsaingeon, D. Vorotnikov, A new optimal transport distance on the space of finite Radon measures. Adv. Differential Equations **21**(11–12), 1117–1164 (2016)
33. H. Lavenant, J. Luckhardt, G. Mordant, B. Schmitzer, L. Tamanini, The Riemannian geometry of Sinkhorn divergences. arXiv:2405.04987 (2024)
34. H. Lavenant, S. Zhang, Y.H. Kim, G. Schiebinger, Towards a mathematical theory of trajectory inference. Ann. Appl. Probab. **34**(1A), 428 (2024)
35. C. Léonard, From the Schrödinger problem to the Monge-Kantorovich problem. J. Funct. Anal. **262**(4), 1879–1920 (2012)
36. C. Léonard, A survey of the Schrödinger problem and some of its connections with optimal transport. Discrete Contin. Dyn. Syst. A **34**(4), 1533–1574 (2014)
37. M. Liero, A. Mielke, G. Savaré, Optimal entropy-transport problems and a new Hellinger-Kantorovich distance between positive measures. Invent. Math. **211**(3), 969–1117 (2018)
38. S. Lisini, Characterization of absolutely continuous curves in Wasserstein spaces. Calc. Var. Partial Differential Equations **28**(1), 85–120 (2007)
39. J. Lott, Some geometric calculations on Wasserstein space. Comm. Math. Phys. **277**, 423–437 (2008)

40. G. Luise, S. Salzo, M. Pontil, C. Ciliberto, Sinkhorn barycenters with free support via Frank–Wolfe algorithm, in *Advances in Neural Information Processing Systems 32 (NeurIPS 2019)* (2019)
41. M. Mauritz, B. Schmitzer, B. Wirth, A Bayesian model for dynamic mass reconstruction from PET listmode data. SIAM J. Math. Anal. **56**(5), 5840–5880 (2024)
42. R.J. McCann, Polar factorization of maps on Riemannian manifolds. Geom. Funct. Anal. **11**(3), 589–608 (2001)
43. G. Peyré, M. Cuturi, *Computational Optimal Transport*, volume 11 of Foundations and Trends in Machine Learning (2019)
44. R.T. Rockafellar, Duality and stability in extremum problems involving convex functions. Pacific J. Math **21**(1), 167–187 (1967)
45. R.T. Rockafellar, Integrals which are convex functionals. Pacific J. Math. **24**(3), 525–539 (1968)
46. W. Rudin, *Real and Complex Analysis*, 3rd edn. (McGraw-Hill Book Company, 1986)
47. L. Rüschendorf, W. Thomsen, Note on the Schrödinger equation and I-projections. Stat. Probab. Lett. **17**, 369–375 (1993)
48. F. Santambrogio, *Optimal Transport for Applied Mathematicians*, volume 87 of Progress in Nonlinear Differential Equations and Their Applications (Birkhäuser Boston, Boston, 2015)
49. B. Schmitzer, K.P. Schäfers, B. Wirth, Dynamic cell imaging in PET with optimal transport regularization. IEEE Trans. Med. Imaging **39**(5), 1626–1635 (2020)
50. C. Villani, *Topics in Optimal Transportation*, Graduate Studies in Mathematics, vol. 58. (American Mathematical Society, Providence, 2003)
51. C. Villani, *Optimal Transport: Old and New*, volume 338 of Grundlehren der mathematischen Wissenschaften (Springer, Berlin, 2009)

A Geometric Perspective on Diffeomorphic and Optimal Transport Flows and Their Applications

François-Xavier Vialard

Abstract

In these notes, we present a geometric point of view on diffeomorphic and optimal transport flows, driven by applications in shape analysis, computational anatomy, and machine learning. This perspective brings tools that facilitate the design, the derivation and the study of geodesic and gradient flows on diffeomorphisms and measures. Along with this geometric perspective, we provide a rigorous treatment of the analysis required to study these flows, particularly the geodesic flows on diffeomorphism groups. The final part studies gradient flows arising in machine learning through the lenses of these geometric and analytic perspectives.

1 Introduction

In these notes, we aim to introduce a unified geometric framework that integrates diffeomorphic flows, optimal transport, and related models. This perspective is inspired by shape analysis and computational anatomy, two fields where the registration problem has garnered significant attention over the past twenty years. It also builds on a large body of work in fluid dynamics, particularly on the line of research initiated by Arnold in the 1970s, and pursued by Ebin and Marsden.

F.-X. Vialard (✉)
Laboratoire d'Informatique Gaspard-Monge, Université Gustave Eiffel, Marne-la-Vallée Cedex 2, France
e-mail: francois-xavier.vialard@univ-eiffel.fr

© The Author(s), under exclusive license to Springer Nature Switzerland AG 2025
W. Li et al., *Variational and Information Flows in Machine Learning and Optimal Transport*, Oberwolfach Seminars 56,
https://doi.org/10.1007/978-3-031-92731-7_2

Being by nature nonlinear spaces, the shape spaces are often equipped with a Riemannian metric, providing us with tools such as geodesics, and parallel transport that can be useful to compare shapes and to solve the registration problem. One approach to endow shape spaces with Riemannian metrics is to leverage group actions combined with appropriately chosen metrics on these groups. Notably, right-invariant metrics on groups paired with left actions systematically induce a Riemannian metric on the action's orbit. This geometric perspective is quite versatile, enabling the development of new models for shape deformation and extending its applicability beyond shape analysis to fields such as optimal transport.

In Sect. 2, we present the vocabulary of group actions as well as simple and important tools, such as the adjoint action, the cotangent action, and the infinitesimal generator. In Sect. 2.6, we present induced Riemannian metrics from group actions, with a simple recipe to build them (see Proposition 33). Several finite dimensional examples are detailed and some infinite dimensional examples are presented. Since right-invariant metrics on groups appear to be a natural tool, Sect. 3 derives the geodesic equations, which are sometimes called Euler-Arnold or Euler-Poincaré equations. We then gives infinite-dimensional examples arising in fluid dynamics.

Obviously, shape spaces are most often infinite dimensional, which makes a key difference with usual Riemannian manifolds. The rigorous mathematical treatment of such objects necessitates the use of analytical tools. Sobolev spaces can sometimes be used to prove the local existence of the flow for these evolution equations. In Sect. 4, we briefly present the ideas together with the main arguments developed by Ebin and Marsden to prove the local well-posedness of the Euler-Arnold equation, of which the incompressible Euler equation is a particular case. We also present the nice property of smoothness propagation of the geodesic flow, called the no loss–no gain Lemma 57.

As presented in Sect. 5, another way to circumvent some analytical issues on the diffeomorphisms group consists of defining deformations by flows of vector fields in a Reproducing Kernel Hilbert Space (RKHS). This point of view has been developed in computation anatomy and it is effective to derive numerical algorithms to implement the boundary value problem associated with geodesics on the diffeomorphisms group. The main result in this section is concerned with the existence of minimizers of the variational problem associated with diffeomorphic registration.

From a right-invariant metric, the induced Riemannian metrics on shape spaces such as groups of points, images or measures can be easily derived. This is the goal of Sect. 6 with a particular emphasis on space of measures. We discuss the case of optimal transport and its unbalanced version. This enables to highlight some key differences between optimal transport and Riemannian metrics on measures induced by a right-invariant metric.

Section 7 is concerned with gradient flows, with a particular focus on gradient flows on measures since it is very relevant in machine learning. We highlight the usefulness of the tools presented in the previous sections on different important examples. We first start with underlining the importance of the co-metric (the inverse of the metric) in gradient flows, in particular in the Polyak-Lojasiewicz inequality, which is a standard tool to prove the global convergence of the gradient flow (see Appendix 5). We then discuss the derivation of different flows of the Kullback-Leibler divergence for the different metrics on the space of measures, making clear the link between Stein Variational Gradient Descent and right-invariant metrics on the diffeomorphisms group. We then address the question of global convergence for two different gradient flows. The first gradient flow is concerned with an objective function which is a discrepancy on probability measures defined by the Coulomb kernel, in the spirit of Maximum Mean Discrepancies in machine learning. The last example is given by residual neural networks (ResNets), which can be modeled (some subparts of ResNets) as diffeomorphic flows. We explain why the optimization landscape may satisfy a local Polyak-Lojasiewicz property.

It is worth noting that the technical level of these notes is relatively elementary. Rather than delving into extensive technical details, we have focused on presenting the main ideas and key techniques in straightforward and accessible settings. Consequently, these notes are concise and can serve as an entry point to the (often extensive) literature on the various topics discussed.

2 Elementary Notions of Group Actions

2.1 Group Actions

This section presents the elementary geometric tools useful for understanding the common points between optimal transport, normalizing flows, and shape analysis from a geometric perspective. We first recall hereafter the notions of group action and its associated constructions. Some readers may find it more efficient to pay attention to examples, which introduce important objects for the next sections. In particular, we provide examples in the infinite-dimensional case, which is our main interest.

Definition 1 Let G_V be a group which is a smooth manifold with a tangent space at the identity Id denoted by V, acting from the *left* on a smooth manifold Q. We denote the action by

$$\Phi : G \times Q \to Q, \quad (g, q) \mapsto g \cdot q := \Phi_g(q). \tag{1}$$

Being a left action means that Φ satisfies the composition law $g_1 \cdot (g_2 \cdot q) = (g_1 g_2) \cdot q$ and $\mathrm{Id} \cdot q = q$ for any $q \in Q$ and $g_1, g_2 \in G$.

Remark 2 At a slightly more abstract level, a group action is a homomorphism between the group and the diffeomorphism group of Q. It also generalizes the notion of flows of time-independent vector fields, since it is a group homomorphism from $\mathbb{R}$ into the diffeomorphisms.

Right actions are also defined in the literature, the difference being the composition $g_1 \cdot (g_2 \cdot x) = (g_2 \cdot g_1) \cdot x$. It is possible to switch between the two actions by precomposition of the action with the inverse map on G. We give hereafter classical examples of finite-dimensional group actions.

Examples 3 The first examples involve group actions of the group on itself. However, we also anticipate the next section and consider infinite-dimensional actions of the diffeomorphisms group.

- Different types of action of a group on itself:
  - Left multiplication:

$$L_g(h) = g \cdot h. \tag{2}$$

  - Right multiplication with the inverse:

$$R_{g^{-1}}(h) = h \cdot g^{-1}. \tag{3}$$

  Indeed, one has $h \cdot g_1^{-1} \cdot g_2^{-1} = h \cdot (g_2 g_1)^{-1}$. If the inverse on g were not present, the action would be a right action instead of a left one.
  - Left and right multiplications commute[1] and the conjugation action is the composition of the two previous ones:

$$\text{Ad}_g(h) = g \cdot h \cdot g^{-1}. \tag{4}$$

- The adjoint action of the group on the tangent space at identity is defined by Ad : $G \times T_{\text{Id}} G \mapsto T_{\text{Id}} G$ by

$$\text{Ad}_g(\xi) = g \cdot \xi \cdot g^{-1}. \tag{5}$$

Note that with a little abuse of notations, we use the same symbol for the conjugation action and the adjoint action. The adjoint action is the differential at the identity of the conjugation action and the notation Ad is usually reserved for the adjoint action.
- The linear group $\text{GL}_d(\mathbb{R})$ acts on $\mathbb{R}^d$ by matrix-vector multiplication.
- The orthogonal group $\text{SO}_d(\mathbb{R})$ acts on the sphere of dimension $d - 1$, $S^{d-1} \subset \mathbb{R}^d$ by matrix multiplication.

---

[1] Prove that the composition of two commuting left actions is a left action.

- A group of diffeomorphic transformations G of $\mathbb{R}^d$ can act on:
  - Groups of n points $(\mathbb{R}^d)^n$ via

  $$(\varphi, (q_i)_{i=1,\ldots,n}) \mapsto (\varphi(q_i))_{i=1,\ldots,n}\,.$$

  - Functions (or 0-forms) from $\mathbb{R}^d$ into $\mathbb{R}$, via

  $$(\varphi, I) \mapsto I \circ \varphi^{-1}\,.$$

In the case of diffeomorphisms, examples of actions are numerous such as the ones given by natural objects in differential geometry, e.g. differential forms. Regarding the two examples on the diffeomorphisms group, note that there is an important difference between the two previous actions with the group of diffeomorphisms: the action on functions is linear with respect to the function, whereas the action on points is nonlinear.

We will make use of semidirect product of groups which can be described as follows:

Definition 4 (Semidirect Product of Groups) Let G be a group acting on the left on another group H by homomorphisms, i.e.

$$(g \cdot h_1)(g \cdot h_2) = g \cdot (h_1 h_2)\,, \tag{6}$$

for $g \in G$ and $h_1, h_2 \in H$. The semidirect product,[2] denoted by $G \ltimes H$, is the Cartesian product $G \times H$ endowed with the group multiplication

$$(g_1, h_1)(g_2, h_2) = (g_1 g_2, h_1(g_1 \cdot h_2))\,. \tag{7}$$

Note that the inverse of (g, h) is (g^{-1}, h^{-1}).

Example 5 The orthogonal group $O(d)$ acts by conjugation on the group of translations of $\mathbb{R}^d$. The group of isometries of $\mathbb{R}^d$ is thus $O(d) \ltimes \mathbb{R}^d$.

Example 6 Consider the group of diffeomorphisms of a closed manifold M denoted by $\mathrm{Diff}(M)$ and $C(M, \mathbb{R}_{>0})$ the space of positive-valued functions. The latter space is a group for the pointwise multiplication. The diffeomorphism group acts on the left by homomorphisms on $C(M, \mathbb{R}_{>0})$ via $(\varphi, \lambda) \in \mathrm{Diff}(M) \times C(M, \mathbb{R}_{>0}) \mapsto \lambda \circ \varphi^{-1}$. Therefore,

---

[2] Note that the cross between the two groups is usually closed next to the group which is acting on the other. In this case, G is acting on H. If H were acting on G, we could have denoted $G \rtimes H$ for their semidirect product.

one can form the semidirect product $\text{Diff}(M) \ltimes C(M, \mathbb{R}_{>0})$. This semidirect product will be used for a generalization of optimal transport to measures of non-equal masses.

2.2 Cotangent Actions

Since an action is nothing else than a map, it implies that the natural operations of differential geometry can be carried out. The simplest is certainly differentiation. Therefore, a left action of a group G on Q induces a left action of G on TQ by $g \cdot (q, \delta q) := (g \cdot q, g \cdot \delta q)$ where the second term means $d[\Phi_g]_{|q}(\delta q)$.

Definition 7 (Cotangent Action) Let G act on the left on Q. The *cotangent action* or *cotangent lifted action* is a left action of G on T^*Q the cotangent bundle of Q defined by

$$g \cdot (q, p_q) \mapsto (g \cdot q, g^{-1*} \cdot (p_q)), \tag{8}$$

where $g^{-1*} \cdot (p_q)$ is a short notation for $d[\Phi_{g^{-1}}]_{|\Phi_g(q)}^\top (p_q)$.

Remark 8 Note that this left action uses g^{-1} instead of g. It is necessary to obtain a left action instead of a right action since taking adjoints (i.e. dual adjoints) reverses the direction of maps.

Examples 9 The cotangent action is an important object for optimization and geodesic flows, thus we give a few examples of cotangent actions.

- The action of $\text{GL}_d(\mathbb{R})$ on $\mathbb{R}^d$ by multiplication is a bilinear map. As a consequence, the cotangent action is

$$M \cdot (x, p) = (Mx, M^{-\top} p), \tag{9}$$

  where $(x, p) \in \mathbb{R}^d \times \mathbb{R}^d$.
- The action of $\text{SO}_d(\mathbb{R})$ on S^{d-1} induces the following cotangent action:

$$O \cdot (x, p) = (Ox, O^{-\top} p), \tag{10}$$

  which is not unexpectedly similar to the previous example.
- The action of diffeomorphisms on groups of points induces the following cotangent action

$$\varphi \cdot (x_i, p_i)_{i=1,\ldots,n} = (\varphi(x_i), [d\varphi^{-1}(\varphi(x_i))]^\top p_i)_{i=1,\ldots,n}. \tag{11}$$

- The dual space of continuous functions is the space of Radon measures. The cotangent action is

$$(\varphi, \mu) \mapsto \varphi_\sharp(\mu), \tag{12}$$

the image measure of μ under φ. Recall that the image measure of μ under $\varphi : X \to Y$ is defined by $\varphi_\sharp(\mu)(B) = \mu(\varphi^{-1}(B))$ where B is a measurable set of Y. We also denote by $\varphi_*(\mu)$ the image measure. It is also defined in duality via

$$\int_Y f(y) d\varphi_*(\mu)(y) = \int_X f(\varphi(x)) d\mu(x). \tag{13}$$

In the case of $X = Y = \mathbb{R}^d$ and a measure μ which has density $\mu = \rho(x)dx$ w.r.t. Lebesgue, the change of variable formula gives $\varphi_*(\mu) = \mathrm{Jac}(\varphi^{-1})(x)\rho(\varphi^{-1}(x))dx$.

- The adjoint action is linear (see remark below) and it has a coadjoint action defined by

$$m \in T^*_{\mathrm{Id}} G \mapsto \mathrm{Ad}^*_{g^{-1}}(m) \in T^*_{\mathrm{Id}} G. \tag{14}$$

Remark 10 For actions of a group on a linear space V which is linear, the cotangent action simplifies to

$$g \cdot (x, v) = (g(x), [g^{-1}]^*(v)), \tag{15}$$

the simplification being that the differential of the action does not depend on the point and it is equal to the action itself. This explains the formula for the coadjoint action of diffeomorphisms on functions in the above examples.

Choosing a reference point q, one can collect all the points that are connected via the group action to q, in other words, the range of the group action when q is fixed. It is called the orbit of q. In a somewhat transverse manner, one can collect all the elements of G that leave a point q fixed. It is called the isotropy subgroup of q.

Definition 11 (Orbit and Isotropy Group, and Others) Consider a group G acting on the left on a space Q. The *orbit* of an element $q \in Q$ is

$$O_q = \{g \cdot q \in Q \,;\, g \in G\}. \tag{16}$$

If there exists q such that $G_q = Q$, the action is said *transitive*. The *isotropy group* of an element $q \in Q$ is the subgroup of G defined by

$$G_q = \{g \,;\, g \cdot q = q \in G\}. \tag{17}$$

The action is said *free* if $G_q = \{\text{Id}\}$ for every $q \in Q$ and *faithful* if for every $g \in G_q$ for every $q \in Q$ implies $g = \text{Id}$. In particular, freeness implies faithfulness.

An important example of a left action is the left multiplication of the group on itself. It is transitive and free. From a control point of view, one can consider the group G as the set of actions and the other copy of G as the set of objects. In this context, it might be natural to consider the question of parametrizing the tangent space at an element g independently of its representation via the left action. For instance, one can write $g = g_1 \cdot h_1 = g_2 \cdot h_2$. A small perturbation of g_1 or g_2 leads to a tangent vector at g, namely $\delta g = \delta g_1 \cdot h_1$ or $\delta g = \delta g_2 \cdot h_2$. Having a parametrization of the tangent space at g which does not depend on such a choice is convenient, if not necessary. To obtain such a parametrization, we compute, by right multiplication of δg with g^{-1} written in the two different forms $g^{-1} = h_1^{-1} \cdot g_1^{-1} = h_2^{-1} \cdot g_2^{-1}$, we get

$$\delta g_1 \cdot h_1 = \delta g_2 \cdot h_2 \implies \delta g_1 \cdot g_1^{-1} = \delta g_2 \cdot g_2^{-1} = \delta g \cdot g^{-1}. \tag{18}$$

Although we used relatively informal notations, it is probably clear that the object $\delta g \cdot g^{-1}$ belongs to the tangent space at identity. The obtained parametrization is thus $\xi \cdot g$ for $\xi \in T_{\text{Id}}$. This map is called right-trivialization and it is defined below.

Definition 12 (Right-Trivialization) Let G be a group and a smooth manifold at the same time, possibly of infinite dimensions, the *right-trivialization* of TG is the bundle isomorphism $\tau : TG \mapsto G \times T_{\text{Id}}G$ defined by $\tau(g, X_h) := (g, dR_{g^{-1}}X_g)$, where $X_g h$ is a tangent vector at point g and $\mathcal{R}_{g^{-1}} : G \to G$ is the right multiplication by g^{-1}, namely, $R_{g^{-1}}(f) = f \cdot g^{-1}$ for all $f \in H$.

Example 13 In fluid dynamics, the right-trivialized tangent vector $dR_{h^{-1}}X_h$ corresponds to the spatial or Eulerian velocity and X_h is the Lagrangian velocity. Importantly, on the group of Sobolev diffeomorphisms, this right-trivialization map is continuous but not differentiable with respect to the variable h. Indeed, right-multiplication R_h is smooth, yet left multiplication is continuous and usually not differentiable, due to a loss of smoothness. Note that we also use the notation $\delta\varphi$ for a tangent vector at point φ and the right translation is $\delta\varphi \cdot \varphi^{-1}$.

Example 14 (The Case of Semidirect Product of Groups) For a semidirect product of groups $G \ltimes H$ (see Definition 4), we get

$$(\delta g, \delta h) \cdot (g, h)^{-1} = (\delta g g^{-1}, \delta h h^{-1} + h((\delta g g^{-1}) \cdot h^{-1})). \tag{19}$$

Remark that we do not require the group in the previous definition to be a Lie group, i.e. a group for which the map $(g, h) \in G^2 \mapsto g^{-1} \cdot h$ is smooth (equivalently, multiplication and inverse are smooth maps). So, why not Lie groups?

2.3 About Lie Groups and Infinite Dimensions

Being a group and a smooth manifold at the same time is often encountered in the literature of Lie groups for which right and left multiplications are smooth maps. As a consequence, the usual examples in finite dimensions are Lie groups. However, in infinite dimensions, it is possible to find groups for which right or left compositions are not smooth operations, and still, the group is a manifold.[3] A prominent example is the Sobolev diffeomorphisms group which is defined and studied in the next sections. In fact, in infinite dimensions, being a Lie group is too restrictive as shown by Omori [52]. Indeed, Omori proved:

Theorem 15 ([52]) *If a connected Banach–Lie group[4] acts faithfully, transitively, and smoothly on a compact manifold, then it is a finite-dimensional Lie group.*

Consider a group of diffeomorphisms of a compact manifold Q and its action on Q defined by $x \mapsto \varphi(x)$, where φ is a diffeomorphism. Then, this action is transitive if the group of diffeomorphisms is rich enough and also smooth and faithful. Hence, Omori's theorem implies that this group of diffeomorphisms cannot be a Banach-Lie group. In conclusion, unless finite-dimensional, the groups we will consider hereafter are not Lie groups, meaning that one must relinquish the smoothness of the right or left multiplications or the inverse map. These weaker structures are called half-Lie groups, see for instance [9].

2.4 Infinitesimal Generator

The introduction of the right-trivialization was motivated by the action of the group G on itself. However, our initial motivation was independent of the space Q on which G acts. From a control point of view, it is a natural idea to obtain a parametrization of the tangent space of an orbit only in terms of G. More precisely, the tangent space of an orbit $O_{g \cdot q}$ can be described via small modifications of the given element $g \in G$. However, this choice again depends a priori on g: if the isotropy subgroup of q is not reduced to identity, there exist $g_1 \neq g_2 \in G$ such that $g_1 \cdot q = g_2 \cdot q$. Here again, it is natural to propose a parametrization (which is encoded at the level of the group) of the tangent space to $O_{g \cdot q}$ which does not depend on the choice of g. The simplest way to do it consists of using right-trivialization which is defined above. The tangent space at an element g is parametrized by $\xi \cdot g$ where $\xi \in T_{\mathrm{Id}}G$. Indeed, assume that $g_1 \cdot q = g_2 \cdot q$ and that $\delta g_1 g_1^{-1} = \delta g_2 g_2^{-1} = \xi$,

---

[3] See also the discussion in [16].

[4] A Banach Lie group is an infinite-dimensional Lie group modeled on a Banach space.

then one gets immediately $\delta g_1 \cdot q = \delta g_2 \cdot q = \xi \cdot g \cdot q$. Here again, the notation is relatively informal since the two dots encode different actions.[5]

Therefore, the tangent space at a point $g \cdot q$ is parametrized via $\xi \in T_{\mathrm{Id}} G$. This operation is called an infinitesimal generator.

Definition 16 (Infinitesimal Generator) The *infinitesimal generator* or *infinitesimal action* of the action corresponding to $\xi \in V$ is the vector field on Q given by

$$\xi_Q(q) := \left. \frac{d}{dt} \right|_{t=0} g(t) \cdot q, \tag{20}$$

where $g(t)$ is any curve such that $g(0) = \mathrm{Id}$ and $g'(0) = \xi$.

Remark 17 The infinitesimal action defines a vector field on Q. As a derivation, it is a linear map from $T_{\mathrm{Id}} G$ to $\chi(Q)$ the space of vector fields on Q. We will often use the notation $\xi \cdot q$ instead of $\xi_Q(q)$.

Examples 18

(1) Consider the action of the group G on itself via left multiplication. The corresponding infinitesimal action is $\xi \cdot g$, which is a map from $T_{\mathrm{Id}} G \to T_g G$. This map is an isomorphism.
(2) Consider the conjugation action of G on itself. Differentiation of $g(t) \cdot h \cdot g(t)^{-1}$ leads to $\xi \cdot h - h \cdot \xi \in T_h G$. This notation may initially be confusing, as the two terms in the last formula can take quite different forms. Consider the diffeomorphisms group of a manifold M as an example for which the group operation is composed. The tangent space at identity is a vector field on M. Then, the infinitesimal action reads

$$\xi \circ \varphi - d\varphi(\xi), \tag{21}$$

where $d\varphi$ denotes the differential of φ and ξ is the vector field.
(3) The adjoint action has an infinitesimal generator which is the Lie bracket between the tangent vector at the identity

$$\mathrm{ad}_\xi(\zeta) = \xi \cdot \zeta - \zeta \cdot \xi, \tag{22}$$

---

[5] We leave to the reader the exercise of making the two different actions involved explicit, see the first example in Examples 18.

which is another tangent vector at identity. It is formally very similar to the previous conjugation action. In the case of the diffeomorphisms group, it reads

$$\mathrm{ad}_\xi(\zeta) = d\xi(\zeta) - d\zeta(\xi) \tag{23}$$

for vector fields $\xi, \zeta \in \chi(Q)$, which is the Lie bracket between the vector fields.

(4) The infinitesimal action of diffeomorphisms on groups of points is simply $(v(x_i))_{i=1,\ldots,n}$ where v is a vector field.
(5) In the case of the linear actions of the group of diffeomorphisms on a function I and a measure which has a density ρ w.r.t. Lebesgue, we have[6]

$$v \cdot I = -\langle \nabla I, v \rangle \tag{24}$$
$$v \cdot \rho = -\mathrm{div}(\rho v), \tag{25}$$

where ∇ and div denote respectively the gradient and the divergence concerning the Lebesgue measure.

(6) The infinitesimal generator of the cotangent action is

$$\xi \cdot (q, p) = (\xi_Q(q), -d\xi_Q(q)^*(p)). \tag{26}$$

The coadjoint action on the Lie algebra is $-\mathrm{ad}_\xi^* \zeta$. In the case of diffeomorphisms and groups of points, it reads

$$v \cdot (q_i, p_i)_{i=1,\ldots,n} = (v(q_i), -dv(q_i)^*(p_i))_{i=1,\ldots,n}. \tag{27}$$

Lie Algebra Homomorphism Recall that a group action can be viewed as a map from G to the diffeomorphism group of Q. By the composition rule of an action, it is a group homomorphism. At the infinitesimal level of the action, this property also holds. Indeed, the infinitesimal action is a linear map from the tangent space at identity and the space of vector fields which are the tangent space of the group of diffeomorphisms at identity. It is of interest to understand the relation between the two. On the group, we have the adjoint action and the group acts on the left on Q, so one can ask if the following formula holds:

$$(\xi \cdot \zeta - \zeta \cdot \xi)_Q = d\xi_Q(\zeta_Q) - d\zeta_Q(\xi_Q). \tag{28}$$

---

[6] We recommend the reader to have done these computations once.

or more compactly:

$$[\mathrm{ad}_\xi(\zeta)]_Q = \mathrm{ad}_{\xi_Q}(\zeta_Q) \,. \tag{29}$$

We leave the proof to the reader.

Remark 19 (Lie Derivative of Vector Fields) Our definition of the adjoint action ad can be related to the Jacobi-Lie bracket (Lie derivative) of vector fields. Namely, one has

$$\mathcal{L}_\xi \zeta = -\,\mathrm{ad}_\xi \zeta \,. \tag{30}$$

The Lie derivative gives the right Lie algebra structure on the space of vector fields rather than the left one. Note that the Lie derivative is defined in coordinates as follows on $\mathbb{R}^d$

$$\mathcal{L}_\xi \zeta = \sum_{i,j=1}^d (\xi^i \partial_j \zeta^i - \zeta^j \partial_j \xi^i)\partial_i \,, \tag{31}$$

and in particular, second-order derivatives do not appear as is well-known. However, this fact will be reused in depth for more general Lie brackets.

So far, we have introduced the notions of left-action and right-trivialization. At least heuristically, that right-trivialization is a natural parametrization of the tangent space in this context. Let us insist again on the following fact, given a curve $g(t)$ and $q(t) = g(t) \cdot q_0$ where q_0 is a chosen point and $\dot{g}(t) = \xi(t) \cdot g(t)$

$$\frac{d}{dt} q(t) = \frac{d}{dt} g(t) \cdot q_0 = \xi(t) \cdot q(t) \,. \tag{32}$$

Let us instantiate this simple fact for adjoint action:

$$\frac{d}{dt} \mathrm{Ad}_{g(t)} = \mathrm{ad}_{\xi(t)} \cdot \mathrm{Ad}_{g(t)} \,. \tag{33}$$

Similarly, for coadjoint action, one has

$$\frac{d}{dt} \mathrm{Ad}^*_{g(t)^{-1}} = -\,\mathrm{ad}^*_{\xi(t)} \cdot \mathrm{Ad}^*_{g(t)^{-1}} \,. \tag{34}$$

In both cases, the dot is simply the composition of linear maps.

2.5 Momentum Maps

Momentum maps are usually defined in the context of Poisson structures. In these notes, we do not need to introduce momentum maps in their full generality since we mainly work on the cotangent space of a manifold. However, for the sake of completeness, we provide the standard definition, even though it slightly extends beyond the scope of the concepts introduced earlier. The following definition is not necessary for understanding the rest of the note.

There are two motivations for introducing momentum maps in these notes. First, it is a conserved quantity for the geodesic flow of a right-invariant metric on a group. Second, in the context of flow optimization, it emerges as a relatively natural object in the computation of the gradient, also known as backpropagation, which can aid in understanding the training flow.

Definition 20 (Momentum Maps) Let G act on a symplectic manifold P via symplectomorphisms (the diffeomorphisms of P that also preserve the symplectic form). The *momentum map* is a map $J : P \mapsto T^*_{\mathrm{Id}}G$ such that $d\langle J(z), \xi\rangle = \omega(\xi_P(z), \cdot)$ where d is the exterior derivative and ω is the symplectic form on P.

Remark 21 Such a momentum map implies that the one-form $\omega(\xi_P(z), \cdot)$ is closed, which is not guaranteed in general. Therefore, the momentum map may not exist.

We are mainly interested in the cotangent action of actions of groups that are symplectic for the canonical symplectic structure on T^*Q. Indeed, one has $\langle g^{-1*} \cdot p, g \cdot \delta q\rangle = \langle p, \delta q\rangle$. Recall that the canonical symplectic structure on T^*Q is given by the exterior derivative of the one form $-\theta(p, q)(\delta p, \delta q) = -\langle p, \delta q\rangle$. Now, applied to the infinitesimal action $\langle \xi \cdot q, -\xi^* \cdot p\rangle$, we get $\langle p, \xi \cdot q\rangle$ which is linear w.r.t. ξ. It is a linear form on $T_{\mathrm{Id}}G$ that we denote $J(p, q) \in T^*_{\mathrm{Id}}G$. Using Cartan's formula and the fact that the action preserves θ,

$$0 = \mathcal{L}_\xi \theta = d\iota_\xi \theta + \iota_\xi d\theta \implies d\iota_\xi \theta = \iota_\xi d[-\theta]. \tag{35}$$

It implies that the momentum map of Definition 20 is equal to $J(p, q)$ since $\iota_\xi \theta = \langle J(p, q), \xi\rangle$.

Corollary 22 (Momentum Map on T^*Q) *Let G act on Q, then the* momentum map of the cotangent action *is given by* $J(p, q)(\xi) = \langle p, \xi \cdot q\rangle$.

The main property of momentum maps is equivariance under the group action. This can be seen at an abstract level, but let us detail the particular case of T^*Q:

$$\langle g^{-1*} \cdot p, \xi \cdot g \cdot q\rangle = \langle p, g^{-1} \cdot \xi \cdot g \cdot q\rangle = \langle p, \mathrm{Ad}_{g^{-1}}(\xi) \cdot q\rangle. \tag{36}$$

This equality means that the cotangent action on T^*Q is seen by the momentum map on $T^*_{\text{Id}}G$ via the coadjoint action.

Proposition 23 *The following equivariance is satisfied*

$$J(g^{-1*} \cdot p, g \cdot q) = \text{Ad}^*_{g^{-1}}(J(p,q)). \tag{37}$$

Examples 24

- The momentum map for the action of $GL_d(\mathbb{R})$ on $\mathbb{R}^d$ is the outer product $p \otimes q = qp^\top$ in vector notations.
- The momentum map for the action of diffeomorphisms on groups of points is[7]

$$J\left((p_i, q_i)_{i=1,\ldots,n}\right) \sum_{i=1}^n p_i \delta_{q_i}, \tag{38}$$

where δ_{q_i} is the Dirac mass at point q_i.
- The momentum map for the action of diffeomorphisms on the space of densities is

$$J(f, \rho) = \rho \nabla f. \tag{39}$$

Dually, the momentum map for the action of diffeomorphisms on the space of functions is

$$J(\rho, f) = -\rho \nabla f. \tag{40}$$

The fact that the two momentum maps differ from a minus sign is a general fact[8] for group actions on linear spaces.

2.6 Riemannian Metrics Induced by Group Actions

Riemannian metrics are used to compute the distances between points on manifolds. Recall that a Riemannian metric g on a Riemannian manifold M gives, at each point $x \in M$, an invertible map $g(x) : T_xM \to T_x^*M$. Comparing objects that live in nonlinear spaces might require the construction of Riemannian metrics. In the case of shapes or objects in Euclidean spaces, Riemannian metrics have been introduced from the point of view of group action, in particular in the field of computational anatomy. The key point for

---

[7] We leave this computation to the reader.

[8] We leave the proof to the reader.

shape spaces is to consider objects up to some reparametrizations. As another example, optimal transport is also a way to induce a Riemannian metric on the space of measures via a variational problem which is formulated by Monge via an action of maps on a given measure. In this section, we give a common framework to construct Riemannian metrics on Q from a group G acting transitively on Q. The result of this construction is that the group action is a Riemannian submersion from the group to Q. Our discussion applies directly to the finite-dimensional case. However, our goal is to derive formally similar results in infinite dimensions.

We first start with the generalization of orthogonal projections to submersions between Riemannian manifolds

Definition 25 (Riemannian Submersion) Let (M, g_M) and (N, g_N) be two Riemannian manifolds and $\pi : M \to N$ a submersion. The map π is said to be a *Riemannian submersion* if for every $x \in M$, the following tangent map is a isometry: $d\pi(x) : H_x \mapsto T_{\pi(x)}N$ where H_x is the *horizontal space* defined as the orthogonal complement w.r.t. g_M to the *vertical space* $\mathrm{Ker}(d\pi(x))$.

There are at least two interesting properties of Riemannian submersions.

Definition 26 (Horizontal Lift) Let $\pi : M \to N$ be a Riemannian submersion and v a vector field on N. The lift of v on M is the vector field $\tilde{v}$ defined at the point x via $[\pi(x)_{|H_x}]^{-1}(v(x))$.

Another formulation of the horizontal lift is variational in the following sense: on the tangent space $T_x M$, solve

$$\arg\min_{w \in T_x M} \frac{1}{2} g(x)(w, w) \text{ s.t. } d\pi_x(w) = v. \tag{41}$$

This formulation shows (after solving the corresponding Lagrange multiplier problem) that the Riemannian submersion property can be expressed using the co-metrics (i.e. the inverse of the metric).

Proposition 27 *A map* $\pi : M \to N$ *is a Riemannian submersion if and only if* $d\pi(x) g_M^{-1}(x) d\pi^*(x) = g_N^{-1}(\pi(x))$ *for all* $x \in M$.

Definition 28 (Horizontal Lift of Geodesics) Let $\pi : M \to N$ be a Riemannian submersion and v a vector field on N. Given a length minimizing curve $y \in C^1([0, 1], N)$ between $y(0)$ and $y(1)$, choose $x_0 \in \pi^{-1}\{y(0)\}$. The horizontal lift of $x(t)$ is the curve such that $\pi(x(t)) = y(t)$ and $d\pi(x(t))(\dot{x}(t)) = \dot{y}(t)$ for all time $t \in [0, 1]$.

To get the existence of the horizontal lift, one can use the horizontal lift of the tangent vector $\dot{y}(t)$ on every point of the fiber $\pi^{-1}(\{y(t)\})$ which defines a vector field on M which is compatible with the one on N. In addition, the horizontal lift has a variational characterization:[9] it is a length-minimizing geodesic between the two fibers, with the given first point.

The most well-known result about Riemannian submersions is O'Neill's formula which says that there is more curvature on N than there is on M:

Theorem 29 (O'Neill's Formula) *Let X, Y be two orthonormal vector fields on N with horizontal lifts $\tilde{X}$ and $\tilde{Y}$, then*

$$K_N(X, Y) = K_M(\tilde{X}, \tilde{Y}) + \frac{3}{4} \| \operatorname{vert}([\tilde{X}, \tilde{Y}]) \|_M^2, \tag{42}$$

where K_M and K_N denote the sectional curvatures of M and N, $\| \cdot \|_M^2$ denotes the squared norm for the metric g_M.

The proof of this theorem relies on the computation of the Levi-Civita connections on M and N. Namely, one can show[10] $\widetilde{\nabla_X Y} = \nabla_{\tilde{X}} \tilde{Y} + \frac{1}{2} \operatorname{Vert}([\tilde{X}, \tilde{Y}])$. Using this formula, O'Neill's result follows from the definition of sectional curvature.

Example 30 Consider on $\mathbb{C}^2$ the sphere parametrized by $S_3 = \{(z_1, z_2) ; |z_1|^2 + |z_2|^2 = 1\}$. On S_3, S_1 acts by isometries by multiplication $e^{i\theta} \cdot (z_1, z_2) = (e^{i\theta} z_1, e^{i\theta} z_2)$. The quotient map $S_3 \mapsto S_3/S_1 = S_2$ is a Riemannian submersion for the usual Riemannian metric on S_2, the sphere of dimension 2.

The example above is a general fact of the principal fiber bundle. The definition of a principal fiber bundle is as follows

Definition 31 Let G be a Lie group acting on the left on P a manifold and $\pi : P \mapsto B$ be a submersion such that locally $\pi^{-1}(U) \approx G \times U$ in such a way that the action of G on $\pi^{-1}(U)$ is the left multiplication on the first factor which is G.

The consequence of the definition is that the action of G is free. Moreover, the submersion π is equal to the quotient map, $\pi : P \mapsto P/G$.

Proposition 32 (Submersion by Isometric Group Action) *Let G, P, B, π be a principal fiber bundle such that P is endowed with a Riemannian metric such that G acts*

---

[9] We leave the proof to the reader.

[10] Proof left to the reader, use the definition of the Levi-Civita connection.

via isometries. Then, there exists a (unique) metric on B such that π is a Riemannian submersion.

Sketch of Proof If there exists a Riemannian metric on B such that π is a Riemannian submersion, it must coincide with the characterization we mentioned above: consider $v \in T_b B$ and any $x \in \pi^{-1}(\{b\})$, $g(b)(v, v) = \min_{w \in T_x P} g(b)(w, w)$. To make it well-posed, we simply need to check that this definition does not depend on the choice of $x \in \pi^{-1}(\{b\})$. It is simply a consequence of the fact that G acts via isometries. □

Note that, as usual, the left action can be replaced with the right action. At first read, this structure may seem to be complicated and a bit rigid, for instance, due to the action of the Lie group via isometries. Our motivation in the next paragraph is to present a recipe to obtain such a structure starting from a left action of G on Q. We first start with the case of a transitive left action.

Proposition 33 *Let G be a Lie group acting transitively on the left on a manifold Q. Choose for each $q \in Q$ a metric on $T_{\mathrm{Id}}G$ denoted by $h(q)(\xi, \xi)$ for $\xi \in T_{\mathrm{Id}}G$. Define the metric on Q as*

$$g(q)(v,v) = \arg \min_{\xi \in T_{\mathrm{Id}}G} h(q)(\xi, \xi) \ s.t. \ \xi \cdot q = v. \tag{43}$$

Then, the map $\pi : G \to Q$ defined by $g \mapsto g \cdot q_0$ is a Riemannian submersion for the Riemannian metric m on G defined by $m(g)(\delta g, \delta g) = h(g \cdot q_0)(\delta g \cdot g^{-1}, \delta g \cdot g^{-1})$.

Moreover, when restricted to the isotropy subgroup G_{q_0}, the metric m is right-invariant.

This proposition can be seen as a particular case of the previous. Indeed, the isotropy subgroup G_{q_0} of q_0 acts via isometries on G by right multiplication. Now, the quotient G/G_{q_0} can be identified to Q. This proposition applies when the choice of the metric $h(q)$ does not depend on q. It is an important special case that has been used to induce metrics on shape spaces.

Corollary 34 (Right-Invariant Metric on G and Left Action Always Induces a Metric on the Orbit) *Let G be a Lie group endowed with a right-invariant metric acting on the left on Q. Choose $q_0 \in Q$, then the map*

$$\pi : G \mapsto Q \tag{44}$$

$$g \mapsto g \cdot q_0. \tag{45}$$

is a Riemannian submersion between G endowed with the right-invariant metric and Q with the induced metric.

The main difference with the previous result is that the metric on G does not depend on the chosen point q_0.

Hereafter, we discuss examples of actions on SPD matrices. Note that SPD matrices can represent different objects in nature: for instance, an SPD matrix can represent the correlation matrix of a multivariate Gaussian. Pushforward of a measure by a measurable map is a natural action. The space of Gaussian measures is preserved by pushforward by a linear map. This pushforward action can be solely read in terms of the correlation matrix. It is the first example of action proposed below. SPD matrices can also encode Riemannian metrics at a given point on a manifold. On Riemannian metrics, a natural action is by pullback, which is a right action. At the matrices level, it corresponds to an action of $GL(d)$ on SPD matrices as $(M, S) \to M^\top S M$.

Example 35 (Induced Metrics on SPD Matrices) Consider the group $GL(d)$ with a Frobenius right-invariant metric acting on the space of positive definite matrices $S_{++}(d)$ via $(M, S_0) \to M S_0 M^\top$. The induced metric on $S_{++}(d)$ is

$$\frac{1}{2} \text{Tr}(\delta S \, \Psi_{S^2}(\delta S)), \tag{46}$$

where $\Psi_{S^2}(\delta S)$ is the unique solution $\Sigma \in S_d$ of the (Riccati) equation

$$\Sigma S^2 + S^2 \Sigma = \delta S. \tag{47}$$

To the best of our knowledge, this metric has not been given a name in the literature.

Example 36 (The Affine Invariant Metric) Interestingly, choosing the Frobenius left-invariant metric on $GL(d)$ also induces a metric on the space of positive definite matrices via the map:

$$M \mapsto MM^\top \tag{48}$$

In this case, the orthogonal group acts by isometries and the fiber bundle result applies.[11] The induced metric is (up to a multiplicative constant) the so-called affine invariant metric

$$\text{Tr}(S^{-1} \delta S S^{-1} \delta S). \tag{49}$$

Example 37 (The Bures-Wasserstein Metric) Consider the non-right invariant metric, standard Frobenius metric on $GL(d)$, and again the map

$$M \mapsto MM^\top. \tag{50}$$

---

[11] We advise the reader to write the details.

The orthogonal group acts on $GL(d)$ via left multiplication by isometries. Using the principal fiber bundle result, one can pass to the quotient and the induced metric on $S_{++}(d)$ is the Bures-Wasserstein metric defined by

$$\text{BW}(S)(\delta S, \delta S) = \text{Tr}(HSH) \text{ s.t. } HS + SH = \delta S, \quad (51)$$

which has a unique solution. Using the recipe above, the metric is defined by minimization

$$\min_{\xi \in M(d)} \text{Tr}(\xi S \xi^\top), \quad (52)$$

under the constraint $\xi MM^\top + MM^\top \xi^\top = \xi S + S\xi^\top = \delta S$ which is the infinitesimal action of the group. After optimization, the optimal ξ is symmetric and one gets the formula proposed above.

Example 38 (Formal Infinite Dimensional Examples) Our main motivation is to induce metrics on infinite dimensional objects such as densities, functions, or groups of points. We can follow the same recipe, however the induced metrics that are constructed are not necessarily well-posed. We address this question in more detail in the next sections.

- For a right-invariant metric and pushforward action on densities, we get

$$\inf_v \|v\|_H^2 \text{ such that } \delta \rho = -\text{div}(\rho v), \quad (53)$$

where $\|\cdot\|_H$ is a Hilbert norm on the space of vector fields. One could a priori use L^2 norms, Sobolev norms, or reproducing kernel Hilbert spaces of vector fields... However, it is not clear that the induced metric is well-behaved. For instance, a non-trivial question is whether or not the induced distance is non-degenerate.

- Similarly, one can induce Riemannian metrics on the finite-dimensional space of groups of (distinct) points.

$$\inf_v \|v\|_H^2 \text{ such that } v_i = v(x_i), \quad (54)$$

where $(x_i)_{1,\ldots,n}$ is the group of distinct points and $(v_i)_{1,\ldots,n}$ are the corresponding tangent vectors. For this particular example, one immediately sees that to make the problem easily well-defined, a simple assumption is that Dirac masses are in the dual of the space of vector fields H.

- The Benamou-Brenier formulation of optimal transport exhibits the following Riemannian metric on the space of densities,

$$\inf_v \|v\|_{L^2(\rho)}^2 \text{ such that } \delta \rho = -\text{div}(\rho v) \quad (55)$$

where ρ is a density and v is a vector field. Here, the metric is L^2 with respect to the current density ρ on the space of vector fields.

As mentioned above, one main interest of Riemannian submersion is to facilitate the understanding of the geodesics downstairs with the help of the geodesics upstairs. We now turn to the derivation of the geodesic equations, which can give some information on the analytical regularity of the flow.

3 Geodesics and the Euler-Arnold-Poincaré Equation

In this section, we derive formulas for the geodesic equations, in particular for right-invariant metrics and optimal transport. These formulas are only derived formally, so we postpone further analysis. Let us recall that for a general Lagrangian $\mathcal{L}$ defined on paths $C^1([0, 1], Q)$ with fixed endpoints,

$$\mathcal{L}(c) = \int_0^1 L(c, \dot{c}) \, dt \tag{56}$$

where Q is a manifold, critical points of this functional satisfy

$$\partial_1 L(c, \dot{c}) - \frac{d}{dt} \partial_2 L(c, \dot{c}) = 0 \,. \tag{57}$$

When this Lagrangian is invariant with respect to a Lie group action denoted by H, that is $L(h \cdot c, h \cdot \dot{c}) = L(c, \dot{c})$ for every $h \in H$. It implies that the infinitesimal action of the group H on $\int_0^1 L(c, \dot{c}) dt$ reads

$$\int_0^1 \partial_1 L(c, \dot{c})(\xi \cdot c) + \partial_2 L(c, \dot{c})(\xi \cdot \dot{c}) dt = 0 \,. \tag{58}$$

Note that $\frac{d}{dt}[\xi \cdot c] = \xi \cdot \dot{c}$ since ξ does not depend on time. In particular, one can integrate by part the last term to get:

$$\int_0^1 \partial_1 L(c, \dot{c})(\xi \cdot c) - \frac{d}{dt} \partial_2 L(c, \dot{c})(\xi \cdot c) dt + [\partial_2 L(c, \dot{c})(\xi \cdot c)]_0^1 = 0 \,. \tag{59}$$

If we assume in addition that c is a solution of the Euler-Lagrange equation (57), it implies that the first two terms cancel. As a consequence, we get

$$[\partial_2 L(c, \dot{c})(\xi \cdot c)]_0^1 = 0 \,, \tag{60}$$

which is Noether's theorem: the quantity $\partial_2 L(c,\dot{c})(\xi \cdot c)$ is preserved along the flow. This quantity is nothing else than the momentum $J(\partial_2 L(c,\dot{c}), c)$. We did not discuss any regularity issue in this result, so we state the following:

Theorem 39 (Noether's Theorem, Informal) *Consider a Lagrangian $L : TQ \mapsto \mathbb{R}$ which is invariant under a left group action G, then any solution $c(t)$ of the Euler-Lagrange equation preserves the momentum (which lives in $T^*_{\mathrm{Id}} H$)*

$$J(\partial_2 L(c(t), \dot{c}(t)), c(t)) = cste. \qquad (61)$$

A right-invariant metric on a Lie group is the case where right multiplications are isometries. From *Noether's theorem*, it implies that a momentum is preserved along the geodesics. Thus, one can expect a particular form of the Euler-Lagrange equation.

3.1 General Euler-Arnold-Poincaré Equation

Let us first derive the Euler-Lagrange equations for

$$\mathcal{L}(g) = \int_0^1 \ell(\xi(t))dt \text{ s.t. } \dot{g}(t) = \xi(t) \cdot g(t). \qquad (62)$$

Since the Lagrangian only depends on $\xi(t)$, looking for an Euler-Lagrange equation written directly on $\xi(t)$ makes sense. Indeed, one has

Theorem 40 (Euler-Arnold-Poincaré Equation) *The Euler-Lagrange equation for a Lagrangian which can be written as*

$$\mathcal{L}(g) = \int_0^1 \ell(\xi(t))dt \text{ s.t. } \dot{g}(t) = \xi(t) \cdot g(t), \qquad (63)$$

reads, with the notation $m = \partial_\xi \ell(\xi(t))$,

$$\dot{m} + \mathrm{ad}^*_{\xi(t)} m = 0. \qquad (64)$$

Proof To prove it, one can first understand how variations of $g(t)$ which have fixed endpoints at times $t = 0, 1$ induce variations of $\xi(t)$. Consider a curve $g(t, s)$ parametrized by $t \in [0, 1]$ and $s \in]-\varepsilon, \varepsilon[$ an open neighborhood of 0 and define $\zeta(t)$ as $\partial_s g(t, s) = \zeta(t, s) \cdot g(t, s)$. Free variations of $g(t, s)$ with fixed endpoints imply "free values" of $\zeta(t, s)$

with fixed endpoints to 0. Differentiating with respect to t and using Schwarz theorem (cross derivatives can be exchanged), we get

$$\partial_{ts}g(t,s) = \partial_t\zeta(t,s) \cdot g(t,s) + \zeta(t,s) \cdot \xi(t,s) \cdot g(t,s)$$
$$= \partial_s\xi(t,s) \cdot g(t,s) + \xi(t,s) \cdot \zeta(t,s) \cdot g(t,s). \tag{65}$$

The previous notation can be misleading if one does not pay attention to the fact that $\zeta(t,s) \cdot \xi(t,s)$ is the infinitesimal action of the right multiplication by $g(t,s)^{-1}$ gives

$$\partial_s\xi(t,s) = \partial_t\zeta + \zeta(t,s) \cdot \xi(t,s) - \xi(t,s) \cdot \zeta(t,s) = \partial_t\zeta - \mathrm{ad}_{\xi(t,s)}\zeta(t,s). \tag{66}$$

It implies that

$$\int_0^1 \partial_t\ell(\xi(t))(\partial_t\zeta(t,0) - \mathrm{ad}_{\xi(t,s)}\zeta(t,0))dt = 0. \tag{67}$$

Integrating by parts and using the adjoint of ad leads to

$$\int_0^1 -\frac{d}{dt}\partial\ell(\xi(t))(\zeta(t,0)) - \mathrm{ad}^*_{\xi(t,s)}(\partial\ell(\xi(t,0)))\,dt = 0. \tag{68}$$

It implies the result

$$\dot{m} + \mathrm{ad}^*_{\xi(t)}m = 0, \tag{69}$$

with $m(t) = \partial\ell(\xi(t))$. □

Example 41 (The Orthogonal Group) On the orthogonal group $O(d)$, we consider the right-invariant metric induced by the Frobenius norm. The Lie algebra is the set of antisymmetric matrices and the Lie bracket on the Lie algebra is $[A, B] = AB - BA$. Therefore, the Euler-Arnold-Poincaré equation takes the form

$$\dot{Z} + [Z^\top, Z] = \dot{Z} = 0, \tag{70}$$

since $Z^\top = -Z$, the Lie bracket vanishes. In this situation, the geodesic equations are remarkably simple, $\dot{Z} = 0$. It is a more general fact for a bi-invariant Riemannian metric on a Lie group, which is the case for the metric considered in this example.

Example 42 (The Orthogonal Group with a Given Inner Product) On the orthogonal group $O(d)$, we consider the right-invariant metric induced by a norm defined by $\mathrm{Tr}(A^\top SA)$ for a positive definite matrix S. The Euler-Arnold-Poincaré equation takes

Diffeomorphic and Optimal Transport Flows

the form

$$\dot{Z} + [S^{-1}Z^\top, Z] = \dot{Z} - [S^{-1}Z, Z] = 0, \tag{71}$$

since $Z^\top = -Z$.

Example 43 (A Left-Invariant Metric on a Lie Group) Consider a left-invariant metric on a Lie group. Since the map $\text{Inv} : g \mapsto g^{-1}$ transforms the left-invariance of the metric into the right-invariance of the metric, one can write the corresponding Euler-Arnold-Poincaré equation through this inverse map. Since the Euler-Arnold-Poincaré equation is the Euler-Lagrange equation on the Lie algebra, it is sufficient to compute the differential of the inverse map. It is given by $d\,\text{Inv}(\text{id})(v) = -v$. Consequently, the Euler-Arnold-Poincaré equation reads

$$-\dot{m} + \text{ad}^*_{-\xi}(-m) = 0, \tag{72}$$

which gives

$$\dot{m} - \text{ad}^*_\xi m = 0. \tag{73}$$

The difference with the right-invariant setting is the minus sign in front of ad^*.

Example 44 (A Bi-invariant Metric on a Lie Group) Consider a bi-invariant metric on a Lie group. It implies that the map $g \mapsto g^{-1}$ is an isometry since, for d the distance on the group, one has $d(g^{-1}, h^{-1}) \underset{\text{r.i.}}{=} d(\text{id}, g^{-1}h) \underset{\text{r.i.}}{=} d(\text{id}, h^{-1}g) \underset{\text{l.i.}}{=} d(h, g)$ where r.i. (resp. l.i.) stands for right-invariance (resp. left-invariance). The Euler-Arnold-Poincaré equation for a left-invariant metric is

$$\dot{m} - \text{ad}^*_\xi m = 0.$$

However, it should also satisfy the Euler-Arnold-Poincaré equation for the right-invariant metric, which implies that $\dot{m} = \text{ad}^*_\xi m = -\text{ad}^*_\xi m = 0$.

Example 45 (On Groups Generated by Vector Fields on $\mathbb{R}^d$) Let L be a positive definite self-adjoint differential operator, define

$$\|v\|^2 := \int_{\mathbb{R}^d} \langle Lv, v \rangle dx. \tag{74}$$

It defines a (pseudo) norm on vector fields and the Euler-Arnold-Poincaré equation reads

$$\partial_t m = -(v \cdot \nabla)m - \text{div}(vm) - m \cdot Dv^\top. \tag{75}$$

3.2 Fluid Dynamic Examples

Arnold's article [3] interpreted the incompressible Euler equation as a geodesic flow on the group of volume-preserving diffeomorphisms. We treat this important example before discussing a few others.

We start by introducing the incompressible Euler equation in its Eulerian formulation on $\mathbb{R}^d$, for a time-dependent vector field $v : \mathbb{R} \times \mathbb{R}^d \to \mathbb{R}^d$,

$$\partial_t v^k + \sum_{i=1}^{d} v^i \partial_i v^k = -\partial_k p \,, \tag{76}$$

with $\text{div}(v) = \sum_{i=1}^{d} \partial_i v^i = 0$ and $p : \mathbb{R} \times \mathbb{R}^d \to \mathbb{R}$ is a function of time and space which is called the pressure. In Lagrangian coordinate (i.e. on the flow map denoted φ), the incompressible Euler equation takes the following simple form

$$\ddot{\varphi} = -\nabla p \circ \varphi \text{ with } \dot{\varphi} = v \circ \varphi \text{ and } \text{div}(v) = 0 \,. \tag{77}$$

A heuristic interpretation of this fluid dynamic model is that the particles of fluid follow straight lines deformed by the pressure field. Indeed, simplifying the equation by setting p to 0, one has $\ddot{\varphi} = 0$, which is a geodesic equation for each particle. From the Lagrangian formulation, the reader knowledgeable in Riemannian geometry might recognize a form of geodesic equation on a sub-manifold in $\mathbb{R}^d$. For a Riemannian sub-manifold M of $\mathbb{R}^d$, the geodesic equation at point $x \in M$ reads

$$\ddot{x} = \mathbf{II}(x)(\dot{x}, \dot{x}) \,, \tag{78}$$

with $\mathbf{II}$ the second fundamental form of M. In particular, a possible ambient metric is the straight L^2 metric on maps that is

$$\int_{\mathbb{R}^d} |\partial_t \varphi(t, x)|^2 dx \text{ under the constraint } \varphi \in \text{SDiff}(\mathbb{R}^d) \,, \tag{79}$$

where the notation $\text{SDiff}(\mathbb{R}^d)$ stands for the group of volume preserving diffeomorphisms. This energy functional is invariant to a change of variable by a volume-preserving diffeomorphism, in particular by φ. As a consequence, this Lagrangian also equals

$$\ell(v) = \int_{\mathbb{R}^d} |v(t, x)|^2 dx \text{ under the constraint } \text{div}(v) = 0 \,. \tag{80}$$

One directly recognizes that this Lagrangian is the right-invariant L^2 metric on $\text{SDiff}(\mathbb{R}^d)$, as put forward in Arnold's article [3]. From these remarks, two different and important directions have been taken in the literature. Developing a calculus of variation approach from the extrinsic formulation (79) to the minimization of the geodesic energy is the point of view taken by Brenier in his seminal article [15]. He was able to propose a tight convex relaxation of the variational problem in dimension greater or equal to 3. In addition, he studied polar factorization and optimal transport as a sort of byproduct of this approach [14]. Ebin and Marsden [28] developed the intrinsic point of view (80) and they proved the short-term existence of solutions by formulating the right-invariant geodesic flow as an ODE in Sobolev spaces. This seminal article has led to a systematic study of the geodesic flow of right-invariant metrics on groups of diffeomorphisms [39,48,49]. At the same time, other fluid dynamic models motivated the study of these geodesic flows, among which the Camassa-Holm equation [21] is famous. The Camassa-Holm (CH) equation as introduced in [21] is a one-dimensional PDE, which is a nonlinear shallow water wave equation and it is usually written as

$$\partial_t u - \partial_{txx} u + 3 \partial_x u\, u - 2 \partial_{xx} u\, \partial_x u - \partial_{xxx} u\, u = 0\,. \tag{81}$$

One of the most important features of the CH equation is that it describes an integrable bi-Hamiltonian system [20] with an infinite number of conserved quantities. Geometrically, it is the geodesic flow for an H^1 Sobolev metric on the diffeomorphisms group on the real line or the circle:

$$\ell(u) = \int_M |u|^2\, d\rho_0 + \frac{1}{4} \int_M |\operatorname{div}(u)|^2\, d\rho_0\,, \tag{82}$$

In dimension one, the divergence is simply the derivative of the vector field. However, this $\dot{H}^1$ metric is a natural generalization of the Camassa-Holm equation.

A last example is the Korteweg-de-Vries equation which corresponds to an L^2 metric on the semidirect product of groups $\operatorname{Diff}(S_1) \ltimes \mathbb{R}$. The corresponding geodesic equation is

$$\partial_t u = -3 u \partial_x u - a \partial_{xxx} u\,. \tag{83}$$

4 Local Existence in Lagrangian Coordinates and Analytical Properties

In this section, we discuss the local existence of the geodesic flow for certain right-invariant Riemannian metrics on diffeomorphisms groups. The results are based on the idea put forward by Ebin and Marsden that the geodesic flow can be written as an ODE in Lagrangian coordinates for smooth enough initial conditions.

4.1 Action on Operators

A key result in Ebin-Marsden's article [28] is that a conjugation action of the group of Sobolev diffeomorphisms on differential operators is smooth. This can then be used to prove that some PDE can be reformulated as ODE on Hilbert spaces. Let us start with the simple example of the Laplacian Δ on $\mathbb{R}^d$ operating on functions.

$$\varphi \mapsto \left(f \mapsto (\Delta(f \circ \varphi^{-1})) \circ \varphi \right) \tag{84}$$

is smooth w.r.t. φ. Before stating and proving a more general result, we first start with a remark. By duality, the last composition by φ can be interpreted as a left action on the dual space. Overall, on the space of function, it is a left action. By duality, it is thus a right action on the Laplacian operator. Consider a standard differential operator denoted by Op and introduce the notation

$$\mathrm{Op}_\varphi := \left(f \mapsto (\mathrm{Op}(f \circ \varphi^{-1})) \circ \varphi \right).$$

A direct computation shows that

$$[\mathrm{Op}_1]_\varphi \circ [\mathrm{Op}_2]_\varphi = [\mathrm{Op}_1 \circ \mathrm{Op}_2]_\varphi. \tag{85}$$

Let us write the infinitesimal action associated with the diffeomorphism group on Op, we get

$$-\mathrm{Op}(\nabla(f \circ \varphi^{-1})\xi) \circ \varphi + \nabla[\mathrm{Op}(f \circ \varphi^{-1})]_{|\varphi}(\xi \circ \varphi) = \left[\nabla_\xi \mathrm{Op} - \mathrm{Op}\, \nabla_\xi \right]_\varphi (f). \tag{86}$$

In conclusion, differentiation with respect to φ representing a tangent vector using right-trivialization $\xi \cdot \varphi$ leads to the commutator of the differential operator and the derivation associated with the vector field ξ. The important observation is that this commutator has the same order as Op if we leave aside the problem of regularity of ξ. In the following, we denote by $\mathcal{B}(H_1, H_2)$ the set of bounded linear operators from H_1 to H_2 two Hilbert spaces.

Theorem 46 (Regularity of Conjugation on Operators) *Let $s > d/2 + 1$ and $0 \leq k \geq d/2 + 1$ and $M = \mathbb{R}^d$ or a d-dimensional closed manifold and $\mathrm{Op} : H^s(M) \mapsto H^{s-k}(M)$ be a differential operator with smooth and bounded coefficients. The map*

$$\mathrm{Diff}_{H^s}(M) \to \mathcal{B}(H^s(M), H^{s-k}(M)), \, \varphi \mapsto \mathrm{Op}_\varphi \tag{87}$$

is C^1.

Sketch of Proof We note that right-trivialization on $T\operatorname{Diff}_{H^s}(M)$ is continuous (see the previous section), i.e. the map $(\delta\varphi, \varphi) \in T\operatorname{Diff}_{H^s}(M) \mapsto \delta\varphi \circ \varphi^{-1} \in H^s(M)$ is continuous. One can thus parameterize continuously the tangent space by (ξ, φ) such that $\delta\varphi = \xi \circ \varphi$.

A direct formal computation gives

$$-\operatorname{Op}(\nabla(f \circ \varphi^{-1})\xi) \circ \varphi + \nabla[\operatorname{Op}(f \circ \varphi^{-1})]_{|\varphi}(\xi \circ \varphi) = \left[\nabla_\xi \operatorname{Op} - \operatorname{Op} \nabla_\xi\right]_\varphi (f). \tag{88}$$

It is important to note that both terms in the commutator are defined as operators from H^s to H^{s-k-1} (if it makes sense), however, their difference is always a well-defined operator from H^s to H^{s-k}.

Instead of proving the general case, we treat the particular case of the simplest differential operator with smooth coefficients on $H^s(\mathbb{R}^d)$ simply written as $\partial^k : H^s(\mathbb{R}^d) \to H^s(\mathbb{R}^d)$ which maps $f \to g\partial^k f$ where g is a smooth and bounded function and ∂^k stands for the k-th derivative in a given coordinate. The commutator reads

$$f \mapsto \xi\partial[h\partial^k f] - h\partial^k[\xi\partial f] = \xi\partial h\partial^k f - \sum_{i=0}^{k-1} \binom{k}{i} h\partial^{i+1} f \partial^{k-i}\xi. \tag{89}$$

In this example, it is clear that the commutator is of order k. The important remaining point is that the norm of the commutator needs to be bounded by the H^s norm of ξ. To prove this fact, we need to use the multipliers' results in Sobolev spaces, recalled in Theorem 47 below. Note that $\partial^{i+1} f \in H^{s-i-1}$ and $\partial^{k-i}\xi \in H^{s+i-k}$. One has $s - i - 1 \geq s - k$ and $s - k + i \geq s - k$ and $s - k + i + s - i - 1 = s + (s - k) - 1 \geq d/2$. The last inequality is true since we assumed $s > d/2 + 1$ and $k \leq s$. Therefore, the operator has its norm bounded (up to a multiplicative constant) by $\|\xi\|_{H^s}$ (as a sum of linear operators satisfying this property). The more general case is similar. We refer the reader to [28] for more details. The resulting map is C^1 since the right-trivialization is continuous. □

Although the theorem below is not stated in this form in [10] for a closed manifold, an argument based on a chart decomposition would give the result.

Theorem 47 (Multiplication in Sobolev Spaces, [10]) *Let $M = \mathbb{R}^d$ or M be a closed manifold of dimension d. The multiplication*

$$H^{s_1}(M) \times H^{s_2}(M) \mapsto H^s(M), (f, g) \mapsto fg, \tag{90}$$

is a continuous bilinear operator if $\min(s_1, s_2) \geq s$, $s \geq 0$, *and* $s_1 + s_2 - s \geq \frac{d}{2}$.

Remark 48 Theorem 46 is stated for differential operators of positive orders. However, using the implicit function theorem in [42], we obtain that their inverses have the same regularity property.

Remark 49 One can re-iterate[12] the argument of C^1 regularity to obtain higher-order regularity. However, C^1 regularity is sufficient to get the local existence of certain PDEs such as incompressible Euler.

4.2 Application to the Incompressible Euler Equation

The article [27] makes very accessible the results in the groundbreaking article [28]. We highly recommend it to the reader, since this paragraph is completely based on it. We have seen that the incompressible Euler equation reads

$$\ddot{\varphi}(t, x) = -\nabla p(t, \varphi(t, x)), \qquad (91)$$

with ∇p corresponding to the Lagrange multiplier associated with the incompressibility constraint.

Theorem 50 (Informal Formulation, See [28]) *The incompressible Euler equation is an ODE when considered on* $\text{Diff}_{H^s}(M)$ *for* $M = \mathbb{R}^d$ *or M a closed d-dimensional Riemannian manifold.*

Ideas of Proof, See [27] We aim to write $\nabla p \circ \varphi$ as a function of $\varphi, \dot\varphi$ to turn it into a second-order ODE on a Sobolev space. We know that $\partial_t v + \nabla_v v = -\nabla p$ in Eulerian coordinates, implying that (taking divergence on both sides)

$$\Delta p = -\operatorname{div}(\nabla_v v).$$

Note that the right-hand side is our data and p is the unknown. This equation defines p up to a constant. Let us write

$$\nabla p = \nabla \Delta^{-1}(-\operatorname{div}(\nabla_v v)). \qquad (92)$$

To apply the conjugation operator result Theorem 46, we need to check that $[\nabla \Delta^{-1} \operatorname{div}(\nabla_v v)] \circ \varphi$ is a conjugation operator. We have

$$[\nabla \Delta^{-1} \operatorname{div}(\nabla_v v)] \circ \varphi = [\nabla \Delta^{-1} \operatorname{div}(\nabla_{\dot\varphi \circ \varphi^{-1}} \dot\varphi \circ \varphi^{-1})] \circ \varphi. \qquad (93)$$

---

[12] We leave it to the reader.

In this formulation, although it is possible to apply the conjugation result, there is a loss of derivative: the operator $\nabla \Delta^{-1}$ div is 0th-order and due to the ∇ inside, there is a loss of one order. The key point is to write this operator itself as a commutator on divergence-free vector fields,

$$\operatorname{div}(\nabla_Z v) = \operatorname{div}(\nabla_Z v) - \nabla_Z \operatorname{div}(v) = [\operatorname{div}, \nabla_Z]. \tag{94}$$

Therefore, the resulting operator is of order 1 which implies that $\nabla \Delta^{-1}[\operatorname{div}, \nabla_v]$ is of order 0. The last point to check is the regularity of

$$(\delta\varphi, \varphi) \mapsto \left(f \mapsto \left(\nabla \Delta^{-1}[\operatorname{div}, \nabla_{\delta\varphi \circ \varphi^{-1}}](f \circ \varphi^{-1}) \right) \circ \varphi \right). \tag{95}$$

By the composition result on operators, it is sufficient to check the regularity of $[\operatorname{div}, \nabla_{\delta\varphi \circ \varphi^{-1}}]_\varphi$ (the only difference with the result above being the dependence on v of δ). For an explicit proof of this fact, we refer to [27], however, let us note that the proof is similar to the smoothness of the operator by introducing $[\operatorname{div}, \nabla.]_{\varphi,2} := ((Z, f) \mapsto [\operatorname{div}, \nabla_{Z \circ \varphi^{-1}}](f \circ \varphi^{-1}))$. This operator is C^1 from H^s into H^{s-1} by similar arguments. $\square$

4.3 Application to Sobolev Type Right-Invariant Riemannian Metrics

Let us start with a quick discussion on Riemannian metrics in infinite dimensions.

Definition 51 (Weak and Strong Riemannian Metrics) Let M be an infinite dimensional manifold modeled on a Hilbert space H. A *weak* Riemannian metric is a map

$$G : TM \times_M TM \to R, \tag{96}$$

such that $G(x)$ is a bilinear map for all $x \in M$ and $G(x)(h, h) \geq 0$ with equality only if $h = 0$. A weak Riemannian metric is said to be *strong* if it satisfies in addition that the topology induced on the tangent space $T_x M$ by G_x coincides with the topology inherited by the manifold structure.

Example 52 The obvious example of a weak Riemannian metric is the study of the L^2 metric on $\operatorname{SDiff}(\mathbb{R}^d)$ the group of volume-preserving diffeomorphisms when studied in the H^s topology for $s > d/2 + 1$.

Example 53 More generally, we are interested in Sobolev right-invariant metrics on the group of diffeomorphisms. We consider a differential operator L such as the one associated with the H^k norm for $k \geq 0$. A typical example of a weak Sobolev metric is to consider an $H^k(M)$ norm on M of dimension d, say for instance $k \leq d/2$, and study it under a stronger topology, the one associated with $\text{Diff}_{H^s}(M)$. In these cases, the H^k right-invariant metric on $\text{Diff}_{H^s}(M)$ is a weak Riemannian metric since the topologies do not coincide. In the case $s = k$, the two topologies coincide and the H^s right-invariant metric on $\text{Diff}_{H^s}(M)$ is strong.

One can write the H^k norm with a differential operator L

$$\int \langle L(\delta\varphi \circ \varphi^{-1}), \delta\varphi \circ \varphi^{-1}\rangle dx \,. \tag{97}$$

This formulation involves the right-trivialization, the term $\delta\varphi \circ \varphi^{-1}$ is only continuous and not differentiable in the second argument. Therefore, one does not expect this Riemannian metric to be smooth, at least in a sufficiently high Sobolev space such as $\text{Diff}_{H^s}(M)$. Perhaps surprisingly we have

Theorem 54 (Regularity of the Sobolev Metrics) *Let* $M = \mathbb{R}^d$ *or a d-dimensional closed Riemannian manifold and $k \geq 0$. The metric tensor*

$$G_{H^k}(\varphi)(\delta\varphi, \delta\varphi) := \|\delta\varphi \circ \varphi^{-1}\|_{H^s}^2 \,, \tag{98}$$

is a smooth metric tensor on $T \text{Diff}_{H^s}(M)$ *for* $s > \min(d/2 + 1, k)$.

Proof Let us write the H^k norm as $\|f\|_{H^s}^2 = \int_x \langle Lf(x), f(x)\rangle dx$ with L a differential operator. Then, the conjugation result shows that $\varphi \mapsto L_\varphi(f)$ is C^1, but iterating the argument, it is smooth. Therefore, with a change of variable by φ^{-1}, recall that $\int f \circ \varphi g dx = \int fg \circ \varphi^{-1} \text{Jac}(\varphi^{-1}) dx$, we get

$$G_{H^k}(\varphi)(\delta\varphi, \delta\varphi) = \int_M \langle L_\varphi(\delta\varphi), \delta\varphi\rangle \text{Jac}(\varphi) dx \,. \tag{99}$$

Now, since $H^s(M)$ is a Hilbert algebra, the map $(\delta\varphi, \varphi) \in T \text{Diff}_{H^s}(M) \mapsto \langle L_\varphi(\delta\varphi), \delta\varphi\rangle \text{Jac}(\varphi) \in H^s(M)$ is smooth. The result follows. □

Corollary 55 (Geodesic Flow of H^k Right-Invariant Metrics) *In the setting of Theorem 54, the geodesic flow associated with the H^k metric is locally well-defined on $\text{Diff}_{H^s}(M)$.*

Proof By the result above, the geodesic flow is a second-order ODE on $\text{Diff}_{H^s}(M)$, and the Banach fixed point theorem can be applied. □

The result is only local and it is not possible to conclude the long-time existence of the geodesic flow for a general right-invariant H^k weak Riemannian metric without any other argument. However, in finite dimensions, on a Lie group with a right-invariant Riemannian metric, the following argument can be applied: local existence at identity can be translated at any given point since the geodesic flow preserves the kinetic energy. This argument cannot be applied for a weak Riemannian metric since the kinetic energy conservation gives a bound on the weaker norm, whereas we would need boundedness in the stronger norm. This is why when the metric is strong, the geodesic flow exists for all times. We have more [17]:

Theorem 56 (Geodesic Completeness of $\text{Diff}_{H^s}(M)$**)** *Let* $s > d/2 + 1$ *where* d *is the dimension of* M *being* $\mathbb{R}^d$ *or a closed manifold. The right-invariant* H^s *metric on* $\text{Diff}_{H^s}(M)$ *is geodesically complete (the geodesic flow exists for all time).*

In the next section, we will see that one can define more general right-invariant metrics on diffeomorphisms groups for which metric completeness will be obtained. Note that geodesic completeness is implied by metric completeness. However, in this construction, we lose the structure of a Hilbert manifold.

We conclude this section by mentioning the no loss–no gain of regularity result on the geodesic flow on $\text{Diff}_{H^s}(M)$. Since the H^k right-invariant metric is smooth on Diff_{H^s} as soon as $s > d/2 + 1$, one can define $T_s^*(\delta\varphi, \varphi)$ the maximal existence time of the geodesic flow with the initial condition $(\delta\varphi, \varphi) \in T \text{Diff}_{H^s}$. Here, there is a choice of regularity under which the solution is studied. It could be possible that the time interval of existence differs between different regularities. However, this is not the case. The following result already present in [28] is nicely explained in [7, Appendix].

Theorem 57 (No Loss–No Gain) *Consider the* H^k *right-invariant metric on* Diff_{H^s}. *Then, one has, for an initial condition* $(\delta\varphi, \varphi) \in \text{Diff}_{H^{s+1}}$

$$T_s^*(\delta\varphi, \varphi) = T_{s+1}^*(\delta\varphi, \varphi). \tag{100}$$

Sketch of Proof Consider the geodesic flow at time t, $(\delta\varphi, \varphi) \to \text{Fl}_t(\delta\varphi, \varphi)$. Assume in addition that $(\delta\varphi, \varphi)$ is in H^{s+1}. By right-invariance of the geodesic flow, one has

$$\text{Fl}_t(\delta\varphi \circ \psi(s), \varphi \circ \psi(s)) = \text{Fl}_t(\delta\varphi, \varphi) \circ \psi(s), \tag{101}$$

for every smooth curve $\psi(s) \in H^s$. The second component of the right-hand side term can be simply written as $\varphi(t) \circ \psi(s)$, which can be differentiated w.r.t. s into $D\varphi(t)(\partial_s\psi(s))$. The left-hand side gives us some information on the right-hand side since $(\delta\varphi \circ \psi(s), \varphi \circ \psi(s))$ can be differentiated in s to obtain $D\text{Fl}_t(\delta\varphi, \varphi)(D\delta\varphi(\partial_s\psi(s)), D\varphi(\partial_s\psi(s)))$. This term is actually in H^s by the smoothness of the geodesic flow, therefore this gives that $D\varphi(t)(\partial_s\psi(s))$ is H^s which implies that $\varphi \in H^{s+1}$. □

This no loss–no gain result is expressed as an initial value problem. From an optimization perspective, the boundary value problem presents more interest, and in this context, it is natural to ask whether a no loss–no gain result holds. This question is positively answered in [35] under mild conditions when M is the flat torus.

5 Diffeomorphisms Groups Generated by Flows of Vector Fields

Instead of starting with an open set of a functional space (e.g. Diff$_{H^s}$), one can directly start with a Hilbert space of vector fields to construct a group just by integrating the flow of a Lipschitz vector field. This was one of the starting points in computational anatomy [26, 59]. In this field, the problem of interest is the registration problem, which consists in finding a deformation of the ambient space that transforms one object into another. These objects can be images, groups of points, or measures.

In this section, we consider a bounded domain $\Omega \subset \mathbb{R}^d$. The construction also works on $\mathbb{R}^d$. In order to have well-defined flows of diffeomorphisms, it is convenient to work with vector fields that belong to $C^1(\Omega, \mathbb{R}^d)$. In what follows, we will denote by V a space of (sufficiently smooth) vector fields. To be precise, we define the notion of *admissibility* for a linear space of vector fields:

Definition 58 (Admissible Space of Vector Fields) A separable Hilbert space of vector fields V defined on Ω is said admissible if

- for any $v \in V$, $v(x) = 0$ and $dv(x) = 0$ if $x \in \partial\Omega$.
- There exists a constant K such that, for all $v \in V$

$$\|v\|_{1,\infty} \leq K \|v\|_V \tag{102}$$

where $\|v\|_{1,\infty}$ denotes the sup norm of v and its first derivative and $\|\cdot\|_V$ denotes the given Hilbert norm on V.

A well-known example of such a space is $H^s(\Omega, \mathbb{R}^d)$ for $s > d/2 + 1$. More generally, the Rellich-Kondrachov theorem states that $W^{j+m,p}(\Omega) \hookrightarrow C^j(\Omega)$ for $mp > d$. We then have:

Proposition 59 *Let $v \in L^1([0, 1], V)$ be a time dependent vector field and consider the Banach space B_0 (where $B_i := C^i(\Omega, \mathbb{R}^d)$ endowed with the sup norm $\|\cdot\|_{i,\infty}$ of the first i derivatives). There exists a unique solution $\varphi \in W^{1,1}([0, 1], B)$ to the flow equation*

$$\begin{cases} \partial_t \varphi(t) = v(t) \circ \varphi(t) \\ \varphi(0) = \mathrm{Id} \,. \end{cases} \tag{103}$$

In addition, $\varphi(\Omega) = \Omega$.

We will also denote $\varphi(t)$ and $v(t)$ by φ_t and v_t.

Proof Since V is admissible, $v \in L^1([0, 1], V)$ implies $v \in L^1([0, 1], B)$ and v is a L^1–Lipschitz function from $I \times B$ to B. The first part of the proposition follows from Theorem 116.

The fact that $\varphi(\Omega) \subset \Omega$ is a direct consequence of uniqueness of solutions: if for some $t \in [0, 1]$, $\varphi_t(x) \in \partial\Omega$, it implies that for all $t \in [0, 1]$, $\varphi_t(x) \in \partial\Omega$. Indeed the first point of the proposition applies not only to a domain but also for any compact sets and thus for $\Omega = \{x\}$. In this case, the solution is $\phi_t(x) = x$ since the boundary condition is $v_t(x) = 0$. □

We can improve on the smoothness of φ.

Theorem 60 *Let $v \in L^1([0, 1], V)$ be a time dependent vector field and $B := C^0(\Omega, \mathbb{R}^d)$ endowed with the sup norm. The solution $\varphi(t)$ is a continuous path in $\mathrm{Diff}^1(\Omega)$, the space of C^1 diffeomorphisms satisfying $\|D\varphi(t)\|_\infty \leq e^{\int_0^t K \|v(t)\| dt}$.*

Proof Let us first assume that $v \in L^1([0, 1], B_2)$, then Theorem 116 implies that φ is in $W^{1,1}([0, 1], B_1)$. Looking at the equation defining $D\varphi$ by differentiating the flow equation, we have:

$$\partial_t D\varphi_t = Dv_t \circ \varphi_t \cdot D\varphi_t \tag{104}$$

This ODE is again a Caratheodory differential equation on $C^0(\Omega, L(\mathbb{R}^d, \mathbb{R}^d))$ endowed with the sup norm.

We now consider an approximation of v_n of $v \in V$ by vector fields in $L^1([0, 1], B_2)$. Using Proposition 59, we get that the solution φ_n is well-defined. Moreover, by a direct application of the dominated convergence theorem $t \to Dv_n(t) \circ \varphi_n(t)$ converges in L^1 to $Dv(t) \circ \varphi$ where φ is the flow of v. Finally, applying Proposition 118, the solution $D\varphi_n$ uniformly converges to $D\varphi$. Since the space B_1 is complete, φ belongs to B_1.

In fact, φ is a diffeomorphism since its inverse is given by the flow of $t \to -v(1 - t)$ and the last inequality is derived from equation (104) and Gronwall's lemma 117. □

Remark 61 We can observe that

(1) The solution $\varphi(t)$ is absolutely continuous.
(2) The previous result can be extended to smoother vector fields. The flow of B_k vector fields will be C^k and the inequality (60) can be generalized to higher derivatives.

One can relax the assumption $v \in B_i$ and replace it with $v \in L^2([0, 1], H^s)$ where H^s is the Sobolev space of order s. It can be proven that for $s > d/2 + 1$, the flow of v is an absolutely continuous path in H^s. The case $s > d/2 + 2$ follows as in Proposition 59, and the other cases require approximation arguments. The proof of this can be found in [17].

We are now in the position of formulating a well-posed variational problem that solves the diffeomorphic image matching problem using time-dependent vector fields:

Theorem 62 *Let $I, J \in L^2(\Omega, \mathbb{R})$ and V be an admissible space of vector fields. The functional*

$$\mathcal{J}(v) = \int_0^1 \|v(t)\|_V^2 dt + \|I \circ \varphi^{-1} - J\|_{L^2}^2 , \tag{105}$$

attains its infimum on $L^2([0, 1], V)$.

Proof The first term $\int_0^1 \|v(t)\|_V^2 dt$ is a norm on a Hilbert space, so that it is lower semi-continuous w.r.t. the weak convergence. Since V is separable, $L^2([0, 1], V)$ is also separable so that bounded balls are compact for the weak topology. The existence of a minimizer for $\mathcal{J}$ follows from the continuity of the flow on B_0 w.r.t. the weak topology, which is proven in the next lemma. Indeed, if I is a Lipschitz function, then $\|I \circ \varphi^n - I \circ \varphi\|_\infty \leq \mathrm{Lip}(I) \|\varphi^n - \varphi\|_\infty$ so that the result is clear. The more general case $I \in L^2$ follows via the application of the dominated convergence theorem using approximations by Lipschitz functions and the fact that $\mathrm{Jac}(\varphi^n)$ is bounded in $L^\infty(\Omega, \mathbb{R})$. □

Remark 63 In fact, the previous theorem is true for any similarity measure that is lower semi-continuous w.r.t. to the weak topology. Using the following lemma, are available similarity measures such as: $\sum_{i=1}^k \|\varphi(q_i) - x_i\|^2$ where q_i and x_i are two given sets of points in Ω.

Lemma 64 *The flow map $\Phi : L^2([0, 1], V) \to B_0$ defined by $\Phi(v) = \varphi(1)$ is continuous for the weak topology on $L^2([0, 1], V)$.*

Proof For any $x \in \Omega$, we have

$$\|\varphi_t^n(x) - \varphi_t(x)\| \leq \left\| \int_0^t v_s^n(\varphi_s^n(x)) - v_s(\varphi_s(x)) ds \right\|$$

$$\leq \int_0^t \|v_s^n(\varphi_s^n(x)) - v_s^n(\varphi_s(x))\| ds + \left\| \int_0^t v_s^n(\varphi_s(x)) - v(\varphi_s(x)) ds \right\| .$$

Remark that the second term can be written as $\|m_x(v_n) - m_x(v)\|$ where

$$m_x(v) := \int_0^t v_s(\varphi_s(x)) ds , \tag{106}$$

which is a continuous linear form on $L^2([0, 1], V)$. In addition, the family m_x are equicontinuous w.r.t. $x \in \Omega$. Indeed,

$$\begin{aligned}
\|m_x(v) - m_y(v)\| &\leq \int_0^t \|v_s(\varphi_s(x)) - v_s(\varphi_s(y))\| ds \\
&\leq \int_0^t K \|v_s\|_V \|\varphi_s(x) - \varphi_s(y)\| ds \\
&\leq K' \int_0^t \|v_s\|_V \|x - y\| ds \\
&\leq K' \sqrt{t} \, \|x - y\| \|v\|_{L^2([0,1],V)}.
\end{aligned}$$

In particular, uniform convergence w.r.t. $x \in \Omega$ is obtained, i.e.

$$\lim_{n \to \infty} \sup_{x \in \Omega} \left\| \int_0^t v_s^n(\varphi_s(x)) - v(\varphi_s(x)) ds \right\| = 0.$$

Going back to the first inequality, we get, since v_n is a bounded sequence,

$$\|\varphi_t^n(x) - \varphi_t(x)\|_\infty \leq \alpha_n(t) + \|\varphi_t^n(x) - \varphi_t(x)\|_\infty \int_0^t K \|v_n(s)\|_V ds,$$

where $\alpha_n(t) = \sup_{x \in \Omega} \|\int_0^t v_s^n(\varphi_s(x)) - v(\varphi_s(x)) ds\|$. The conclusion follows by application of Gronwall's lemma 117. □

A consequence of the previous lemma is the following theorem that can be found in [59] or [63],

Theorem 65 (Trouvé) *The image of the map Φ (i.e. all the flows at time 1) is a group of C^1 diffeomorphisms. The distance defined on this group denoted by G_V by*

$$d(\varphi, \psi) := \inf\{\|v\|_{L^2([0,1],V)} \mid \Phi(v) \circ \varphi = \psi\} \tag{107}$$

makes the group a complete metric space and there exists a minimizing vector field realizing the distance between two given φ and ψ.

Proof There is only one important point in proving that d is a distance:

$$d(\varphi, \psi) = 0 \implies \varphi = \psi.$$

Let us consider $\varphi \neq \psi$ then there exists $x \in \Omega$ such that $\varphi(x) \neq \psi(x)$ and consider a path φ_s joining φ to ψ. We thus have

$$\|\varphi(x) - \psi(x)\| \leq \int_0^1 \|v_s(\varphi_s(x))\| ds \leq \int_0^1 K \|v_s\|_V ds \leq \|v\|_{L^2([0,1],V)} \tag{108}$$

and it follows that $d(\varphi, \psi) \geq \frac{1}{K} \|\varphi(x) - \psi(x)\| > 0$.

The existence of a minimizer follows as in the previous theorem: if $v_n \in L^2([0,1], V)$ is a minimizing sequence satisfying the boundary conditions $\Phi(v_n) \circ \varphi = \psi$, by weak compactness of bounded balls, there exists a subsequence of v_n that weakly converges to v. By lower semi-continuity of the norm, v is a minimizer provided that $\Phi(v) \circ \varphi = \psi$ which is true by the previous lemma.

Let us show that G_V is a complete metric space: Consider φ_n a Cauchy sequence such that $d(\varphi_n, \varphi_{n+1}) \leq 1/2^n$, it is thus possible to concatenate the minimizing vector fields u_n between φ_n and φ_{n+1} to obtain a vector field $u_\infty \in L^2([0,1], V)$. It is easy to prove that its flow φ_∞ is the limit of φ_n. For a general Cauchy sequence, there exists a subsequence satisfying the above condition. This subsequence converges and consequently, the sequence as well. □

Remark 66 If the metric is not strong enough then this lower bound may vanish. For instance, the L^2 right-invariant metric gives a degenerate distance on the group of diffeomorphisms [19, 47]. In fact, on S_1 the H^s right-invariant distance is known to be degenerate for $s \leq 1/2$, which is the critical index: If $s > 1/2$, the distance is not degenerate [8]. In addition, it is not degenerate for $s \geq 1$ in any dimension [47]. In higher dimensions, the critical value is $s = 1$, if $s \geq 1$, the metric is not degenerate. For more details on this question, we refer the reader to [6, 38].

The following result is based on the previous results and [29], it improves a bit Theorem 9.1 in [50]:

Theorem 67 *For $s > d/2 + 1$, $M = \mathbb{R}^d$ or a d-dimensional closed Riemannian manifold, the group $\mathrm{Diff}_{H^s}(M)$ is an infinite dimensional Riemannian manifold modeled on H^s, which is metrically complete and geodesically complete. In addition, between any two diffeomorphisms in Diff_{H^s}, there exists a minimizing geodesic. This minimizing geodesic is unique on a G_δ dense subset.*

Proof See [17]. □

Question 68 *Are there other interesting examples of diffeomorphism groups generated by flows of vector fields in a reproducing kernel Hilbert space that are also (infinite dimensional) manifolds?*

6 Induced Metrics and the Case of Measures

As we have seen in Sect. 2.6, right-invariant metrics on groups induce metrics on the spaces on which they act. Since we have defined the diffeomorphism group in more detail, we look at examples and focus on the case of measures, which is an infinite-dimensional case. In infinite dimensions, the smoothness of the Riemannian submersion setting is sometimes lost. However, the metric properties might still be preserved. To capture such situations, Berestovskii introduced the notion of submetries in [12].

Definition 69 (Submetries) A map $\pi : M \to N$ between two metric spaces is a *submetry* if for every $r > 0$

$$\forall x \in M, \ \pi(B(x,r)) = B(\pi(x), r). \tag{109}$$

Examples of submetries include quotient maps of isometric group actions, possibly in infinite dimensions. A nice example is optimal transport.

Example 70 (Optimal Transport as a Submetry) Consider a Polish space (M, d) and a reference probability measure with no atoms $\eta \in \mathcal{P}(M)$. Let $L^2(M, M; \eta)$ be the set of measurable maps f from M to M such that $\int_M d(x_0, f(x))^2 d\eta(x)$ is finite where $x_0 \in M$ is any reference point. Then, the map

$$L^2(M, M) \to \mathcal{P}(M), \ \varphi \mapsto \varphi_\sharp(\eta), \tag{110}$$

is a submetry between $L^2(M, M; \eta)$ and $\mathcal{P}(M)$ endowed with the Wasserstein L^2 metric. To check that this is the case, we remark that the set of measurable maps $L_\eta : M \to M$ that preserve the measure η act by isometries by right-multiplication: namely, $\|\varphi_1 \circ L_\eta - \varphi_2 \circ L_\eta\|^2 = \|\varphi_1 - \varphi_2\|^2$ by the change of variable formula. Now, by [61], we know that optimal transport maximizes the correlation between two probabilities, in other words,

$$W_2(\mu, \nu)^2 = \min_{\varphi_1, \varphi_2} \int_M d(\varphi_1(x), \varphi_2(x))^2 d\eta(x), \ \text{s.t.} \ [\varphi_1]_\sharp(\eta) = \mu \text{ and } [\varphi_2]_\sharp(\eta) = \nu. \tag{111}$$

The optimization set is nonempty since η has no atoms. From the Kantorovich formulation of optimal transport, the infimum is attained, see [61]: Let $\gamma \in \mathcal{P}(M \times M)$ be an optimal transport plan between μ and ν. Then, since η has no atom, there exists a measurable map $(\varphi_1, \varphi_2) : M \to M \times M$ such that $(\varphi_1, \varphi_2)_\sharp \eta = \gamma$. Now, given $\varphi_1 : M \to M$ such that $[\varphi_1]_\sharp(\eta) = \mu_1$, one can consider a coupling plan

6.1 Induced Metrics on Groups of Points (a.k.a. Space of Landmarks)

Let us consider the action of a diffeomorphism group Diff_V (equipped with a right-invariant metric determined by a reproducing kernel Hilbert space V) on the set of n-distinct points in a domain $\Omega \subset \mathbb{R}^d$. This set of points is often called the space of landmarks. The manifold of landmarks is an open set of Ω^n defined by

$$\mathcal{L}_n := \{(q_1, \ldots, q_n) \in \Omega^n \mid q_i \neq q_j \text{ for } i \neq j\}.$$

It is a connected manifold if $d \geq 2$. Recall that the action $\text{Diff}_V \times \mathcal{L}_n \to \mathcal{L}_n$ is defined by

$$(g, (x_1, \ldots, x_n)) \mapsto (g(x_1), \ldots, g(x_n))$$

and the infinitesimal action is thus

$$v \cdot (x_1, \ldots, x_n) \mapsto (v(x_1), \ldots, v(x_n)).$$

The momentum map reads in coordinates

$$(x_1, \ldots, x_n), (p_1, \ldots, p_n) \in T^*\mathcal{L}_n \mapsto \sum_{i=1}^n \delta_{x_i}^{p_i} \in V^*. \tag{112}$$

Since V is an admissible space of vector fields, this application is well-defined since the Dirac operators belong to V^*. The tangent action reads

$$g \cdot (x_1, \ldots, x_n), (v_1, \ldots, v_n) \mapsto (g(x_1), \ldots, g(x_n)), \left(Tg_{x_1}(v_1), \ldots, Tg_{x_n}(v_n)\right).$$

and the co-tangent action is:

$$g \cdot (x_1, \ldots, x_n), (p_1, \ldots, p_n) \mapsto (g(x_1), \ldots, g(x_n)), \left(Tg_{x_1}^{-1*}(p_1), \ldots, Tg_{x_n}^{-1*}(p_n)\right).$$

Since V is a reproducing kernel Hilbert space (see Appendix 2), the kernel k satisfies $(\delta_x^p, v)_{V^*, V} = \langle k(., x)p, v \rangle_V$. To compute the metric on $T\mathcal{L}_n$, it suffices to find the minimal norm vector field $v \in V$ that interpolates the values $v_i \in T_{x_i}M$. By the representer theorem for reproducing kernel Hilbert spaces, this minimal norm vector field can be written as $v(x) = \sum_{j=1}^n k(x, x_j) p_j$ for $p = (p_1, \ldots, p_n)$ such that $\sum_j k(x_i, x_j) p_j = v_i$. Since the kernel matrix $[k(x_i, x_j)]_{i,j}$ is invertible, the solution exists and is unique. In particular, we have (by definition of the norm in V), for $q = (x_1, \ldots, x_n)$ and $p = (p_1, \ldots, p_n)$ corresponding to the minimal norm vector field v,

$$\|v\|^2 = \frac{1}{2}\|J(p, q)\|_{V^*}^2 = \frac{1}{2} \sum_{i,j=1}^n \langle p_i, k(x_i, x_j) p_j \rangle. \tag{113}$$

The geodesic flow associated with this Riemannian metric on the landmark space can be written in Hamiltonian formulation

$$\begin{cases} \dot{p}_i = -\partial_{q_i} H(p,q) = -\sum_{j=1}^n \langle p_i, \partial_1 k(q_i, q_j) p_j \rangle = -dv^*(q_i)(p_i) \\ \dot{q}_i = \partial_{p_i} H(p,q) = \sum_{j=1}^n k(q_i, q_j) p_j = v(q_i), \end{cases} \quad (114)$$

where

$$H(p,q) := \frac{1}{2} \sum_{i,j=1}^n \langle p_i, k(x_i, x_j) p_j \rangle.$$

Note that $\partial_p H(p,q) = \sum_{j=1}^n k(q_i, q_j) p_j$ which is simply the evaluation at point q_i of the vector field $v(x) = \sum_{j=1}^n k(x, q_j) p_j$. This Hamiltonian formulation can be used for numerical purposes, the forward model being the Hamiltonian flow.

Proposition 71 *If V is admissible, the landmark space $\mathcal{L}_n$ is a complete Riemannian manifold. In other words, solutions to System (114) are defined for all time.*

Proof Denote by φ the horizontal lift of (p,q) to Diff$_V$, i.e. the flow of the time-dependent vector field $k \star J(p(t), q(t))$ (this notation means the integral kernel operation on $\sum_{i=1}^n \delta_{x_i}^{p_i}$ and k is the kernel associated with V). Since the Hamiltonian function is constant along a solution, it implies that $d^2(\text{Id}, \varphi(t)) \leq cste\, t$. Using the bounds on the flow (60), it implies that (p,q) stays in a compact set of $T^*\mathcal{L}_n$ for t bounded. Let us underline that the important point is to lower bound the minimum distance $|x_i - x_j|$ which is possible since they are the images of the initial points by the diffeomorphism $\varphi(t)$. Thus, the solutions can be extended for all time. □

Remark 72 (Peakons) The above result does not apply to the kernel associated with the H^1 norm $k(x, y) = e^{-\|x-y\|/\sigma}$ Id (in one dimension). Indeed, H^1 cannot be continuously embedded in C^1 endowed with the sup norm, whatever the dimension. Solutions to System (114) for this particular kernel are called peakons [36]. The Riemannian submersion approach still applies but the normal metric does not make $\mathcal{L}_n$ a complete Riemannian manifold. In general, conditions on the kernel can be derived to obtain the geodesic completeness of the space of landmarks.

6.2 Induced Metrics on Measures

We are now interested in $\mathcal{P}(\Omega)$, the space of probability measures (it could be more generally measures). We consider the group Diff$_V$ acting by pushforward on $\mathcal{P}(\Omega)$: $\varphi \cdot \mu := \varphi_\sharp \mu$. The action of Diff$_V$ is not transitive on $\mathcal{P}(\Omega)$ since diffeomorphisms preserve

Dirac masses and densities. More generally, Diff$_{H^s}$ acts on H^{s-1}-densities, preserving their H^{s-1} regularity. This can be seen from the definition of the action when written on densities:

$$\varphi \cdot \rho = \mathrm{Jac}(\varphi^{-1})\rho(\varphi^{-1}) \,. \tag{115}$$

This situation is in sharp contrast with optimal transport in which one can always transport between two probabilities. The induced metric of diffeomorphisms groups on probabilities is only a pseudo-metric since the metric is only defined between two elements that belong to the same group orbit. However, let us write the formal geodesic equations in Hamiltonian form which is very similar to the landmark space:

$$\begin{cases} \dot{p} + \langle \nabla p, v \rangle = 0 \\ \dot{\rho} + \mathrm{div}(\rho v) = 0 \\ v(x) = \int_\Omega k(x,y) \nabla p(y) \rho(y) dy \,. \end{cases} \tag{116}$$

In the previous equations, p is a function, and ρ is a density. The Hamiltonian reads

$$H(p,\rho) = \frac{1}{2} \int_{\Omega^2} \langle \nabla p(x), k(x,y) \nabla p(y) \rangle \rho(x) \rho(y) dx dy \,. \tag{117}$$

We only stay at a formal level and do not discuss the regularity issues associated with the momentum $\rho \nabla p$. The first equation of System (116) is an advection equation on the values of p. The second equation is the continuity equation, the transport action on mass.

Remark 73 In computational anatomy, the case of images is important and the geodesic flow in Hamiltonian form is similar to System (116) except that the state and momentum equations are exchanged. These equations can be used in practical implementations [60].

6.3 The Optimal Transport Metric via the Benamou–Brenier Formulation

In this section, we introduce the Benamou-Brenier formulation [11] of the optimal transport problem (see also [61] for more details and [62] for the reader interested in a broad perspective on optimal transport). Formally, the L^2 metric on the diffeomorphisms group descends to a metric on the space of probability measures. This is made rigorous analytically by the Benamou–Brenier formula in its convex formulation.

Definition 74 (Benamou-Brenier—Non-convex Formulation) Consider a Riemannian manifold (M, g),

$$W_2(\rho_0, \rho_1)^2 = \inf_{\rho, v} \int_0^1 \int_M \frac{1}{2} g(v(t,x), v(t,x)) d\rho(t,x) dt, \tag{118}$$

under the continuity equation constraint $\partial_t \rho(t,x) + \mathrm{div}(\rho(t,x) v(t,x)) = 0$ and time boundary constraints $\rho(0) = \rho_0, \rho(1) = \rho_1$.

Introducing the flow φ_t of the vector field $v(t,x)$ when well-defined, and using the time-dependent change of variable $y = \varphi(t,x)$, we get

$$\int_0^1 \int_M \frac{1}{2} g(\varphi(t,x))(\partial_t \varphi(t,x), \partial_t \varphi(t,x)) d\rho_0(x) dt$$

$$= \frac{1}{2} \int_0^1 \int_M g(y)(v(t,y), v(t,y)) d\rho_t(x) dt \tag{119}$$

by definition of the image measure. One easily recognizes the straight (in opposition to right-invariant) L^2 metric of the group of diffeomorphisms. We directly get the inequality

$$\int_M \frac{1}{2} d^2(x, \varphi_1(x)) d\mu(x) \leq \int_0^1 \int_M \frac{1}{2} g(\varphi(t,x))(\partial_t \varphi(t,x), \partial_t \varphi(t,x)) d\rho_0(x) dt, \tag{120}$$

which is an equality if for every x, $t \mapsto \varphi(t,x)$ is a geodesic on M. In the case of the Euclidean cost, the equality is achieved if and only if $\varphi_t = \mathrm{Id} + t(\nabla \varphi_1 - \mathrm{Id})$. Minimizing the quantity $\int_M \frac{1}{2} d^2(x, \varphi_1(x)) d\mu(x)$ under the pushforward constraint $[\varphi_1]_\sharp(\rho_0) = \rho_1$ is the Monge formulation of optimal transport, for which the convex relaxation is called the Kantorovich formulation. However, the kinetic formulation (118) is nonconvex. It is one of the key contributions by Benamou and Brenier to propose a convex reformulation amenable to non-smooth convex optimization:

Definition 75 (Benamou–Brenier—Convex Formulation) Consider two given measures $\rho_0, \rho_1 \in \mathcal{P}(M)$ and the optimization set $m \in \mathcal{M}([0,1] \times M, TM)$ i.e. measures taking values in the tangent space of M and $\rho(t,x) \in \mathcal{M}([0,1] \times M)$.

$$\frac{1}{2} \inf_{\rho, m} \int_0^1 \int_M \frac{m(t,x)^2}{\rho(t,x)} d\rho(t,x) dt, \tag{121}$$

under the linear constraint $\partial_t \rho(t, x) + \text{div}(m) = 0$ and same time boundary constraints on ρ. The linear constraint is satisfied in the following weak sense, for every $f(t, x) \in C^1([0, 1] \times M)$,

$$-\langle \partial_t f, \rho \rangle - \langle \nabla f, m \rangle = \langle f(t = 1), \rho_1 \rangle - \langle f(t = 0), \rho_0 \rangle, \tag{122}$$

where the pairing is the dual pairing between measures and continuous functions. On a closed Riemannian manifold, there are no boundary conditions but on a bounded convex set in $\mathbb{R}^d$, this weak constraint also encodes the homogeneous Neumann boundary condition (zero flux).

Remark 76 The advantage of this reformulation is to facilitate the proof of the existence of minimizers (by the application of Fenchel–Rockafellar theorem for instance). Moreover, one can use non-smooth convex optimization methods to solve it in practice [55].

We are now interested in an actual formulation of the Riemannian(-like) tensor of the Wasserstein metric at a given (sufficiently smooth) density ρ (not any measure). Following the point of view of induced Riemannian metrics, we are interested in the minimization of the kinetic functional

$$G_{W_2}(\rho)(\delta\rho, \delta\rho) = \inf_v \frac{1}{2} \int |v(t, x)|^2 d\rho(t, x) \text{ under the constraint } -\text{div}(\rho v) = \delta\rho. \tag{123}$$

Using Lagrange multipliers, the vector field v is necessarily the gradient of the Lagrange multiplier. Therefore, one can replace the space of vector fields by the space of function p, however, in this case, the constraint reads

$$\Delta_\rho(p) := -\text{div}(\rho \nabla p) = \delta \rho. \tag{124}$$

On a closed (connected) Riemannian manifold and if ρ has sufficient (mild) regularity, this is an elliptic equation whose solution is unique up to a constant. Now, one can write the Riemannian(-like) metric tensor of the Wasserstein space, at least for sufficiently smooth densities ρ:

Proposition 77 (Informal, Wasserstein Metric Tensor) *The Riemannian(-like) metric tensor at a density ρ is*

$$G_{W_2}(\rho)(\delta\rho, \delta\rho) = \int_M \|\nabla \Delta_\rho^{-1} \delta\rho\|^2 \, d\rho(x) = \int_M \Delta_\rho^{-1}(\delta\rho) \delta\rho \, dx. \tag{125}$$

The order of the Wasserstein metric is -1 since one first integrates twice and then differentiates once in the above formula. It is thus similar to an H^{-1} metric that depends on the current density. In contrast, the induced Riemannian metric by a right-invariant Sobolev metric of order s is $s - 1$. Comparing the Wasserstein metric and a form of H^{-1} metric (the formula above in which ρ is taken to be constant as the reference volume measure, e.g. Lebesgue in Euclidean space) has been studied in the literature since it is potentially useful in numerical simulation of gradient flows.

To compare with the metrics induced by right-invariant metrics on diffeomorphisms groups, we finish with the geodesic flow which are the Hamilton–Jacobi equations

$$\begin{cases} \dot{p} + \frac{1}{2}\|\nabla p\|^2 = 0 \\ \dot{\rho} + \text{div}(\rho \nabla p) = 0. \end{cases} \quad (126)$$

Note that there is no dependency in ρ in the equation driving the potential p (the dual variable to ρ), in contrast to System (116).

Remark 78 When applying Fenchel–Rockafellar theorem, the first equation of the geodesic flow in the sense of convex analysis needs to be re-written as

$$\dot{p} + \frac{1}{2}\|\nabla p\|^2 \leq 0. \quad (127)$$

We refer the reader to [23] for more details.

To trace back this metric and the corresponding Hamiltonian equations in the literature, one of the earliest articles presenting them is arguably the work of Lafferty [41] on the Schrödinger equation, even before the seminal works of Otto [54] and Brenier [14].

6.4 The Wasserstein-Fisher-Rao Metric on Positive Densities

Wasserstein L^2 optimal transport is concerned with probability measures. To deal with positive measures of different total mass, it is possible to renormalize each measure by its total volume. Another natural idea consists of allowing the creation and deletion of mass at every point in space, as well as displacement. Displacement only is the subject of the Wasserstein metric. Change in mass only is the subject of L^2 like metrics on measures. Among them, the Hellinger distance squared is one-homogeneous w.r.t. the mass and is Riemannian-like, similar to the Wasserstein metric. Recall that the Hellinger distance is the squared L^2 norm between the square roots of the measures and is defined by

$$\text{Hellinger}(\mu_1, \mu_2)^2 = \mu_1(M) + \mu_2(M) - 2 \int_M \sqrt{\frac{d\mu_1}{d\lambda} \frac{d\mu_2}{d\lambda}} d\lambda, \quad (128)$$

where λ is a common dominating measure (the result does not depend on this choice by one-homogeneity). The corresponding metric tensor (sometimes called Fisher-Rao when restricted to probability measures) can be written in the following way

$$G_{\text{Hellinger}}(\rho)(\delta\rho, \delta\rho) = \int_M \left(\frac{\delta\rho}{\rho}\right)^2 \rho. \tag{129}$$

It is convex and one-homogeneous in $(\delta\rho, \rho)$, similar to the Benamou-Brenier formulation. Interestingly, these two properties characterize the metric tensor of the Hellinger distance at least locally: it must be quadratic in $\delta\rho$ and one-homogeneous in $(\delta\rho, \rho)$ so pointwisely it is colinear to $\left(\frac{\delta\rho}{\rho}\right)^2 \rho$.

One way to mix the two metrics simply consists of adding up the two metric tensors. However, by doing so, we would lose the interpretation of minimizing a Lagrangian of an over-parametrized action on an object and the corresponding geometric picture of the submersion result. Moreover, there are certain directions that are not admissible in the Wasserstein space (i.e. changing mass). To circumvent these two issues, we propose to use a minimizing approach, also called infimal convolution (in convex analysis): Let $\delta > 0$ be a parameter, define

$$G_{\text{WFR}}(\rho)(\delta\rho, \delta\rho) = \inf_{v,\alpha} \int_M |v(t,x)|^2 + \delta^2 \alpha(t,x)^2 d\rho(t,x), \tag{130}$$

under the constraint of the generalized continuity equation

$$\partial_t \rho + \text{div}(\rho v) = \alpha\rho. \tag{131}$$

This Lagrangian involves two control variables (that are redundant in certain directions), the vector field $v(t,x)$ and the growth rate $\alpha(t,x)$. The resulting tensor is again one-homogeneous w.r.t. $(\delta\rho, \rho)$, by construction. With this Lagrangian, the extension of the Benamou-Brenier formulation is as follows, minimize, under Eq. (131), the action

$$\inf_{\rho,v,\alpha} \int_0^1 \int_M \frac{1}{2}(\|v(t,x)\|^2 + \frac{\delta^2}{2}\alpha(t,x)^2) \, d\rho(t,x) \, dt, \tag{132}$$

in which we underline the dependence of the control variable on time and space. This slight modification of Benamou-Brenier leads to the following dual problem: Maximize $\int_M \varphi_1(y) d\rho_1(y) - \int_M \varphi_0(y) d\rho_0(y)$ under the constraint that $\varphi \in C^1([0,1], M)$ such that

$$\partial_t \varphi + \frac{1}{2}(\|\nabla\varphi\|^2 + \varphi^2) \leq 0. \tag{133}$$

Note that, as it is the case in standard OT, this equation does not depend on the current density $\rho(t)$. It implies that given one $\varphi(t)$ that realizes the equality (instead of the inequality), one can use it to solve the generalized continuity equation and thus obtain a path of minimizing energy.

Interestingly, this optimization problem is a slight modification of the Benamou-Brenier formulation and the same numerical framework can be used to solve the problem. Again, using the language of convex analysis, the new metric tensor is obtained as the infimal convolution of the Wasserstein metric tensor and the Fisher-Rao metric tensor, whence its name Wasserstein-Fisher-Rao. It has been named Hellinger-Kantorovich in [43] since at the level of distances, it interpolates between Kantorovich and Hellinger distances. We refer the reader to [23, 24, 43] for more details on the Wasserstein-Fisher-Rao metric.

From the Eulerian to the Lagrangian Formulations For the Wasserstein-Fisher-Rao metric, the group $\text{Diff}(M)$ is replaced with the semidirect product of groups between $\text{Diff}(M)$ and the space of positive functions on M which is a group under pointwise multiplication. On this semidirect product of groups (see Example 6), the composition law is defined such that the map π_1 given by

$$\pi_1 : (\text{Diff}(M) \ltimes C(M, \mathbb{R}_{>0})) \times \text{Dens}(M) \mapsto \text{Dens}(M)$$

$$\pi_1((\varphi, \lambda), \rho) := \varphi_*(\lambda \rho)$$

is a left-action of the group $\text{Diff}(M) \ltimes C(M, \mathbb{R}_{>0})$ on the space of densities. Similarly to the optimal transport case, this action is a Riemannian submersion between $L^2(M, \mathcal{C}(M); \rho_0)$ and $\text{Dens}(M)$ endowed with the WFR metric. Note that the L^2 metric is defined by a density (the initial density) on M and the cone metric on $\mathcal{C}(M)$ defined in (see [30] for more details). On the cone, the Riemannian metric reads

$$g_{\mathcal{C}(M)}((x, m))(\delta x, \delta m) = m g_x(\delta x, \delta x) + \delta^2 \frac{\delta m^2}{m}.$$

Up to the change of variable $m = r^2$, we find that the metric can be rewritten as

$$g_{\mathcal{C}(M)}((x, m))(\delta x, \delta r) = r^2 g_x(\delta x, \delta x) + 4\delta^2 (\delta r^2), \tag{134}$$

which is called a cone metric. Under this change of variables, the action is changed into

$$\pi_\rho : (\text{Diff}(M) \ltimes C(M, \mathbb{R}_{>0})) \times \text{Dens}(M) \to \text{Dens}(M)$$

$$\pi_\rho((\varphi, \lambda), \rho) := \varphi_*(\lambda^2 \rho),$$

and the metric on $\mathcal{C}(M)$ is the cone metric (134). As done in [31] we can identify this semidirect product of groups with the automorphism group of the cone $\mathcal{C}(M)$ (since it has a multiplicative group structure in the $\mathbb{R}_{>0}$ component). Thus, to shorten the notations, we use $\text{Aut}(\mathcal{C}(M))$ instead of $\text{Diff}(M) \ltimes C(M, \mathbb{R}_{>0})$. We now state the (formal) Riemannian submersion result obtained in [31].

Proposition 79 *Let $\rho_0 \in \text{Dens}(M)$ be a positive density and π be the map*

$$\pi : \text{Aut}(\mathcal{C}(M)) \to \text{Dens}(M)$$
$$\pi(\varphi, \lambda) = \varphi_*(\lambda^2 \rho_0) \,. \tag{135}$$

Then, π is a Riemannian submersion between $\text{Aut}(\mathcal{C}(M))$ endowed with the metric $L^2(M, C(M); \rho_0)$ and $\text{Dens}(M)$ with the WFR metric.

This proposition follows from the application of Proposition 33. This can be used to formulate a Monge formulation of the Wasserstein-Fisher-Rao metric as done in [33]. In addition, this formal Riemannian submersion picture can be made rigorous by using the weaker notion of submetries (see Definition 69).

Exercise 80 *Write the formal metric tensor of the Wasserstein-Fisher-Rao metric, similar to Proposition 77 for standard optimal transport.*

Remark 81 (The Camassa-Holm Equation on the Isotropy Subgroup) By Proposition 33, the induced Riemannian metric on the isotropy group of the action is a right-invariant metric. The isotropy group can be parametrized by $\text{Diff}(M)$ since $\pi(\varphi, \lambda) = \varphi_*(\lambda^2 \rho_0) = \rho_0$ implies $\text{Jac}(\varphi^{-1})\lambda^2 \circ \varphi^{-1} = 1$, which gives $\lambda = \sqrt{\text{Jac}(\varphi)}$. One can simply evaluate the metric at identity to get the H^1 metric as presented in Eq. (82). This fact can be used to derive the natural relaxation of the variational problem associated with the minimizing geodesics (as done by Brenier for the incompressible Euler equation), see [32]. Using this geometrical insight, the Camassa-Holm equation can be rewritten as an incompressible Euler equation on the cone, see [51] for more details and some generalizations.

7 A Geometric Point of View on Gradient Flows on Measures

Gradient descent methods and their continuous-time counterparts, gradient flows, are widely used in machine learning. In optimization, one is often concerned with the global convergence of the gradient flow. This global convergence is guaranteed under some convexity conditions of the function f, or weaker conditions such as the Polyak-Lojasiewicz condition (see Appendix 5).

Arising from machine learning, hereafter are some motivating examples for gradient flows on measures.

Sampling The Langevin diffusion equation can be seen as a gradient flow of the relative entropy in the Wasserstein geometry, as shown in the seminal work of Otto [54].

Generative Modeling In several situations, the data are modeled by parametrized densities and accessed via samples, i.e. empirical measures. In generative modeling, the data is modeled as an empirical measure, and one wants to find a parametric map that pushes forward a reference measure such as a Gaussian measure onto the observed data. In this situation, the objective functional is defined on measures and pulled back on the parameter space.

Mean Field Training Dynamic for Optimization of Neural Networks Gradient descent and stochastic gradient descent are routinely used to optimize nonlinear functions obtained via neural networks. To understand the training dynamics, a particular regime has been put forward, the mean-field limit: the neurons are replaced with a measure, and the training dynamic is studied directly on this measure. The gain is to pass from a nonlinear parametrization (with neurons) to a linear parametrization (with the measure). The downside is to pass from a finite-dimensional optimization problem to an infinite one. It is an intermediate relaxation before the standard convex relaxation of nonlinear optimization: solving $\min_{x \in D} f(x)$ is equivalent to solving $\min_{\mu \in \mathcal{P}(D)} \int f(x) d\mu(x)$.

7.1 The Importance of the Co-metric in Gradient Flows

First, recall how to compute a gradient in the case of a function on the Euclidean space with respect to a metric defined by a symmetric positive matrix $g(x)$. It has the following variational formulation (check using first-order optimality condition):

$$\nabla f(x) = \arg\min_{w} \frac{1}{2} \langle w, g(x)w \rangle - df_x(w),$$

where the symbol $\langle \cdot, \cdot \rangle$ denotes the dual pairing. It gives

$$\nabla f(x) = g(x)^{-1}(df_x). \tag{136}$$

Let us insist that what matters in the previous formula is the co-metric $g(x)^{-1}$ which is a map from T_x^*M to T_xM (since $g(x) : T_x \to T_x^*$). The gradient flow associated with a function f is the ODE

$$\dot{x} = -\nabla f(x) = -g(x)^{-1}(df_x). \tag{137}$$

Note that changing the metric changes the gradient, and thus the curves of the gradient flow. However, changing the metric obviously does not change the landscape of the function (local/global minima for instance). Since we have highlighted the role of the co-metric in the gradient flow, we can deduce

Proposition 82 *Let $\pi : M \to N$ be a Riemannian submersion between two Riemannian manifolds and $f : N \to \mathbb{R}$ a smooth function. The gradient flow of f on N is the image of the gradient flow of $f \circ \pi$ on M.*

Proof The two gradient flow equations are, with $y = \pi(x)$,

$$\dot{x} = -g_M(x)^{-1}(d\pi^*(df_{\pi(x)})) \tag{138}$$

$$\dot{y} = -g_N(y)^{-1}(df_y) \tag{139}$$

since $d[f \circ \pi]_x) = d\pi^*(df_{\pi(x)})$. Starting from a gradient flow on M, one can differentiate in time its image $\pi(x)$. By composing the first equation with $d\pi_x$, we have

$$\dot{y} = d\pi_x(\dot{x}) = -[d\pi_x g_M(x)^{-1} d\pi^*](df_{\pi(x)}). \tag{140}$$

By the property of Riemannian submersions, the co-metrics satisfy the following relation

$$d\pi(x) g_M^{-1}(x) d\pi^*(x) = g_N^{-1}(\pi(x)).$$

Therefore, the last equation reads $\dot{y} = -g_N(y)^{-1}(df_y)$ which is the flow equation on N.
□

Remark 83 The converse implication is also true in the following formulation: if for every function f and every point $x \in M$, the gradient flows (on M and on N) coincide through the submersion π, it implies that π is a Riemannian submersion.

As mentioned previously, the Polyak-Lojasiewicz condition is an elementary condition that guarantees exponential convergence of the gradient flow. It is useful to think of it in terms of the co-metric.

Proposition 84 (PL Inequality for Infimal Convolution of Riemannian Metrics) *Let M be a manifold, g_1, g_2 two Riemannian metrics. Let $f : M \to \mathbb{R}$ be a function satisfying a PL inequality for the metric g_1. Then, f satisfies a PL inequality for the infimal convolution (see Formula (187) for the definition) of the metrics $g_1 \boxplus g_2$.*

Diffeomorphic and Optimal Transport Flows

Proof One has $(\frac{1}{2}(g_1 \boxplus g_2))^{-1} = \frac{1}{2}(g_1^{-1} + g_2^{-1})$. Then, the PL inequality (see Appendix 5) implies (with $x_\star$ a minimizer of f)

$$f(x) - f(x_\star) \leq C \langle g_1^{-1}(df_x), df_x \rangle \leq C \langle (g_1^{-1} + g_2^{-1})(df_x), df_x \rangle. \tag{141}$$

It is the PL inequality for $g_1 \boxplus g_2$. □

Remark 85 The PL inequality is preserved under the infimal convolution operation on metrics, which stands in sharp contrast to convexity. Indeed, a function f which is strongly convex in the geometry of g_1 might not be convex in the geometry of $g_1 \boxplus g_2$. An example is given below, with the Wasserstein metric and the Wasserstein-Fisher-Rao metric.

These two facts underline the importance of the co-metric in gradient flows.

7.2 A Geometric Perspective on Different Entropy Flows, Including Stein Variational Gradient Descent

Let us start with a comparative discussion of the gradient flow of the Kullback-Leibler divergence with respect to two different Riemannian-like metrics on the space of measures, optimal transport, and induced right-invariant metric. The Kullback-Leibler divergence is a particular case of Csizàr divergences, which we now present.

Definition 86 An *entropy function* $F : \mathbb{R} \to [0, +\infty]$ is a convex, lower semi-continuous, nonnegative function such that $F(1) = 0$ and $F(x) = +\infty$ if $x < 0$. Its *recession constant* is defined as $F'_\infty = \lim_{r \to +\infty} \frac{F(r)}{r}$.

An entropy function is minimal at 1, it is on purpose to compare ratios. From an entropy function, one can use the corresponding *perspective function* to obtain a convex function on $\tilde{F} : \mathbb{R} \times \mathbb{R}_{>0} \to \mathbb{R}$ defined by the formula $(x, t) \to tF(x/t)$. To obtain a divergence between measures, this perspective function is evaluated on $\frac{d\mu}{d\nu}$ and integrated in space, taking care of infinite values with the recession function.

Definition 87 Let F be an entropy function and μ, ν be Radon measures on $\mathbb{R}^d$. The *Csiszàr divergence* associated with F is

$$D_F(\mu, \nu) = \int_{\mathbb{R}^d} F\left(\frac{d\mu(x)}{d\nu(x)}\right) d\nu(x) + F'_\infty \int_{\mathbb{R}^d} d\mu^\perp,$$

where $\mu^\perp$ is the orthogonal part of the Lebesgue decomposition of μ with respect to ν.

For $F(x) = x\log(x) - x + 1$, D_F is the *Kullback–Leibler divergence* or *relative entropy*, and it reads

$$\text{KL}(\mu, \nu) = \int_{\mathbb{R}^d} \frac{d\mu}{d\nu} \log\left(\frac{d\mu}{d\nu}\right) d\nu + \int_{\mathbb{R}^d} d\nu - \int_{\mathbb{R}^d} d\mu. \tag{142}$$

Note that this divergence is defined on the whole space of positive Radon measures. However, we will restrict it to probability measures. Let us assume that both measures μ, ν have density w.r.t. the Lebesgue measure respectively denoted by ρ, ρ_∞. Moreover, we assume that $\rho_\infty = e^{-V(x)}$ for a function $V : \mathbb{R}^d \to R$. In this case, one can write (with a slight abuse of notation for the arguments)

$$\text{KL}(\rho, \rho_\infty) = \int_{\mathbb{R}^d} \rho \log(\rho) dx + \int_{\mathbb{R}^d} V(x) \rho(x) dx. \tag{143}$$

One may want to optimize this functional with respect to ρ, most often on a subset of the space of probability measures. For instance, this subset can be a parametrized family or an orbit of a group of diffeomorphisms. It is instructive to look at the gradient flow of this functional from the group action viewpoint: we act on densities using horizontal (pushforward by a vector field) or vertical (multiplication by a pointwise multiplication) variations. The infinitesimal action of this semidirect product of groups (see Sect. 6.4) on the space of positive Radon measures is

$$(v, \alpha) \cdot \rho = -\text{div}(\rho v) + \alpha \rho, \tag{144}$$

if a total volume constraint is included to stay in the space of probability measures, an additional constraint has to be put on α, namely $\int_{\mathbb{R}^d} \alpha(x)\rho(x)dx = 0$. As developed in Sect. 2.6, one can induce a Riemannian(-like) metric on the orbits of the measure from a Riemannian(-like) metric on the group. Then, one can compute the formal expression for the gradient flow for different metrics:

The Wasserstein Metric

$$\partial_t \rho = \Delta \rho + \text{div}(\rho \nabla V). \tag{145}$$

The reader might recognize the Fokker-Planck equation, which is the PDE that arises from the law of the process associated with the Langevin dynamic:

$$dX(t) = -\nabla V(X(t)) + \sqrt{2} dB(t). \tag{146}$$

The law of $X(t)$ denoted by $\rho(t)$ evolves accordingly to Formula (145). Note that ρ_∞ is the unique steady state of this PDE. Indeed, $0 = \text{div}(\rho \nabla V) + \Delta \rho = \text{div}(\rho(\nabla V + \nabla \log \rho)$

which implies $\rho = e^{-V}$ (the multiplicative constant is 1). Importantly, this process is used in generative modeling to sample from ρ_∞. The key point in simulating (146) is that one does not need to have access to the renormalizing constant $\int e^{-V(x)}dx$, which can be very costly to estimate. Indeed, the flow equation is invariant to adding a constant to the potential.

The convergence of $\rho(t)$ to ρ_∞ is ensured by functional inequalities: if ρ_∞ satisfies a Poincaré or a log-Sobolev inequality, then exponential convergence holds. Indeed, the log-Sobolev inequality reads

$$\mathrm{KL}(\rho, \rho_\infty) \leq C \int_{\mathbb{R}^d} |\nabla \log(\rho/\rho_\infty)|^2 \rho(x)dx = C \|\nabla_\rho \mathrm{KL}(\rho, \rho_\infty)\|_{W_2}^2, \qquad (147)$$

which is a Polyak-Lojasiewicz (PL) inequality for the Kullback-Leibler divergence under the Wasserstein metric. The PL inequality guarantees exponential convergence of the gradient flow to the set of minimizers (see Appendix 5). In this case, the minimizer is unique ρ_∞.

The Wasserstein-Fisher-Rao Metric

$$\partial_t \rho = \Delta \rho + \mathrm{div}(\rho \nabla V) - \frac{1}{\delta^2}(V + \log(\rho))\rho, \qquad (148)$$

where δ is the weight multiplying the squared norm of α. In sharp contrast with the Wasserstein metric, the entropy is not convex in this geometry (see [44]). However, the long time behavior of the gradient flow of the entropy is still preserved: To guarantee the convergence of the gradient flow, the Polyak-Lojasiewicz inequality (see Appendix 5) is somehow weaker than convexity conditions. A simple fact proven in Proposition 84 is that if a function $\mathcal{F}$ satisfies a PL inequality for the Wasserstein metric, then it also satisfies a PL inequality for the WFR metric. Indeed, at least formally,

$$\|\nabla \mathcal{F}(\rho)\|_{\mathrm{WFR}}^2 = \int \left(\left|\nabla \frac{\delta \mathcal{F}}{\delta \rho}\right|^2 + \frac{1}{\delta^2} \left|\frac{\delta \mathcal{F}}{\delta \rho}\right|^2 \right) \rho(x)dx \geq \|\nabla \mathcal{F}(\rho)\|_{W_2}^2 \geq C\mathcal{F}(\rho), \qquad (149)$$

where C is a positive constant. Similar results are used for the Spherical-Hellinger Kantorovich (WFR restricted to probability measures) gradient flows in [40]. From a practical point of view, the presence of the $\rho \log(\rho)$ term makes it difficult to simulate, see [45] for a possible numerical algorithm in the case of the Fisher-Rao metric which makes use of kernel estimation of the density.

An Induced Metric by a Right-Invariant Metric (a.k.a. Stein Variational Gradient Descent)
For a right-invariant metric on the semidirect product of groups, i.e. for a metric on the

vector field v and α that does not depend on the current density ρ, we get:

$$\partial_t \rho = +\operatorname{div}(\rho v) - \alpha \rho. \tag{150}$$

where

$$\begin{cases} v = K_1 \star (\rho(\nabla \log(\rho) + \nabla V)), \\ \alpha = \frac{1}{\delta^2} K_2 \star (\rho \log(\rho) + \rho V)), \end{cases} \tag{151}$$

and K_1, K_2 are respectively the inverses of the differential operators L_1, L_2 defining the norm

$$\frac{1}{2} \int_{\mathbb{R}^d} \langle L_1 v, v \rangle + \delta^2 \langle L_2 \alpha, \alpha \rangle \, dx. \tag{152}$$

As such, these formulas are not enough to attract interest. Let us explain why these equations have drawn much interest in machine learning and analysis. Let us focus on the case of a right-invariant metric on the group of diffeomorphisms, i.e. $\alpha = 0$. The evolution equation that is obtained on ρ reads

$$\partial_t \rho = \operatorname{div}(\rho K_1 \star \rho \nabla \log(\rho) + \rho \nabla V) = \operatorname{div}(\rho K_1 \star \nabla \rho + \rho \nabla V). \tag{153}$$

Remark that $K_1 \star \nabla \rho = K_1 \log(\rho)$ so that one can easily compute the right-hand side

$$\partial_t \rho = \operatorname{div}(\rho \nabla K_1 \star \rho + \rho \nabla V). \tag{154}$$

It somehow amounts to a kernelized estimation of $\nabla \rho$. This equation was first introduced under the name of Stein variational gradient descent in [46]. The interest of this equation is that it can be instantiated for empirical measures (which are preserved by the flow of this PDE) $\rho(t) = \frac{1}{n} \sum_{i=1}^{n} \delta_{x_i}$ for $x_i \in \mathbb{R}^d$. It reads as follows on the position of the Dirac masses

$$\dot{x}_i = -\frac{1}{n} \sum_{j=1}^{d} \left(\nabla K_1(x_i, x_j) - K_1(x_i, x_j) \nabla V(x_j) \right). \tag{155}$$

In sharp contrast with the Wasserstein metric, there are no guarantees of exponential convergence in the mean-field limit. For the practical case of a finite number of particles, we refer the reader to [58] as an entry point in the rapidly growing literature about Stein variation gradient descent, see also [25] for a slightly different presentation of this geometric point of view.

Although one could also consider from a theoretical perspective the flow induced by a metric in α, its practical implementation needs more work since one would need to use an estimation of $\rho \log(\rho)$, which is not directly accessible.

7.3 Wasserstein Gradient Flow of the Coulomb Discrepancy

Recall that the squared norm of the MMD discrepancy is given by

$$\|\mu - \nu\|_K^2 = \langle K \star (\mu - \nu), \mu - \nu \rangle. \tag{156}$$

One can consider the functional $\mathcal{E}_\nu(\mu) := \frac{1}{2}\|\mu - \nu\|_K^2$. One can assume that μ is parameterized by $\theta \in \Theta$ and one can perform gradient descent on $\mu(\theta)$. Note that optimizing this function directly on μ is obvious since it is a quadratic functional which attains its unique minimum at $\mu = \nu$. An important model is the case where μ is pushforwarded infinitesimally by a vector field v and this vector field is measured by its $L^2(\mu)$ norm. This corresponds to a Wasserstein gradient flow which reads

$$\partial_t \mu = \operatorname{div}(\mu \nabla K \star (\mu - \nu)). \tag{157}$$

For this system, it is natural to check Polyak-Lojasiewicz inequality which reads

$$\|\varphi_{\mu-\nu}\|_{K^{-1}}^2 \leq \int_M \|\nabla \varphi_{\mu-\nu}(x)\|^2 d\mu(x). \tag{158}$$

Let us discuss at the order of the operator in this inequality. The inverse of the kernel can be thought as a differential operator of order $2k$ and the right-hand side is an operator of second-order. Thus, for this inequality to hold, it is natural to first look at the case $k = 1$. This is the case of the Coulomb kernel Δ^{-1}, which we now discuss. In the case $K^{-1} = -\Delta$, the PL inequality reads

$$\int_M \|\nabla \varphi_{\mu-\nu}(x)\|^2 d\operatorname{vol}(x) \leq \int_M \|\nabla \varphi_{\mu-\nu}(x)\|^2 d\mu(x), \tag{159}$$

where vol is the volume measure of the closed Riemannian manifold M. We will only treat the case of a closed manifold since we need compactness for the next inequality: Assume that μ is bounded below by a positive constant denoted $\underline{\mu}$, then

$$\int_M \|\nabla \varphi_{\mu-\nu}(x)\|^2 d\operatorname{vol}(x) \leq \frac{1}{\underline{\mu}} \int_M \|\nabla \varphi_{\mu-\nu}(x)\|^2 d\mu(x), \tag{160}$$

which is the desired PL inequality. This PL inequality is not global since the PL coefficient depends on the measure μ. A priori, such an inequality is not enough to conclude to global convergence. However, it is sufficient to guarantee that this inequality holds along the flow. Indeed, we have

$$\partial_t \mu = \langle \nabla \mu, \nabla K \star (\mu - \nu) \rangle + \mu \Delta K \star (\mu - \nu) \,. \tag{161}$$

Let us evaluate this equation at a critical point of μ, $\nabla \mu(x) = 0$, we have since $\Delta K = -\operatorname{Id}$,

$$\partial_t \mu(x) = -\mu(x)(\mu(x) - \nu(x)) \,. \tag{162}$$

This evolution equation implies that the minima and the maxima of μ will stay in a compact set in $\mathbb{R}_{>0}$ if ν is lower and upper bounded. Therefore, we have shown formally that under a regularity assumption on ν and the fact that ν is lower bounded, the PL inequality is preserved along the flow. This implies that the optimization question of convergence of the flow is reduced to a question of regularity of the flow. Note that the local existence of the flow in Sobolev regularity $s > d/2 + 1$ can be deduced by similar arguments as in Theorem 50. Indeed, this equation can be rewritten using the flow of the corresponding vector field, and in this new variable, the PDE is transformed into an ODE in Sobolev spaces. Then, the remaining key question is global existence, which we proved in [13] using standard but fine inequalities in potential theory:

Theorem 88 (See [13]) *Let μ_0 and ν be Hölder continuous densities with compact support on M, both bounded away from 0. Then, the gradient flow equation admits a unique global solution μ_t that is Hölder continuous at all times and $\mu(t)$ converges exponentially to ν for the Wasserstein-2 distance.*

Question 89 *Are there other kernels for which similar results can be derived?*

A good candidate for this question is the energy distance kernel, which coincides with the Coulomb kernel in dimension 1, and in higher dimensions has the form: $k(x, y) = -\|x - y\|$ in Euclidean space. Note that this kernel is only conditionally positive definite, i.e. positive definite on the subspace of zero total mass (signed) measures, which makes the MMD functional a well-behaved loss function. The energy distance kernel is repulsive, and numerical simulations tend to showgood behavior for global convergence.

7.4 Infinite Width/Infinite Depth Neural Networks

Infinite Width Let us start with a single-hidden layer (SHL) function, which takes the following form:

$$f_{(\theta_1,\theta_2)} : \mathbb{R}^d \to \mathbb{R} \,, x \mapsto \theta_1 \sigma(\theta_2(x)) \,, \tag{163}$$

where $\sigma : \mathbb{R} \to \mathbb{R}$ is a nonlinear function such as ReLU : $x \mapsto \max(0, x)$ Consider a loss function which is the L^2 norm of $\|f_{(\theta_1,\theta_2)} - g\|^2_{L^2(\rho)}$ with ρ representing the data density (ρ is an empirical measure in practice). Optimizing on the parameters θ_1, θ_2 is non-convex, however, the relaxation, as proposed in [5] by interpreting these parameters as an empirical measure makes it convex. This relaxation consists in associating to a measure $\mu \in \mathcal{P}(\Theta)$ where Θ is the space of parameters, the function on $\mathbb{R}^d$ given by $f_\mu(x) := \int_\Theta f(\theta, x) d\mu(\theta)$. This relaxation is also called *infinite width limit* since the number of neurons may be infinite. However, there is no free lunch here, since this is a convex optimization problem in <u>infinite</u> dimensions. The functional on probability measures on the parameter space $\mu \in \mathcal{P}(\Theta)$ reads

$$\int_\Theta \int_\Theta \int_x (f(\theta, x) - g(x))(f(\theta', x) - g(x)) d\rho(x) d\mu(\theta) d\mu(\theta') \geq 0. \qquad (164)$$

This functional is "vertically" convex, that is under the usual affine structure on probability measures. Indeed, it is quadratic and nonnegative. However, as studied in [22], the gradient descent on the parameters corresponds to a Wasserstein gradient flow on the probability measure μ. Therefore, studying the convergence of a gradient descent scheme in the mean field limit starts with the study of the gradient flow of (164) in the Wasserstein geometry. We refer to [22] for more details.

Infinite Depth One of the keys to the success of deep learning is the use of skip connection when stacking several layers of neural network functions. It consists in introducing, inside the architecture of the neural network, the iterative scheme, for an initial data x_0,

$$x_{k+1} = x_k + f_{\theta_k}(x_k) \quad k = 0, \ldots, n-1 \qquad (165)$$

This architecture is called *residual* since f_{θ_k} encode the difference $x_{k+1} - x_k$. In the literature, such networks are called *ResNets* and they have been extremely useful to improve the performances. We think that this performance boost is partly due to the landscape of functional and partly due to a simpler initialization of the network weights. Indeed, the first remark is that composing linear networks corresponds to a product of matrices, say $\Pi_{i=1}^n A_i$. If n is not too small, each matrix A_i need to be close to Id. So it is natural to write $A_i = \text{Id} + \frac{1}{n} R_i$. More importantly, this expansion shows that diagonal terms and extra-diagonal terms do have the same behavior in the gradient descent. Elaborating on this remark, one can explain, in this simple case, the observed phenomenon of *vanishing gradients*, which is completely resolved by using residual connections.

In addition, we now show the landscape of the functional is rather better behaved, in particular in the case of infinite width for the residual functions f_{θ_k}. The first simplification we make is to pass to an infinite number of timesteps to obtain a neural ODE in the following sense

$$\dot{x} = f_{\theta(t)}(x). \qquad (166)$$

Then, one is left with optimization on the time-dependent function $\theta : [0, 1] \to \Theta$. We make another simplification by assuming that the vector fields $f_{\theta(t)}$ are linear with respect to $\theta(t)$. The simplest case is to consider that $f_{\theta(t)}$ belongs to an <u>admissible</u> Reproducing Kernel Hilbert Space V. The functional of interest is now:

$$\mathcal{L}(v) := \frac{1}{N} \sum_{i=1}^{N} \ell(\varphi(q_i), y_i), \qquad (167)$$

where ℓ is a loss such as the quadratic loss, $q_i \in \mathbb{R}^d$ are the initial data and $y_i \in \mathbb{R}^d$ are the target data. The variable φ is computed using the flow constraint equation

$$\begin{cases} \partial_t \varphi(t, x) = v(t, \varphi(t, x)) \\ \varphi(0, x) = x \, . \end{cases} \qquad (168)$$

We now need to perform a gradient flow under some geometry which is naturally $L^2([0, 1], V)$. Remark that, when V is a finite-dimensional space of vector fields, the metric on V can be written as a straight Euclidean metric on an orthonormal basis. Thus, it corresponds to a usual gradient flow in the parameter space.

Consequently, one can use the geometric tools developed in the previous sections to study the PL condition.

Proposition 90 *The gradient of $\mathcal{L}$ is given by*

$$D\mathcal{L}(v)(\eta) = \int_0^1 \langle J(p(t), q(t)), \eta(t) \rangle dt, \qquad (169)$$

where p, q are obtained by the differential equation

$$\begin{cases} \dot{p}_i = -dv^*(q_i)(p_i) \\ \dot{q}_i = v(q_i) \, , \end{cases} \qquad (170)$$

with initial conditions $p_i(1) = -\partial_q \ell(q_i(1), y_i)$, for $i = 1, \ldots, N$.

Recall that J denotes the momentum map and is an element of the dual space V^*. Equations (170) are the adjoint equations for computing the gradient, which are computed by the backpropagation algorithm. Assuming that one has computed the flow $\varphi(t)$ associated to $v(t)$, one has

$$J(p(t), q(t)) = J(\varphi \cdot p, \varphi \cdot q), \qquad (171)$$

To prove a local PL inequality, we have to compare the information in the squared gradient norm $\int_0^1 \|J(p(t), q(t))\|_{V^*}^2 dt$ with the loss. Let us assume that the loss ℓ satisfies a global PL inequality then

$$\frac{1}{N}\sum_{i=1}^N \ell(\varphi(q_i), y_i) \leq C \frac{1}{N} \sum_{i=1}^N \|\partial_q \ell(q_i, y_i)\|^2 = C \frac{1}{N} \sum_{i=1}^N \|p_i(1)\|^2. \tag{172}$$

The key point is to compare the usual L^2 norm of p and the norm of $J(p, q)$. More precisely, we need to have the existence of a positive constant $C(q)$ such that

$$\sum_{i=1}^N \langle p_i, k(q_i, q_j) p_j \rangle \geq C(q) \sum_{i=1}^N \|p_i\|^2. \tag{173}$$

Since the kernel is positive definite, it is always true for a fixed number N. However, assuming that the kernel is translation invariant, the constant depends on the separation distance defined by $\delta_q := \min_{i \neq j} \|q_i - q_j\|$. The smaller this quantity, the less invertible the kernel matrix $k(q_i, q_j)$ and the smaller the constant $C(\delta_q)$. The second point then is to control the evolution of $p(t)$ along the flow. Recall that $\varphi \cdot p = ([d\varphi^{-1}(\varphi(x_i))]^\top p_i))$. Therefore, $\|p\| \leq \|d\varphi\|_\infty \|\varphi(t) \cdot p\|$ and

$$\|J(\varphi(t) \cdot p, \varphi(t) \cdot q)\| \geq [\inf_t C(\delta_{q(t)})][\sup_t \|d\varphi(t)\|_\infty]\|J(p(0), q(0))\| \tag{174}$$

Now, the two quantities $\delta_{q(t)}$ and $[\sup_t \|d\varphi(t)\|_\infty]$ are well-defined since the path $\varphi(t)$ is continuous in time in the diffeomorphism group (see Theorem 60). The quantity $\delta_{q(t)}$ never vanishes since φ is a diffeomorphism. Yet, both quantities $\inf_t C(\delta_{q(t)})$ and $[\sup_t \|d\varphi(t)\|_\infty]$ have an exponential dependence in terms of $v \in L^2([0, 1], V)$ since it is the case for $\|\varphi\|_{1,\infty}$ and $\|\varphi^{-1}\|_{1,\infty}$ (see again Theorem 60). However, it is sufficient to obtain a local PL inequality.

Exercise 91 *Prove the following inequality*

$$\|J(\varphi \cdot p, \varphi \cdot q)\|_{V^*} \leq M \max(\|\varphi\|_{1,\infty}, \|\varphi^{-1}\|_{1,\infty}) \|J(p, q)\|_{V^*}. \tag{175}$$

As a direct consequence of the discussion above and the previous exercise, we obtain

Proposition 92 *Let $R > 0$, there exists c, C two positive constants that depends exponentially on R such that for all $v \in L^2([0, 1], V)$ such that $\|v\| \leq R$,*

$$\begin{cases} c(R)\mathcal{L}(v) \leq \|\nabla \mathcal{L}(v)\|^2 \\ \|\nabla \mathcal{L}(v)\|^2 \leq C(R)\mathcal{L}(v). \end{cases} \tag{176}$$

As a direct consequence of the first inequality above, [57] implies the convergence provided that the initial loss is sufficiently small.

Corollary 93 *If $v_0 \in L^2([0, 1], V)$ is such that $4\mathcal{L}(v_0) \leq R^2 c(R + \|v_0\|)$, then the gradient flow initialized at v_0 converges exponentially to a global minimizer.*

The general case of an infinite depth and finite or infinite width is treated in [4]. In this paper, we introduce the metric structure on the parameter space, which is called conditional optimal transport, roughly $L^2([0, 1], \mathcal{P}_2(\Theta))$ where the Wasserstein metric is used on $\mathcal{P}_2(\Theta)$. Since the case of a neural ODE with residual blocks, which are standard neural networks, is contained in this setting, the PL inequality cannot hold over the entire space of probability measures. However, it can hold on a subset (e.g. overparametrized neural blocks) thereby ensuring a local convergence result.

Appendix 1: Sobolev Spaces

In this section, our goal is to collect the necessary tools that are needed to understand the fact that the incompressible Euler equation (among others) can be formulated as a standard ordinary differential equation on Hilbert spaces and some Riemannian metrics on Sobolev diffeomorphism groups leading to smooth geodesic flow. We do not prove the results (apart from the easy case of dimension 1) and instead refer to textbooks such as [1].

Let us start with the usual definition:

Definition 94 Let M be a compact Riemannian manifold or $\mathbb{R}^d$ the Euclidean space, the Sobolev space of order $q \in \mathbb{N}$ denoted by $H^q(M, \mathbb{R})$ is given by the space of L^2 integrable functions $f : M \to \mathbb{R}$ such that its distributional derivative ∂^i up to order q are square integrable, that is

$$\|f\|_q^2 := \int_M \sum_{|i| \leq q} |\partial^i f(x)|^2 d\mathrm{vol}(x), \tag{177}$$

where $i = (i_1, \ldots, i_d) \in \mathbb{N}^d$ is a multi-index and $|i| = \sum_{k=1}^d i_k$.

For example, when $M = \mathbb{R}$, the space $H^1(\mathbb{R}, \mathbb{R})$ is the space of $L^2(\mathbb{R})$ functions such that their first derivative is also in $L^2(\mathbb{R})$. By the fundamental theorem of calculus $\int_a^b f'(x)dx = f(b) - f(a)$, this easily implies that f is bounded and that f is continuous.[13] Embedding Sobolev spaces in a space of smooth functions is a general fact:

---

[13] Using Cauchy-Schwarz inequality, it is $1/2$-Hölder continuous.

Theorem 95 (Sobolev Embedding) *If $q > d/2 + k$, then $H^q \hookrightarrow C(M, \mathbb{R})$, where d is the dimension of M.*

Consider $f, g \in H^1(\mathbb{R}, \mathbb{R})$, then the product $fg \in H^1(\mathbb{R}, \mathbb{R})$. Indeed, $f'g + fg' \in L^2(\mathbb{R})$ since f, g are bounded and $f', g' \in L^2(\mathbb{R})$. It is also a general fact.

Theorem 96 *If $k > d/2$ where d is the dimension of the manifold M and $0 \leq l \leq k$, then the map*

$$H^k(M) \times H^l(M) \to H^l(M) \tag{178}$$

$$(f, g) \mapsto fg \tag{179}$$

is a continuous bilinear map.

In particular, $H^k(M)$ is a Hilbert algebra for $k > d/2$. In this case, it implies that functions defined on $H^k(M)$ such as $f \mapsto f^\alpha$ with $\alpha \in \mathbb{N}$ is a smooth map on $H^k(M)$. We now consider the set of diffeomorphisms on M or $\mathbb{R}^d$. The notation $\mathrm{Diff}(M)$ stands for the set of invertible C^1 maps with C^1 inverses.

Definition 97 (Sobolev Diffeomorphisms) For $k > d/2 + 1$, we also denote by

$$\mathrm{Diff}_{H^k}(\mathbb{R}^d) = \{\varphi \in \mathrm{Diff}_+(\mathbb{R}^d)\,;\, \varphi - \mathrm{Id} \in H^k(\mathbb{R}^d)\}, \tag{180}$$

the set of Sobolev (of order k) diffeomorphisms, where $\mathrm{Diff}_+(\mathbb{R}^d)$ denotes the space of orientation preserving C^1 diffeomorphisms.

As a consequence of the Sobolev embedding theorem, $\mathrm{Diff}_{H^k}(\mathbb{R}^d) \subset \mathrm{Diff}(\mathbb{R}^d)$. For example, $\mathrm{Diff}_{H^2}(\mathbb{R})$ coincides with the set of $\mathrm{id} + f$ where $f \in H^2$ and $f' > -1$. Since f' is continuous, the set of $f \in H^2(\mathbb{R})$ such that $\mathrm{id} + f$ is a diffeomorphism is an open set of $H^2(\mathbb{R})$. This is a general fact since one has (see [16])

$$\mathrm{Diff}_{H^k}(\mathbb{R}^d) = \{\varphi\,;\, \varphi - \mathrm{id} \in H^k(\mathbb{R}^d) \text{ and } \det(D\varphi(x)) > 0\, \forall x \in \mathbb{R}^d\}.$$

It is not clear that the set of diffeomorphisms defined above is stable by taking inverses. Let us remark that if $\varphi = \mathrm{id} + f$ is such that $\det(d\varphi)(x) > 0$, then it is a local C^1 diffeomorphism. In addition, $\lim_{\|x\| \to \infty} \varphi(x) = x$, the image of φ is closed and open so φ is surjective. Since $\mathbb{R}^d$ is simply connected, φ is a diffeomorphism by the Hadamard theorem. Now, one needs to show that $\varphi^{-1} - \mathrm{id} \in H^k(\mathbb{R}^d)$ which is proven in [37]. To give some insight into the proof, we present the main arguments in dimension 1 for $H^2(\mathbb{R})$: First, we show that $\varphi^{-1} - \mathrm{id} \in L^2$. One needs to bound, using a change of variable

$$\int_{\mathbb{R}} |\varphi^{-1} - \mathrm{id}|^2 dx = \int_{\mathbb{R}} |\mathrm{id} - \varphi|^2 \mathrm{Jac}(\varphi) dx. \tag{181}$$

If $\mathrm{Jac}(\varphi) = \varphi' \in L^\infty$, then the result is clear since $\mathrm{id} - \varphi \in L^2$. This fact is satisfied since $1 - \varphi' \in H^1$ so that φ' has a limit in $\pm \infty$ equal to 1. In particular, φ' is bounded on $\mathbb{R}$. We are now interested in showing that $(\varphi^{-1})' - 1 = \frac{1}{\varphi'(\varphi^{-1}(x))} - 1$ is in L^2:

$$\int_{\mathbb{R}} |\frac{1}{\varphi'(\varphi^{-1}(x))} - 1|^2 dx = \int_{\mathbb{R}} |1 - \varphi'(x)|^2 \frac{1}{\varphi'(x)} dx. \tag{182}$$

Therefore, the result follows if $\varphi'(x)$ is bounded by below by a positive constant. This is the case since φ' is continuous, strictly positive, and has the limit 1 in $\pm \infty$. Last, the second derivative $(\varphi^{-1})''$ has to be in L^2 to conclude the proof. Since $(\varphi^{-1})'' = \frac{\varphi''(\varphi^{-1}(x))}{(\varphi'(\varphi^{-1}))^3}$, we use again a change of variable to get

$$\int_{\mathbb{R}} |\frac{\varphi''(\varphi^{-1}(x))}{(\varphi'(\varphi^{-1}))^3}|^2 dx = \int_{\mathbb{R}} |\frac{\varphi''(x)}{\varphi'(x)^3}|^2 \varphi'(x) dx. \tag{183}$$

Since φ' is bounded below by a positive constant the result follows.

These elementary one-dimensional computations can be extended in higher dimensions. Passing by, these computations rely on a simple change of variable which transforms the quantities of interest on φ^{-1} into rational functions in terms of φ and its derivatives. This remark helps understand other properties of the Sobolev diffeomorphisms group as a Riemannian manifold. More details can be found in [37], from which we cite:

Theorem 98 (Th 1.2 [37]) *Let M be a closed oriented manifold of dimension d or $\mathbb{R}^d$, N be a smooth manifold and an integer $s > d/2 + 1$, then for any $k \geq 0$ integer, the two following maps are C^k:*

$$\text{Composition}: H^{s+k}(M, N) \times \mathrm{Diff}_{H^s(M)} \to H^s(M, N), \, (f, \varphi) \mapsto f \circ \varphi,$$

$$\text{Inverse}: \mathrm{Diff}_{H^{s+k}(M)} \to \mathrm{Diff}_{H^s(M)}, \, \varphi \mapsto \varphi^{-1}.$$

Remark 99 (Smoothness of Composition w.r.t. the First Argument) One important point is that the composition map is actually smooth with respect to its first argument f. Indeed, it is simply a linear map. In other words, the map $f \in H^s \mapsto f \circ \varphi \in H^s$ for a diffeomorphism $\varphi \in \mathrm{Diff}_{H^s}(M)$ is smooth.

What is not smooth is the composition on the right with the diffeomorphism. To get C^k regularity w.r.t. both arguments, one must have k-higher regularity on the argument f.

Example 100 (Right-Trivialization on $\text{Diff}(M)$**)** The right-trivialization on $T\,\text{Diff}(M)$ is only continuous on $\text{Diff}_{H^s}(M)$, i.e. $(\delta\varphi, \varphi) \mapsto \delta\varphi \circ \varphi^{-1}$.

Due to the remark above, it might come as a surprise that the right-invariant H^s metric is a smooth Riemannian metric!

Other important spaces for the study of flows are Hölder spaces, which present some interesting regularity results on the composition that sometimes compare advantageously with Sobolev spaces.

Appendix 2: Reproducing Kernel Hilbert Spaces

Definition 101 A reproducing kernel Hilbert space (of vector fields) is a Hilbert space V of functions from Ω to $\mathbb{R}^d$ such that the pointwise evaluation maps denoted by $\delta_x : f \in H \mapsto f(x) \in \mathbb{R}^d$ are continuous. Denoting $K : V^* \mapsto V$ the Riesz isomorphism between V^* (the dual of V) and V, the reproducing kernel associated with the space V is defined by $k(x, y) = \langle \delta_x, K\delta_y \rangle \in L(\mathbb{R}^d, \mathbb{R}^d)$, where the bracket $(\cdot, \cdot)$ denotes the dual pairing.

The kernel completely specifies the reproducing kernel Hilbert space V: we refer the reader to [56] for more information on RKHS, but we can cite:

Definition 102 A positive kernel of dimension d on a set X is a map $k : S \times S \mapsto L(\mathbb{R}^d)$ such that

(1) for all $x, y \in S$, $k(x, y) = k(y, x)^*$
(2) for any $n \in \mathbb{N}$, $x_1, \ldots, x_n \in S$ and $p_1, \ldots, p_n \in \mathbb{R}^d$

$$\sum_{i,j=1}^n \langle p_i, k(x_i, x_j) p_j \rangle \geq 0. \tag{184}$$

The kernel is said to be strictly positive if inequality (184) is strict whenever there exists i such that $p_i \neq 0$.

Theorem 103 *The kernel associated with an RKHS is positive. To each positive kernel corresponds a unique RKHS of functions defined on S with values in $\mathbb{R}^d$.*

Proof The first assertion comes from the fact that

$$\left\| \sum_{i=1}^n \delta_{x_i}^{p_i} \right\|_{V^*}^2 = \sum_{i,j=1}^n \langle p_i, k(x_i, x_j) p_j \rangle. \tag{185}$$

The second assertion is obtained by defining

$$V = \overline{\mathrm{Span}\left\{k(.,x)p \mid x \in S \text{ and } p \in \mathbb{R}^d\right\}}$$

where the closure is taken with respect to the (semi) norm defined by Formula (184). Note that V is still a space of functions. Indeed, the injection

$$\mathrm{Span}\left\{k(.,x)p \mid x \in S \text{ and } p \in \mathbb{R}^d\right\} \mapsto \mathcal{F}(S, \mathbb{R}^d)$$

is continuous for the topology of pointwise convergence on $\mathcal{F}(S, \mathbb{R}^d)$. We now prove that the positivity of the semi-norm. Let $x \in S$ and consider $\langle f(x), p \rangle = \langle \delta_x^p, f \rangle \leq \|\delta_x^p\| \|f\| = 0$, using the Cauchy-Schwarz inequality. Therefore, $f(x) = 0$ for any $x \in S$ and thus $f = 0$. □

We will be particularly interested by kernels defined on $\mathbb{R}^d$. If the kernel is smooth, then the RKHS is composed of smooth functions of the same smoothness. When the kernel is translation invariant, the kernel $k(x,y)$ is in fact a function of the variable $x - y$ that we will denote with a little abuse of notation $k(x - y)$.

Theorem 104 (Bochner) *Let $k \in L^1(\mathbb{R}^d, \mathbb{R}^d \times \mathbb{R}^d)$ with Fourier transform $\hat{k}$ also in $L^1(\mathbb{R}^d, \mathbb{R}^d \times \mathbb{R}^d)$. Then $k(x - y)$ is a positive kernel if and only if $\hat{k}(\xi)$ is a self-adjoint positive matrix definite matrix for all $\xi \in \mathbb{R}^d$.*

We can be also interested in kernels that are invariant by rotations. A strict subset of those are of the form $k_d(x, y) = k(\|x - y\|_{\mathbb{R}^d})$ Id and among those, functions k such that k_d is a kernel for any $d \in \mathbb{N}^*$ are characterized by a theorem of Schoenberg:

Theorem 105 (Schoenberg) *A function $k : \mathbb{R}_+ \mapsto \mathbb{R}$ defines a kernel $k(\|x - y\|_{\mathbb{R}^d})$ Id for any $d \in \mathbb{N}^*$ if and only if it satisfies one of the two equivalent properties:*

(1) $f : r \to k(\sqrt{r})$ is a completely monotonic function, i.e. for any $n \in \mathbb{N}$,

$$(-1)^n f^{(n)}(t) \geq 0.$$

(2) There exists a finite Radon measure μ such that

$$k(r) = \int_0^{+\infty} e^{-r^2 u^2} d\mu(u).$$

How to Build Kernels? In practice, Gaussian kernels are widely used, however a wide range of kernels are available. Moreover, various operations combine kernels to produce new ones. The set of kernels is a cone, i.e. stable under addition and multiplication by a positive scalar. It is also stable by multiplication for scalar kernels. The addition of kernels has a nice variational interpretation: Let us consider a finite set of admissible Hilbert spaces H_i with kernels k_i and Riesz isomorphisms K_i between H_i^* and H_i for $i = 1, \ldots, n$. Denoting $H = H_1 + \ldots + H_n$, the space of all functions of the form $v_1 + \ldots + v_n$ with $v_i \in H_i$, the following norm can be defined on H:

$$\|v\|_H^2 = \inf \left\{ \sum_{i=1}^n \|v_i\|_{H_i}^2 \,\Big|\, \sum_{i=1}^n v_i = v \right\}. \tag{186}$$

The minimum is achieved for a unique n-tuple of vector fields and the space H, endowed with the norm defined by (186), is complete. This statement is simply a particular case of the involution of the Legendre-Fenchel transform which maps sums onto infimal convolutions. For the reader interested, the infimal convolution of two convex functions f, g is defined by

$$f \boxplus g(z) = \inf_{x+y=z} f(x) + g(y). \tag{187}$$

Then, one has for f, g two convex, lower semicontinuous, and proper (that is it takes at least one finite value and it never attains $-\infty$) functions

$$(f \boxplus g)^* = f^* + g^*, \tag{188}$$

where the Legendre-Fenchel transform is defined on $f : \mathbb{R}^d \to \mathbb{R}$ by

$$f^*(p) = \sup_x \langle p, x \rangle - f(x). \tag{189}$$

Proposition 106 *The formula (186) induces a scalar product on H which makes H a RKHS, and its associated kernel is $k := \sum_{i=1}^n k_i$, where k_i denotes the kernel of the space H_i.*

In applications, the choice of the kernel is crucial to avoid poor local minima, as shown in [18], and also generate more plausible deformations. It is also possible to build kernels that are associated with spaces of divergence-free or curl-free vector fields and thus put more information in the variational formulation.

Appendix 3: Bochner Integral

In this section, we give a brief introduction to the Bochner integral which is a generalization of the Lebesgue integral on real-valued functions to Banach-valued functions.

Let B denote a real Banach space and $U \subset B$ be an open set. If $A \subset B$ is a set, the notation $\mathbf{1}_A$ is the indicator function of A defined by $\mathbf{1}_A(x) = 1$ if $x \in A$ and $\mathbf{1}_A(x) = 0$ otherwise.

Definition 107 Let $(\Omega, \mathcal{A}, \mu)$ be a complete σ-finite measure space. A function $f : \Omega \to E$ is called a *step function* if it can be written as

$$f(x) = \sum_{i=1}^{n} \mathbf{1}_{E_i}(x) a_i, \tag{190}$$

where $n \geq 1$ and $E_i \in \mathcal{A}$ are measurable sets of finite measure and $a_i \in B$.

The integral of f is defined by

$$\int_\Omega f d\mu := \sum_{i=1}^{n} \mu(E_i) a_i$$

Definition 108 A function $f : \Omega \to B$ is called μ−*measurable* if it is almost everywhere (μ−a.e.) the pointwise limit of step functions, i.e. there exists a sequence $(f_n)_{n \in \mathbb{N}}$ of step functions such that μ−a.e. $\lim_{n \to \infty} \|f_n(t) - f(t)\| = 0$.

Note that this definition implies the convergence μ−a.e. of the norms by using the triangle inequality $\lim_{n \to \infty} |\|f_n(t)\| - \|f(t)\|| = 0$. Since $\|f_n(x)\| = \sum_{i=1}^{n} \mathbf{1}_{E_i}(x) \|a_i\|$ for a step function, this implies that $\|f_n\|$ is Lebesgue measurable. In particular, using Fatou's lemma, we have $\int_\Omega \|f\| d\mu \leq \liminf_{n \to \infty} \int \|f_n\| d\mu$.

Remark 109 This definition implies that for every $e \in B^*$, the dual space of B, the real function $e(f)$ is measurable.

Definition 110 A measurable function $f : \Omega \to B$ is called μ-*integrable* if there exists a sequence of step functions g_n converging μ-a.e. on Ω to f such that $\lim_{n \to \infty} \int_\Omega \|g_n - f\| d\mu = 0$. The integral of f is defined by

$$\int_\Omega f d\mu := \lim_{n \to \infty} \int_\Omega g_n d\mu.$$

It can be checked easily that the limit does not depend on the chosen converging sequence g_n. Let us prove the following equivalence,

Proposition 111 *A measurable function f is μ-integrable if and only if $\int_\Omega \|f\| d\mu < \infty$.*

Proof Let us prove the first implication: By the triangle inequality, the definition implies that $|\int_\Omega \|f_m\| d\mu - \int_\Omega \|f_n\| d\mu| \leq \int_\Omega \|f_m - f\| d\mu + \int_\Omega \|f_n - f\| d\mu$, which shows that $\int_\Omega \|f_m\| d\mu$ is a Cauchy sequence. Its limit is $\int_\Omega \|f\| d\mu$ (again by triangle inequality).

We now prove the reverse implication in the case $\mu(\Omega) < \infty$ (the case where μ is only σ-finite follows easily). In addition, by restriction to the measurable set $\{x \in \Omega \mid \|f(x)\| > 0\}$, we can assume $\|f(x)\| > 0$ for almost all $x \in \Omega$.

Let f be a measurable function such that $\int_\Omega \|f\| d\mu < \infty$. Since f is measurable, there exists a sequence f_n of step functions converging pointwisely to f. For such a given sequence, it is not true a priori that $\lim_{n\to\infty} \int_\Omega \|f - f_n\| d\mu = 0$, but this is true for a modification of f_n. Consider $\varepsilon > 0$,

$$A_{n,\varepsilon} := \{x \in \Omega \mid \|f_n(x) - f(x)\| > \varepsilon \|f(x)\|\}$$

and $B_{N,\varepsilon} = \cup_{n \geq N} A_{n,\varepsilon}$. By pointwise convergence of f_n, $\lim_{n\to\infty} \mu(B_{N,\varepsilon}) = 0$. Thus, for $\varepsilon = 1/k$, there exists $N(k)$ such that $\mu(B_{N(k),1/k}) \leq 1/k^2$. Let us define the following step function

$$g_k := \frac{1}{1 + 1/k} \mathbf{1}_{\Omega \setminus B_{N(k),1/k}} f_{N(k)} .$$

First, remark that $\|g_k\| \leq \|f\|$ since on $\Omega \setminus B_{N(k),1/k}$, we have

$$\|f_{N(k)}(x)\| - \|f(x)\| \leq \frac{1}{k} \|f(x)\|,$$

so that

$$\|f_{N(k)}(x)\| \leq (1 + \frac{1}{k}) \|f(x)\| .$$

In addition, $g_k(x)$ converges to $f(x)$ for almost all $x \in \Omega$: Indeed, we have

$$\|f(x) - g_k(x)\| \leq \|f(x) - f_{N(k)}(x)\| + (1 - \frac{1}{1+n}) \|f_{N(k)}(x)\| \leq \frac{2}{k}$$

on $\Omega \setminus \cup_{k \geq n} B_{N(k),1/k}$. Since $\mu(\cup_{k \geq n} B_{N(k),1/k}) \leq \sum_{k=n}^\infty \frac{1}{k^2} \xrightarrow[n\to\infty]{} 0$, the result ensues. □

Definition 112 Let $p \geq 1$, the space $L^p(\Omega, \mu, B)$ is the Banach space of μ-measurable functions (equivalence classes for the equivalence relation $f \sim g$ if $g = f$ μ-a.e.) such that $\|f\| \in L^p(\Omega, \mu)$.

Hereafter are some direct consequences and other useful remarks:

- Continuous functions on an interval I, $C^0(I, B)$ are Bochner integrable.
- Compositions of continuous maps are Bochner integrable.
- Let $I \subset \mathbb{R}$ be an interval and H a Hilbert space, then $L^2(I, H)$ is a Hilbert space.
- For $p \geq 1$, $W^{1,p}(I, H)$ defined as primitive integrals of elements in $L^p(I, B)$ is a Banach space. Note that of $W^{1,p}(I, B) \subset C^0(I, B)$.
- $W^{1,2}(I, H)$ is a Hilbert space, and it is separable if H is separable.

We end this section with a proposition that will be used in the appendix on ODE:

Proposition 113 *Let I be an interval and $f : I \times B \to B$ a function that is Bochner measurable when the second variable is fixed and continuous in the second variable for almost every $t \in I$. Then, if $x : I \to B$ is continuous, the composition $t \to f(t, x(t))$ is Bochner measurable.*

Proof The proof is left to the reader as an exercise. □

Appendix 4: ODE on Banach Spaces

This section follows [53].

Integration of ODE

In this section, we are concerned with the initial value problem of ordinary differential equations (ODE):

Let I be an interval, B be a Banach space and $f : I \times B \to B$ be a function. Find a function $x : I \to B$ such that

$$\begin{cases} \dot{x}(t) = f(t, x(t)) \\ x(0) = a \in B. \end{cases} \tag{191}$$

In what follows, we define the analytical framework suitable for our purpose. The class of functions f we want to work with is called *Caratheodory functions*.

Definition 114 A function $f : I \times B \to B$ is an L^p–Caratheodory function if

(1) The map $y \to f(t, y)$ is continuous for almost all $t \in I$,
(2) For all $y \in B$, the map $t \to f(t, y)$ is Bochner measurable,
(3) For every $r > 0$, there exists $h_r \in L^p(I, \mathbb{R})$ such that if $\|y\| \le r$ then $|f(t, y)| \le h_r(t)$ a.e. on I.

Definition 115 A function $f : I \times B \to B$ is said to be L^p-Lipschitz if there exists $\alpha \in L^p(I, \mathbb{R})$ such that for all $x, y \in B$

$$\|f(t, x) - f(t, y)\| \le \alpha(t)\|x - y\|,$$

for almost all $t \in I$.

Theorem 116 *Let $f : [0, T] \times B \to B$ be a L^p-Caratheodory function and L^p-Lipschitz. Then, there exists a unique $x \in W^{1,p}([0, T], B)$ solving (191).*

Proof We denote $I := [0, T]$. Define $A(t) = \int_0^t \alpha(s)ds$, so that a.e. $A'(t) = \alpha(t)$. We introduce the norm on $C^0(I, B)$, $\|y\|_A = \sup_{x \in I} \|y(t)e^{-A(t)}\|$. This norm is equivalent to the standard sup norm since $e^{-A(t)}$ is bounded below and above by positive real numbers. Thus the space $(C^0(I, B), \|\cdot\|_A)$ is a Banach space.

The map $F : C^0(I, B) \to C^0(I, B)$ defined by

$$F(y)(t) = \int_0^t f(s, y(s))ds,$$

is well-defined since $s \to f(s, y(s))$ is Bochner measurable by proposition 113 and it is integrable by Proposition 111: its norm is integrable using the Lipschitz property. Using a standard property of Bochner integral and the fact that f is L^p-Lipschitz, we get

$$\|F(x)(t) - F(y)(t)\| \le \int_0^t \|f(s, x(s)) - f(s, y(s))\|ds$$
$$\le \int_0^t \alpha(s)\|x(s) - y(s)\|ds.$$

Multiplying the previous inequality by $e^{-A(t)}$, we get:

$$e^{-A(t)}\|F(x)(t) - F(y)(t)\| \le e^{-A(t)} \int_0^t \|f(s, x(s)) - f(s, y(s))\|ds$$
$$\le e^{-A(t)} \int_0^t \alpha(s)e^{A(s)}e^{-A(s)}\|f(s, x(s)) - f(s, y(s))\|ds$$

$$\leq e^{-A(t)} \|x - y\|_A \int_0^T \alpha(s) e^{A(s)} \, ds$$

$$\leq (1 - e^{-A(t)}) \|x - y\|_A \, .$$

Noting that $k := \sup_{t \in I} 1 - e^{-A(t)} < 1$, it implies that F is a contraction

$$\|F(x) - F(y)\|_A \leq k \|x - y\|_A \, , \tag{192}$$

and existence and uniqueness of the solution in $C^0(I, B)$. Let x be this continuous solution. Since f is L^p−Lipschitz, the map $t \to f(t, x(t))$ is $L^p(I, B)$, so that $x \in W^{1,p}(I, B)$. □

Continuity of Solutions w.r.t Parameters

We also state without proof Gronwall's lemma (see [63] for a proof):

Lemma 117 (Gronwall's Lemma) *Let f, a, b be three measurable positive real functions defined on the interval $[0, T]$ for $T > 0$. If*

$$f(t) \leq a(t) + \int_0^t b(s) f(s) \, ds \, , \tag{193}$$

then, for all $t \in [0, T]$

$$f(t) \leq a(t) + \int_0^t a(s) b(s) e^{\int_0^s b(u) du} \, ds \, . \tag{194}$$

A consequence of this lemma is the continuity of the solutions with respect to the initial condition. Another interesting perturbation that will be used is the following:

Proposition 118 *Let f_n be a bounded sequence of L^1−Lipschitz functions such that on any bounded sets in B, there exists a sequence of L^1 functions α_n s.t. $\|f_n(t, x) - f(t, x)\| \leq \alpha_n(t)$ with $\lim_{n \to \infty} \|\alpha_n\|_{L^1} = 0$ ($(f_n)_{n \in \mathbb{N}}$ converges uniformly to f on bounded sets) then $\lim_{n \to \infty} \sup_{t \in [0,1]} \|x_n(t) - x(t)\| = 0$.*

Proof First, remark that all the solutions $x_n(t)$ and $x(t)$ are bounded on $[0, 1]$. We have,

$$\|x_n(t) - x(t)\| \leq \int_0^t \|f_n(s, x_n) - f(s, x)\| ds$$

$$\leq \int_0^t \|f_n(s, x_n(s)) - f(s, x_n(s))\| ds + \int_0^t \|f(s, x_n(s)) - f(s, x(s))\| ds$$

$$\leq \int_0^t \alpha_n(s) ds + \int_0^t M(s) \|x_n(s) - x(s)\| ds.$$

Applying Grönwall's lemma, we obtain, denoting $\int_0^t \alpha_n(s) ds = A_n(t)$

$$\|x_n(t) - x(t)\| \leq A_n(t) + \int_0^t A_n(s) M(s) e^{\int_0^s M(u) du} ds$$

$$\leq \sup_{t \in [0,1]} A_n(t) \left(1 + \int_0^t M(s) e^{\int_0^s M(u) du} ds\right).$$

This gives the result since $\|\alpha_n\|_{L^1} \to_{n \to \infty} 0$. □

Appendix 5: Polyak-Lojasiewicz Inequality

In optimization, a question of interest is the global convergence of gradient flow (continuous time) or gradient descent (discrete time). Gradient descent algorithms find, at best, critical points of the optimized function, which turn out to be a global minimizer if the function is convex. Outside the convex setting, obtaining global convergence might seem difficult. A simple condition to obtain global convergence is the Polyak-Lojasiewicz (PL) condition. It is defined for a C^1 function (the regularity can be weakened, see [34]) $f : X \to \mathbb{R}$ with minimum value f_* as follows:

$$\lambda(f(x) - f_*) \leq \frac{1}{2} \|\nabla f(x)\|^2, \tag{195}$$

for a positive real λ. In this formulation, one only needs to make sense of the norm of the gradient (and not necessarily the gradient itself). Clearly, the terms in this inequality are defined on Riemannian manifolds but the formulation is more general and it can be applied to spaces of measures endowed with the Wasserstein metric, for instance. Note that the factor $1/2$ is here to have $\frac{1}{2}\|x\|^2$ is PL($\lambda = 1$). We are interested in the convergence of the gradient flow $\dot{x} = -\nabla f(x)$. Let us assume that $f \in C^{1,1}$ for the gradient flow to be well-defined (here again, the regularity assumption can be weakened). Note that we get

$$\frac{d}{dt}(f(x) - f_*) = -\|\nabla f(x)\|^2 \leq 2\lambda(f(x) - f_*). \tag{196}$$

Therefore,
$$f(x(t)) - f_* \leq (f(x(0)) - f_*)e^{-2\lambda t}, \tag{197}$$

which is a linear rate of convergence towards the minimum value.

However, at this point, nothing is said about the boundedness, nor convergence of $x(t)$. To prove the convergence of $x(t)$, it is sufficient to bound the length of the curve by introducing $g(x) = 2\sqrt{f(x) - f_*}$. We compute

$$\begin{aligned} g(x(0)) - g(x(t)) &= \int_0^t \langle \nabla g(x(s)), \dot{x}(s) \rangle ds \\ &= \int_0^t \left\langle \frac{\nabla f(x(s))}{\sqrt{f(x(s)) - f_*}}, \dot{x}(s) \right\rangle ds \geq \lambda \|x(t) - x(0)\|. \end{aligned} \tag{198}$$

The two terms in the scalar product are colinear, thus this term can be lower bounded by $\mu \|\dot{x}(s)\|$ due to the PL inequality. More generally, it proves, for $t' < t$ $\|x(t) - x(t')\| \leq g(x(t')) - g(x(t))$. This inequality implies that $x(t)$ converges to some x_∞ for which $f(x_\infty) = f_*$ due to the continuity of f. Unlike convexity, the PL condition is stable under the composition of the function f with a sufficiently regular diffeomorphism.

Proposition 119 (Stability of PL) *Let $\varphi : \Omega \to \Omega$ be a C^1 diffeomorphism of the definition domain of f, then $\varphi^* f(y) \triangleq f \circ \varphi(y)$ satisfies $PL(\lambda/M^2)$ if f satisfies $PL(\lambda)$ for $M = \sup_{x \in \Omega} \|d\varphi(x)^{-1}\|$.*

Indeed, one has $\nabla[f \circ \varphi](x) = d\varphi(x)^\top (\nabla f(\varphi(x)))$ and the result follows by computing its norm. So if $M(x)$ is the minimal singular value of $d\varphi(x)$, one has to minimize $M(x)$ over Ω. Not every function can be PL obviously, this condition implies the following growth condition.

Proposition 120 *The PL condition implies quadratic growth measured from the set of minimizers of the function. That is, given x such that its gradient flow converges to $x_* \in \arg\min f$, one has*

$$f(x) - f(x_*) \geq \frac{\lambda}{2} \|x - x_*\|^2, \tag{199}$$

for a function f that satisfies $PL(\lambda)$.

Proof Introduce $h(x) \triangleq \sqrt{f(x) - f(x_*)}$ then the PL condition reads

$$\|\nabla h(x)\|^2 \geq \frac{\lambda}{2} \tag{200}$$

for x that is not a minimizer (equivalently a minimizer). It is, in fact, the Kurdika-Lojasiewicz inequality. One gets the result by integrating along unit speed curves defined by the gradient flow. □

In fact, the strategy of proof gives a reparametrization of the level lines of the function into a sort of quadratic map from the minimizers. Let us assume for instance that x_* is the unique minimizer of f, then for each $x \neq x_*$ one can associate the length of the curve described by the gradient flow $l(x)$, it gives a homeomorphism (which is actually smooth but at x_*) from $\mathbb{R}^d$ into $\mathbb{R}^d$ defined by $x \mapsto l(x)^2 \frac{x}{\|x\|}$ that maps f to $\frac{\lambda}{2}\|x - x_*\|^2$. The topology of the set of minimizers is also constrained for a PL function:

Proposition 121 *If a C^2 function f is λ-PL, then its gradient flow gives a deformation retract from $\mathbb{R}^d$ to the set of its minimizers. Therefore, its set of minimizers is homotopically equivalent to $\mathbb{R}^d$.*

Proof The retraction can be constructed as $T : \mathbb{R}^d \to \mathbb{R}^d$ by $T(x) = \lim_{t \to \infty} x(t)$ with $x(t)$ the gradient flow of f starting at x. It is the identity on the set of minimizers since they are critical points. Consider the flow map at time t defined by $\mathrm{Fl}_t(x) = x(t)$. It is continuous due to the regularity assumption on f. Due to the inequality (198), the flow maps $(\mathrm{Fl}_n)_{n \in \mathbb{N}}$ uniformly converge on every compact set to Fl_∞.[14] Therefore, the limit is continuous. Up to time-reparametrization, the flow Fl gives a strong deformation retract from $\mathbb{R}^d$ to the set of minimizers. □

Strong Convexity Implies PL In practice, obtaining the PL property can be difficult. It holds for strongly convex functions for instance. Recall that a function f is μ strongly convex if $f(x) - \frac{\mu}{2}\|x\|^2$ is convex or equivalently,

$$f(y) \geq f(x) + \langle y - x, \nabla f(x) \rangle + \frac{\mu}{2}\|y - x\|^2. \tag{201}$$

The PL condition can be obtained directly from this estimate: minimize over the variable y to get $\nabla f(x) + \mu(y - x) = 0$ so that

$$f(x) - f(x_\star) \leq \frac{1}{2\mu}\|\nabla f(x)\|^2. \tag{202}$$

By adding to this inequality the same one exchanging y, x, one gets the second inequality

$$\langle y - x, \nabla f(y) - \nabla f(x) \rangle \geq \mu \|y - x\|^2. \tag{203}$$

---

[14] We leave the proof to the reader.

This condition is equivalent to strong convexity and it implies that the distance to a minimizer can be controlled as

$$\|\nabla f(x)\| \geq \mu \|x - x_*\|. \tag{204}$$

Under the additional assumption that f has a L Lipschitz gradient, the last inequality implies PL since this smoothness condition implies

$$f(y) \leq f(x) + \langle \nabla f(x), y - x \rangle + \frac{L}{2}\|y - x\|^2, \tag{205}$$

and by setting $x = x_*$ a minimizer, we get PL($\frac{\mu}{L}$),

$$f(x) - f(x_*) \leq \frac{L}{2}\|x - x_*\|^2 \leq \frac{L}{2\mu}\|\nabla f(x)\|^2. \tag{206}$$

Thus, Eq. (204) and L-smoothness implies PL (with a function f not necessarily convex[15]).

Generalization of Gradient Flows to Metric Spaces The notion of a gradient flow has been generalized to metric spaces, see [2]. The starting point is the gradient flow of a smooth functions f in Euclidean spaces (or Hilbert spaces), $\dot{x} = -\varepsilon \nabla f(x)$, by remarking[16] that such curves $x(t)$ minimize

$$\frac{1}{2}\int_0^T |\dot{x}|(t)^2 + \varepsilon^2 \left(|\nabla f|(x(t))\right)^2 dt. \tag{207}$$

Now, in a metric space, the *metric speed* of a curve $x(t)$ can be defined $|\dot{x}|(t) = \lim_{s \to t} \frac{d(x(s), x(t))}{|s - t|}$ and the quantity $|\nabla f(x)|$ can be replaced with the *descending slope* defined by:

$$|D^- f|(x) := \lim_{y \mapsto x} \frac{\max(-(f(y) - f(x)), 0)}{d(y, x)}. \tag{208}$$

Thus, one can give a sense to the right-hand side of Formula (207) in a metric space for functions that have a well-defined descending slope. Now, several definitions of gradient flow curves have been given in the literature. We refer the reader to [57] for more details on the PL property on metric spaces.

---

[15] The proof above applies, since one only needs Formula (205) at a minimizer.

[16] In a Hilbert space, the equality $\dot{x} = \nabla f(x)$ can be recast as an equality case in the inequality $\langle \dot{x}, \nabla f(x) \rangle \leq \frac{1}{2}(|\dot{x}|^2 + |\nabla f(x)|^2)$.

Definition 122 (Local PL Property) Let (X, d) be a complete metric space and $f : X \to \mathbb{R}$ be a function such that the descending slope (defined above) is well-defined at every point $x \in X$. We say that f satisfies a *local PL property*, if there exists x_0 and for every $R > 0$, there exists a positive constant $\lambda(R)$ such that

$$\lambda(R)(f(x) - f_*) \leq \frac{1}{2}(|D^- f|(x))^2 \tag{209}$$

for all $x \in X$ such that $d(x, x_0) \leq R$.

Remark 123 This condition is weaker than the global one since the constant $\lambda(R)$ can increase with R. Yet, the property implies that every critical point is a global minimizer. In general, this property is insufficient to conclude the gradient flow's global convergence. Indeed, the function $x \to e^x$ does satisfy this condition and the gradient flow converges to $-\infty$.

A simple result is the following:

Corollary 124 (Bounded Flow Implies Global Convergence in Value) *If a function f satisfies a local PL condition and the trajectory $x(t)$ of the gradient flow of f stays in a bounded domain, then global convergence holds.*

References

1. R.A. Adams, *Sobolev Spaces*, 2nd edn. (Academic Press, New York, 2003)
2. L. Ambrosio, N. Gigli, G. Savaré, *Gradient Flows: In Metric Spaces and in the Space of Probability Measures* (Springer Science & Business Media, New York, 2005)
3. V. Arnold, Sur la géom'etrie différentielle des groupes de Lie de dimension infinie et ses applications 'a l'hydrodynamique des fluides parfaits, in *Ann. Inst. Fourier (Grenoble)*, vol. 16, fasc. 1 (1966), pp. 319–361. ISSN: 0373-0956
4. R. Barboni, G. Peyré, F.X. Vialard, Understanding the training of infinitely deep and wide ResNets with Conditional Optimal Transport (2024). arXiv: 2403.12887 [cs.LG]. https://arxiv.org/abs/2403.12887
5. A.R. Barron, Universal approximation bounds for superpositions of a sigmoidal function. IEEE Trans. Inf. Theory **39**(3), 930–945 (1993)
6. M. Bauer, C. Maor, Can we run to infinity? The diameter of the diffeomorphism group with respect to right-invariant Sobolev metrics. Preprint. arXiv:1910.04834 (2019)
7. M. Bauer, K. Modin, Semi-invariant Riemannian metrics in hydrodynamics. Calc. Var. Partial Differential Equations **59**(2) (2020). ISSN: 1432-0835. https://doi.org/10.1007/s00526-020-1722-x
8. M. Bauer, M. Bruveris, P. Harms, P.W. Michor, Geodesic distance for right invariant Sobolev metrics of fractional order on the diffeomorphism group. Ann. Global Anal. Geom. **44**(1), 5–21 (2013). ISSN: 0232-704X. https://doi.org/10.1007/s10455-012-9353-x
9. M. Bauer, P. Harms, P.W. Michor, Regularity and completeness of half-Lie groups (2023). arXiv: 2302.01631 [math.DG]. https://arxiv.org/abs/2302.01631

10. A. Behzadan, M. Holst, Multiplication in Sobolev spaces, revisited (2021). arXiv: 1512.07379 [math.AP]. https://arxiv.org/abs/1512.07379
11. J.D. Benamou, Y. Brenier, A computational fluid mechanics solution to the Monge-Kantorovich mass transfer problem. Numer. Math. **84**(3), 375–393 (2000)
12. V.N. Berestovskiĭ, "Submetries" of three-dimensional forms of nonnegative curvature. Sibirsk. Mat. Zh. **28**(4), 44–56, 224 (1987). ISSN: 0037-4474
13. S. Boufadène, F.X. Vialard, On the global convergence of Wasserstein gradient flow of the Coulomb discrepancy (2024). arXiv: 2312.00800 [math.AP]. https://arxiv.org/abs/2312.00800
14. Y. Brenier, Polar factorization and monotone rearrangement of vector-valued functions. Comm. Pure Appl. Math. **44**(4), 375–417 (1991). ISSN: 0010-3640. https://doi.org/10.1002/cpa.3160440402
15. Y. Brenier, The dual least action problem for an ideal, incompressible fluid. Arch. Rational Mech. Anal. **122**(4), 323–351 (1993). ISSN: 1432-0673. https://doi.org/10.1007/BF00375139
16. M. Bruveris, Riemannian geometry for shape analysis and computational anatomy (2018). arXiv: 1807.11290 [math.DG]
17. M. Bruveris, F.X. Vialard, On completeness of groups of diffeomorphisms. J. Eur. Math. Soc. (JEMS) **19**(5), 1507–1544 (2017). ISSN: 1435-9855. https://doi.org/10.4171/JEMS/698
18. M. Bruveris, L. Risser, F.X. Vialard, Mixture of kernels and iterated semidirect product of diffeomorphisms groups. Multiscale Model. Simul. **10**(4), 1344–1368 (2012). ISSN: 1540-3459. https://doi.org/10.1137/110846324
19. M. Bauer, M. Bruveris, P. Harms, P.W. Michor, Vanishing geodesic distance for the Riemannian metric with geodesic equation the KdV-equation. Ann. Global Anal. Geom. **41**(4), 461–472 (2012). ISSN: 0232-704X. https://doi.org/10.1007/s10455-011-9294-9
20. R. Camassa, D.D. Holm, An integrable shallow water equation with peaked solitons. Phys. Rev. Lett. **71**(11), 1661–1664 (1993). ISSN: 0031-9007. https://doi.org/10.1103/PhysRevLett.71.1661
21. R. Camassa, D.D. Holm, An integrable shallow water equation with peaked solitons (English). Phys. Rev. Lett. **71**(11), 1661–1664 (1993). ISSN: 0031-9007; 1079-7114/e. https://doi.org/10.1103/PhysRevLett.71.1661
22. L. Chizat, F. Bach, On the global convergence of gradient descent for over-parameterized models using optimal transport, in *Advances in Neural Information Processing Systems*, vol. 31 (2018)
23. L. Chizat, G. Peyré, B. Schmitzer, F-X. Vialard, Unbalanced optimal transport: dynamic and Kantorovich formulations. J. Funct. Anal. **274**(11), 3090–3123 (2018). https://doi.org/10.1016/j.jfa.2018.03.008
24. L. Chizat, G. Peyré, B. Schmitzer, F.X. Vialard, An interpolating distance between optimal transport and Fisher–Rao metrics. Found. Comput. Math. **18**, 1–44 (2018)
25. A. Duncan, N. Nusken, L. Szpruch, On the geometry of Stein variational gradient descent. J. Mach. Learn. Res. **24**(56), 1–39 (2023). http://jmlr.org/papers/v24/20-602.html
26. P. Dupuis, U. Grenander, M.I. Miller, Variational problems on flows of diffeomorphisms for image matching. Q. Appl. Math. **56**, 587–600 (1998)
27. D.G. Ebin, Groups of diffeomorphisms and fluid motion: reprise, in *Geometry, Mechanics, and Dynamics: The Legacy of Jerry Marsden*, ed. by D.E. Chang, D.D. Holm, G. Patrick, T. Ratiu (Springer New York, New York, 2015), pp. 99–105. ISBN: 978-1-4939-2441-7. https://doi.org/10.1007/978-1-4939-2441-7_6
28. D.G. Ebin, J. Marsden, Groups of diffeomorphisms and the motion of an incompressible fluid. Ann. Math. (2) **92**, 102–163 (1970). ISSN: 0003-486X
29. I. Ekeland, The Hopf-Rinow theorem in infinite dimension. J. Differential Geometry **13**(2), 287–301 (1978). ISSN: 0022-040X. http://projecteuclid.org.proxy.lib.fsu.edu/euclid.jdg/1214434494
30. D.S. Freed, D. Groisser, The basic geometry of the manifold of Riemannian metrics and of its quotient by the diffeomorphism group. Michigan Math. J. **36**(3), 323–344 (1989)

31. T. Gallouët, F.X. Vialard, The Camassa–Holm equation as an incompressible Euler equation: a geometric point of view. J. Differential Equations **264**(7), 4199–4234 (2018). ISSN: 0022-0396. https://doi.org/10.1016/j.jde.2017.12.008. http://www.sciencedirect.com/science/article/pii/S0022039617306435
32. T.O. Gallouët, A. Natale, F.X. Vialard, Generalized compressible flows and solutions of the $H(\mathrm{div})$ geodesic problem. Arch. Ration. Mech. Anal. **235**(3), 1707–1762 (2020). ISSN: 0003-9527,1432-0673. https://doi.org/10.1007/s00205-019-01453-x
33. T. Gallouët, R. Ghezzi, F.X. Vialard, Regularity theory and geometry of unbalanced optimal transport. First version, comments welcome (2021). https://hal.archives-ouvertes.fr/hal-03498098
34. G. Garrigos, Square distance functions are Polyak-Łojasiewicz and vice-versa (2023). arXiv: 2301.10332 [math.OC]. https://arxiv.org/abs/2301.10332
35. P. Heslin, Two-point boundary value problems on diffeomorphism groups (2022). arXiv: 2110.12278 [math.DG]. https://arxiv.org/abs/2110.12278
36. D.D. Holm, Peakons (2009). arXiv: 0908.4351 [nlin.SI]. https://arxiv.org/abs/0908.4351
37. H. Inci, T. Kappeler, P. Topalov, On the regularity of the composition of diffeomorphisms. Mem. Amer. Math. Soc. **226**(1062), vi+60 (2013). ISSN: 0065-9266. https://doi.org/10.1090/S0065-9266-2013-00676-4
38. R.L. Jerrard, C. Maor, Geodesic distance for right-invariant metrics on diffeomorphism groups: critical Sobolev exponents. Ann. Glob. Anal. Geom. **56**(2), 351–360 (2019)
39. B. Khesin, J. Lenells, G. Misiołek, Generalized Hunter-Saxton equation and the geometry of the group of circle diffeomorphisms. Math. Ann. **342**(3), 617–656 (2008). ISSN: 0025-5831. https://doi.org/10.1007/s00208-008-0250-3
40. S. Kondratyev, D. Vorotnikov, Spherical Hellinger-Kantorovich gradient flows (2019). arXiv: 1809.03430 [math.FA]. https://arxiv.org/abs/1809.03430
41. J.D. Lafferty, The density manifold and configuration space quantization. Trans. Am. Math. Soc. **305**(2), 699–741 (1988). ISSN: 00029947. http://www.jstor.org/stable/2000885 (visited on 01/06/2025)
42. S. Lang, *Fundamentals of Differential Geometry*, vol. 191. Graduate Texts in Mathematics (Springer, New York, 1999), pp. xviii+535. ISBN: 0-387-98593-X. https://doi.org/10.1007/978-1-4612-0541-8
43. M. Liero, A. Mielke, G. Savaré, Optimal entropy-transport problems and a new Hellinger–Kantorovich distance between positive measures. Invent. Math. **247**(6), 1–149 (2015)
44. M. Liero, A. Mielke, G. Savaré, Fine properties of geodesics and geodesic λ-convexity for the Hellinger–Kantorovich distance. Arch. Rational Mech. Anal. **247**(6), 112 (2023). https://doi.org/10.1007/s00205-023-01941-1
45. Y. Lu, D. Slepčev, L. Wang, Birth–death dynamics for sampling: global convergence, approximations and their asymptotics. Nonlinearity **36**, 5731–5772 (2023). https://doi.org/10.1088/1361-6544/acf988
46. Q. Liu, D. Wang, Stein variational gradient descent: a general purpose bayesian inference algorithm (2019). arXiv: 1608.04471 [stat.ML]. https://arxiv.org/abs/1608.04471
47. P.W. Michor, D. Mumford, Vanishing geodesic distance on spaces of submanifolds and diffeomorphisms. Doc. Math. **10**, 217–245 (2005). ISSN: 1431-0635
48. G. Misiołek, Stability of flows of ideal fluids and the geometry of the group of diffeomorphisms. Indiana Univ. Math. J. **42**(1), 215–235 (1993). ISSN: 0022-2518. https://doi.org/10.1512/iumj.1993.42.42011
49. G. Misiołek, A shallow water equation as a geodesic flow on the Bott-Virasoro group. J. Geom. Phys. **24**(3), 203–208 (1998). ISSN: 0393-0440. https://doi.org/10.1016/S0393-0440(97)00010-7

50. G. Misiolek, S.C. Preston, Fredholm properties of Riemannian exponential maps on diffeomorphism groups (English). Invent. Math. **179**(1), 191–227 (2010). ISSN: 0020-9910. https://doi.org/10.1007/s00222-009-0217-3
51. A. Natale, F.X. Vialard, Embedding Camassa-Holm equations in incompressible Euler. J. Geom. Mech. **11**(2), 205–223 (2019). ISSN: 1941-4889, 1941-4897. https://doi.org/10.3934/jgm.2019011
52. H. Omori, On Banach-Lie groups acting on finite dimensional manifolds. Tôhoku Math. J. **30**(2), 223–250 (1978)
53. D. O'Regan, *Existence Theory for Nonlinear Ordinary Differential Equations* (Springer, Dordrecht, 1997)
54. F. Otto, The geometry of dissipative evolution equations: the porous medium equation. Commun. Partial Differential Equations **26**(1–2), 101–174 (2001)
55. N. Papadakis, G. Peyré, E. Oudet, Optimal transport with proximal splitting. SIAM J. Imaging Sci. **7**(1), 212–238 (2014). https://doi.org/10.1137/130920058. http://hal.archives-ouvertes.fr/hal-00816211/
56. S. Saitoh, *Theory of Reproducing Kernels and its Applications*. Pitman Research Notes in Mathematics (Longman Scientific & Technical, Harlow, 1988)
57. L.D. Schiavo, J. Maas, F. Pedrotti, Local conditions for global convergence of gradient flows and proximal point sequences in metric spaces (2023). arXiv: 2304.05239 [math.OC]
58. J. Shi, L. Mackey, A finite-particle convergence rate for stein variational gradient descent (2023). arXiv: 2211.09721 [cs.LG]. https://arxiv.org/abs/2211.09721
59. A. Trouvé, Infinite dimensional group action and pattern recognition. Tech. rep. Unpublished. DMI, Ecole Normale Supérieure, 1995
60. F.X. Vialard, L. Risser, D. Rueckert, C.J. Cotter, Diffeomorphic 3D image registration via geodesic shooting using an efficient adjoint calculation. Int. J. Comput. Vision **97**, 229–241 (2012)
61. C. Villani, *Topics in Optimal Transportation*. Graduate Studies in Mathematics, vol. 58 (American Mathematical Society, Providence, 2003), pp. xvi+370. ISBN: 0-8218-3312-X. https://doi.org/10.1090/gsm/058
62. C. Villani, *Optimal Transport: Old and New*, vol. 338. Grundlehren der mathematischen Wissenschaften (Springer, Berlin, 2009)
63. L. Younes, *Shapes and Diffeomorphisms* (Springer, Berlin, 2010)

Generalized Wasserstein Dynamics in Mathematical Data Sciences

Wuchen Li

Abstract

There are several well-established books in the area of optimal transport (Villani, Optimal transport: old and new. Springer, 2009), gradient flows (Ambrosio et al., Gradient flows in metric spaces and in the space of probability measures. Birkhauser Verlag, 2008), and mean field control and games (Cardaliaguet et al., The master equation and the convergence problem in mean-field games, 2015) with computations and applications. Why do we need this lecture note for the Oberwolfach seminar 2023? We present a different angle or a shortcut from current books. The note follows a modern applied mathematics angle toward dynamical optimal transport and its generalizations. It presents some recent developments in related areas, written at an informal level. The note targets readers of high-year undergraduate students, graduate students, postdocs, and early career faculties. They may find some interesting problems and motivations to work in this area, where the technical part is left in the above-mentioned famous books. The note also presents some mathematics formulations and computational methods. This explains why optimal transport-related equations or algorithms are essential in mathematical data sciences. Towards this goal, we particularly prepare readers to digest some terminologies in modern data sciences,

---

Wuchen Li is partially supported by AFOSR YIP award No. FA9550-23-1-0087, NSF DMS-2245097, and NSF RTG: 2038080.

W. Li (✉)
Department of Mathematics, University of South Carolina, Columbia, SC, USA
e-mail: wuchen@mailbox.sc.edu

including tangent space, co-tangent space, Wasserstein metrics, Jordan–Kinderlehrer–Otto (JKO) schemes, generative models, neural ODEs, Deep JKO methods, and so on.

1 Introduction

In recent years, optimal transport dynamics and their generalizations [25, 26, 34, 35] have been widely studied in a variety of communities in pure and applied mathematics, such as finite and infinite dimensional differential geometry, probability, functional inequalities, spectral graph theory, mathematics physics, and scientific computing. This area has also become a research focus in data science, in particular, generative artificial intelligence [14, 22].

Nowadays, dynamical optimal transport formulations, analysis, approximations, and computations are core areas in the mathematical foundations of data sciences (MDS). The main goals of MDS lie in the selection of models and optimizations in the probability density space. We note that the model represents a computational method that approximates probability density functions. Approximating the probability density function can be formulated explicitly or implicitly, depending on how the probability density function is constructed. This has been investigated in the optimal transport problem, in which we explore the variational problem in both Eulerian (explicit models) and Lagrangian (implicit models) coordinates. Optimization methods in MDS often refer to solvers, which can find solutions in an efficient way to represent the observed data sets from selected models from samples or particle evolutions. Here, two components are essential in this area: objective functionals and optimization schemes, such as gradient descent directions, in the probability density space. Again, these concepts are well-established directions in optimal transport, including Lyapunov functionals and gradient flows. The Lyapunov functional are often chosen as distances or divergences to measure the closeness between two different probability density functions. Typical examples include the Kullback–Leibler divergence or Wasserstein distance, etc. At the same time, the minimization problem is considered between the current density function and the target distribution, which is often given as an empirical distribution provided by data samples. In this procedure, the gradient descent in Wassertein-2 space provides a descent direction in the probability space. The algorithm will converge to the target distribution in probability density space by the iterative procedure in discrete time. Of course, there are essential issues in implementing these gradient flows in probability densities regarding samples or particles in the sample space [24].

In this note, we mainly divide the area of generalized optimal transport into three components, including original dynamical optimal transport problems, optimal transport gradient flows, and optimal transport Hamiltonian flows.

(i) Optimal transport problems: We illustrate the formulation of optimal transport, starting from Wasserstein distances and Benamou-Brenier formulas [4].
(ii) Optimal transport Hamiltonian flows: We present the optimal control formulation of generalized optimal transport problems, namely mean field control problems. We briefly write down their minimizer systems and discuss their Hamiltonian structures [5, 25, 26].
(iii) Optimal transport gradient flows: We reformulate gradient-drift Fokker-Planck equations as gradient flows. One important example is that the gradient-drift Fokker-Planck equation is the gradient flow of relative entropies (also named Kullback–Leibler divergence) in Wasserstein-2 spaces [16]. The generalized Wasserstein-type gradient flow formulations are also discussed [26, 30].

We organize this note as follows. We briefly discuss the basics of Wasserstein dynamics in Sect. 2. We then present a machine learning-related computational scheme for Wasserstein dynamics in Sect. 3. Numerical examples are presented in Sect. 3.7. In Sect. 4, we study the accelerated optimization methods in the probability density space. A discussion section on future directions of computational Wasserstein dynamics is presented in Sect. 5.

2 Basics of Wasserstein Dynamics

In this section, we first review the dynamical optimal transport from the Benamou-Brenier formula. We then provide the generalized optimal transport problems based on generalized Benamou-Brenier formulas. From this viewpoint, we discuss gradient flows and Hamiltonian flows in generalized Wasserstein spaces, see related studies in [21].

2.1 Review of Dynamics in Euclidean Space

Consider an optimization problem in the Euclidean space:

$$\min_{x \in \mathbb{R}^d} f(x),$$

where the function $f \colon \mathbb{R}^d \to \mathbb{R}$ is a given smooth objective function. To find the minimizer of function f, consider an initial value dynamical system as

$$\frac{dx(t)}{dt} = -g(x(t))^{-1} \nabla f(x(t)), \quad x(0) = x_0, \tag{1}$$

where $g \colon \mathbb{R}^d \to \mathbb{R}^{d \times d}$ is a given matrix function. In practice, there are several natural choices of matrix functions g.

(i) If $g(x) = \mathbb{I}$, where $\mathbb{I}$ is an identity matrix. Dynamic (1) is the gradient flow in Euclidean space.
(ii) If $g(x) = \nabla^2 f(x)$, where ∇^2 is the Euclidean Hessian operator. Dynamic (1) satisfies the Newton's flow in Euclidean space.

Assume that the matrix function g is positive definite. We observe that the objective function f decays along with dynamic (1). In other words,

$$\frac{d}{dt}f(x(t)) = \nabla f(x(t))^\mathsf{T}\frac{dx(t)}{dt} = -\nabla f(x(t))^\mathsf{T} g(x(t))^{-1}\nabla f(x(t)) \leq 0. \tag{2}$$

Find a function along which the dynamics decay is known as a Lyapunov method, in which the objective function is a "natural" Lyapunov function for Eq. (1). In literature, dynamic (1) can be viewed as a gradient flow in the metric space $(\mathbb{R}^d, g)$. And the matrix function g is often called the metric tensor. Sometimes, it is also called the matrix operator or the preconditioner matrix in the optimization community.

2.1.1 Hamiltonian Flows

In this metric space $(\mathbb{R}^d, g)$, one often considers the following variational problem. Denote $k \colon \mathbb{R}^d \to \mathbb{R}$ as a given smooth potential function and formulate $L \colon \mathbb{R}^d \times \mathbb{R}^d \to \mathbb{R}$ a Lagrangian function as

$$L(x, v) = \frac{1}{2}v^\mathsf{T} g(x)v - k(x).$$

Consider

$$\frac{1}{2}\mathrm{D}(x_0, x_1)^2 := \inf_{x \colon [0,1] \to \mathbb{R}^d} \int_0^1 L(x(s), \frac{dx(s)}{ds})ds, \tag{3}$$

where the infimum is taken among all smooth path functions $x \colon [0, 1] \to \mathbb{R}^d$ with fixed initial and terminal values x_0, x_1. The Euler-Lagrange equation of the variation problem (3) can be formulated into a Hamiltonian flow. Write a Hamiltonian function $H \colon \mathbb{R}^d \times \mathbb{R}^d \to \mathbb{R}$ as

$$H(x, p) = \frac{1}{2}p^\mathsf{T} g(x)^{-1} p + k(x). \tag{4}$$

where $x \in \mathbb{R}^d$ represents the state variable and $p \in \mathbb{R}^d$ is the momentum variable. The minimizer of variational problem (3) satisfies $x(0) = x_0, x(1) = x_1$ with

$$\begin{cases} \dfrac{dx(s)}{ds} = \nabla_p H(x(s), p(s)), \\ \dfrac{dp(s)}{ds} = -\nabla_x H(x(s), p(s)). \end{cases}$$

This can be rewritten as

$$\begin{cases} \dfrac{dx(s)}{ds} = g(x(s))^{-1} p(s), \\ \dfrac{dp(s)}{ds} = -\dfrac{1}{2}\nabla_x(p(s)^\mathsf{T} g(x(s))^{-1} p(s)) - \nabla_x k(x(s)). \end{cases}$$

In particular, if $k = 0$, the Hamiltonian flow (4) is called the geodesic equation in the metric space $(\mathbb{R}^d, g)$, in which $D(x_0, x_1)$ is also named the distance function. Besides, the above Hamiltonian flow is a characteristic of the Hamilton-Jacobi equation. In other words, consider a value function $U \colon [0, \infty) \times \mathbb{R}^d \to \mathbb{R}$, such that $p(t) = \nabla_x U(t, x(t))$, then

$$\partial_t U(t, x) + H(x, \nabla_x U(t, x)) = 0.$$

2.1.2 Hamiltonian Formalism of Gradient Flows and Variational Time Discretizations

We remark that both the gradient flow (1) and the Hamiltonian flow (4) are connected with each other. Formally speaking, the gradient flow (1) can be written by the Hamiltonian formalism below:

$$\begin{cases} \dfrac{dx}{dt} = \nabla_p H(x, p), \\ p = -\nabla_x f(x), \end{cases}$$

where H is a quadratic Hamiltonian function defined in (4) with $k = 0$. The above Hamiltonian formalism is useful in designing the variational time discretization for the gradient flow. Denote a time stepsize by $\Delta t > 0$. One can construct a sequence $\{x_k\}_{k=1}^\infty$ as follows. Consider an iterative variational sequence as

$$\inf_{x \colon [t_k, t_{k+1}] \to \mathbb{R}^d} \left\{ \int_{t_k}^{t_{k+1}} L\!\left(x(t), \dfrac{dx(t)}{dt}\right) dt + f(x(t_{k+1})) \colon x(t_k) = x_k \right\}. \tag{5}$$

Write

$$x_{k+1} = x(t_{k+1}),$$

where $x(t_{k+1})$ is the minimizer of variational problem (5). We observe that the following claim holds.

Claim

$$\{x_k\}_{k=1}^\infty \text{ is a first-order time discretization of gradient flow (1)}.$$

Proof of Claim To see this fact, one can solve the minimization problem (5) and observe that

$$\begin{cases} \dfrac{dx}{ds} = \nabla_p H(x, p), \quad \dfrac{dp}{ds} = -\nabla_x H(x, p) \\ x(t_k) = x_k, \quad p(t_{k+1}) = -\nabla_x f(x(t_{k+1})). \end{cases}$$

It is clear that our update is

$$\begin{aligned} x_{k+1} &= x(t_{k+1}) = x_k + \int_{t_k}^{t_{k+1}} \frac{dx(t)}{dt} dt \\ &= x_k + \int_{t_k}^{t_{k+1}} \nabla_p H(x(t), p(t)) dt \\ &= x_k + (t_{k+1} - t_k) \nabla_p H(x(t), p(t))|_{t=h} + o(h) \\ &= x_k - h g(x_{k+1})^{-1} \nabla_x f(x_{k+1}) + o(h). \end{aligned}$$

In other words,

$$\frac{x_{k+1} - x_k}{h} = -g(x_{k+1})^{-1} \nabla_x f(x_{k+1}) + o(h).$$

The above formulation is a backward Euler time discretization up to a small order perturbation. A known fact is that the backward Euler method is a first order method w.r.t. a time stepsize $h > 0$. □

2.2 Mean-Field Information Metric Spaces with Gradient Flows and Hamiltonian Flows

In this section, we first review the mean-field information metric on the space of positive densities; see [2,25]. We next review both gradient flows and Hamiltonian flows in positive density metric space.

In this subsection, we review some known facts about reaction–diffusion equations in term of optimal transport type gradient flow formalisms. See related studies of diffusion equations in [31], and reaction diffusion equations with Onsager principles in [25,26]. We can construct metrics and Hamiltonian formulations for some nonlinear reaction diffusion equations by designed Lyapunov functionals. Consider a scale value nonlinear reaction diffusion equation by

$$\partial_t u(t, x) = \Delta F(u(t, x)) + R(u(t, x)), \tag{6}$$

where $x \in \Omega$, $\Omega \subset \mathbb{R}^d$ is a compact convex spatial domain, $u \in \mathcal{M}(\Omega) = \{u \in C^\infty(\Omega) : u \geq 0\}$, and Δ is the Euclidean Laplacian operator. For the variable u, we assume either periodic boundary conditions or zero flux conditions on the boundary of the spatial domain Ω. We also assume that $F, R \colon \mathbb{R}_+ \to \mathbb{R}_+$ are given smooth functions.

We next construct a Lyapunov functional $\mathcal{G} \colon \mathcal{M}(\Omega) \to \mathbb{R}$ to study Eq. (6). Consider

$$\mathcal{G}(u) = \int G(u(x)) dx,$$

where $G \colon \mathbb{R} \to \mathbb{R}$ is a given convex function with $G''(u) > 0$. In this case, along the reaction diffusion equation (6), we observe that

$$\frac{d}{dt} \mathcal{G}(u(t, \cdot))$$
$$= \int G'(u(t,x)) \cdot \partial_t u(t,x) dx$$
$$= \int G'(u(t,x))(\Delta F(u(t,x)) + R(u(t,x))) dx$$
$$= -\int \Big(\nabla G'(u(t,x)), \nabla F(u(t,x))\Big) dx + \int G'(u(t,x)) R(u(t,x)) dx$$
$$= -\int \Big(\nabla G'(u(t,x)), \nabla u(t,x)\Big) F'(u(t,x)) dx + \int G'(u(t,x)) R(u(t,x)) dx$$
$$= -\int \Big(\nabla G'(u(t,x)), \nabla u(t,x)\Big) G''(u(t,x)) \frac{F'(u(t,x))}{G''(u(t,x))} dx$$
$$\quad + \int G'(u(t,x))^2 \frac{R(u(t,x))}{G'(u(t,x))} dx$$
$$= -\int \Big(\nabla G'(u(t,x)), \nabla G'(u(t,x))\Big) \frac{F'(u(t,x))}{G''(u(t,x))} dx + \int G'(u(t,x))^2 \frac{R(u(t,x))}{G'(u(t,x))} dx,$$

where we apply integration by parts in the third equality and $\nabla G'(u) = G''(u) \nabla u$ in the last equality.

Here we assume that $R \in C^1(\Omega)$ is a given function with $-\frac{R}{G'} > 0$, and $F'(u) > 0$ for $u > 0$. Under these assumptions, it is clear that

$$\frac{d}{dt} \mathcal{G}(u) \leq 0.$$

This indicates that functional $\mathcal{G}(u)$ is non-increasing along the dynamics.

In fact, the above decay behavior indicates a gradient flow formulation for dynamics (6). To see that, we introduce the following notations. Denote a weighted elliptic operator $g\colon C^\infty(\Omega) \to C^\infty(\Omega)$ by

$$g(u) := \Big(-\nabla\cdot(\frac{F'(u)}{G''(u)}\nabla) - \frac{R(u)}{G'(u)}\Big)^{-1}.$$

Using the elliptic operator $g(u)$, we simply write

$$\begin{aligned}\partial_t u &= -g(u)^{-1}\frac{\delta}{\delta u}\mathcal{G}(u)\\ &= -\Big(-\nabla\cdot(\frac{F'(u)}{G''(u)}\nabla) - \frac{R(u)}{G'(u)}\Big)\frac{\delta}{\delta u}\mathcal{G}(u)\\ &= \nabla\cdot(\frac{F'(u)}{G''(u)}\nabla G'(u)) + \frac{R(u)}{G'(u)}G'(u)\\ &= \Delta F(u) + R(u).\end{aligned}$$

where $\frac{\delta}{\delta u}$ represents the L^2 first variation w.r.t. $u \in \mathcal{M}(\Omega)$ in the first equality. In the above notation, the dissipation of Lyapunov functional $\mathcal{G}$ along the dynamics (6) can be formulated below

$$\frac{d}{dt}\mathcal{G}(u) = -\int \Big(\frac{\delta}{\delta u}\mathcal{G}(u), g(u)^{-1}\frac{\delta}{\delta u}\mathcal{G}(u)\Big)dx \leq 0.$$

Clearly, our assumptions on F, R are sufficient conditions to describe the fact that $g(u)$ is a "positive definite" operator.

2.2.1 Metric Spaces and Gradient Flows

In this subsection, we illustrate a formal definition of metric space and gradient flows. See details in [2, 26].

Denote the smooth positive probability space by

$$\mathcal{M} = \Big\{u \in C^\infty(\Omega)\colon u > 0\Big\}.$$

Given $F, G\colon \mathbb{R} \to \mathbb{R}$ satisfying $F'(u) > 0$ if $u > 0$, and $G''(u) > 0$. Denote

$$V_1(u) = \frac{F'(u)}{G''(u)}, \qquad V_2(u) = -\frac{R(u)}{G'(u)}.$$

Denote the tangent space of $\mathcal{M}$ at $u \in \mathcal{M}$ by

$$T_u \mathcal{M} = \Big\{ \sigma \in C^\infty(\Omega) \Big\}.$$

We define the F, G, R dependent metric tensor in the positive density space below.

Definition 1 (Mean-Field Information Metric) The inner product $g(u): T_u \mathcal{M} \times T_u \mathcal{M} \to \mathbb{R}$ is given below. For any $\sigma_1, \sigma_2 \in T_u \mathcal{M}$, define

$$g(u)(\sigma_1, \sigma_2) = \int_\Omega \sigma_1 \Big(-\nabla \cdot (V_1(u)\nabla) + V_2(u) \Big)^{-1} \sigma_2 dx,$$

where

$$\Big(-\nabla \cdot (V_1(u)\nabla) + V_2(u) \Big)^{-1} : T_u \mathcal{M} \to T_u \mathcal{M}$$

denotes the inverse operator of weighted elliptic operator $-\nabla \cdot (V_1(u)\nabla) + V_2(u)$. The other formulation of metric is given below. Denote $\Phi_i \in C^\infty(\Omega)$, such that

$$\sigma_i = -\nabla \cdot (V_1(u)\nabla \Phi_i) + V_2(u)\Phi_i, \quad i = 1, 2.$$

Hence the metric satisfies

$$g(u)(\sigma_1, \sigma_2) = \int_\Omega (\nabla \Phi_1, \nabla \Phi_2) V_1(u) dx + \int_\Omega \Phi_1 \Phi_2 V_2(u) dx.$$

Remark 1 We remark that the above metric contains the definition of Wasserstein metric, which is well-studied in optimal transport. It corresponds to $V_1 = u$, $V_2 = 0$. It also contains the Fisher-Rao metric, which is important in information geometry [1]. It corresponds to $V_1 = 0$, $V_2 = u$. There are several interactive studies of them in unbalanced optimal transport $V_1 = V_2 = u$ and unnormalized optimal transport $V_1 = u$, $V_2 = 1$. They are different choices of metric operators $g(u)$, depending on the selected Lyapunov functional.

We are now ready to formulate gradient flows in $(\mathcal{M}, g)$.

Proposition 2 (Mean-Field Information Gradient Flow) *Given an energy functional $\mathcal{F}: \mathcal{M} \to \mathbb{R}$, the gradient flow of $\mathcal{F}$ in $(\mathcal{M}(\Omega), g)$ satisfies*

$$\partial_t u(t, x) = \nabla \cdot (V_1(u) \nabla \frac{\delta}{\delta u} \mathcal{F}(u))(t, x) - V_2(u) \frac{\delta}{\delta u} \mathcal{F}(u)(t, x).$$

Proof The proof follows the definition. The gradient operator in $(\mathcal{M}(\Omega), g)$ is defined by

$$g(\sigma, \text{grad}\mathcal{F}(u)) = \int \frac{\delta}{\delta u(x)}\mathcal{F}(u) \cdot \sigma(x)dx, \quad \text{for any } \sigma(x) \in T_u\mathcal{M}. \tag{7}$$

In other words,

$$\begin{aligned}
\text{grad}\mathcal{F}(u) &= g(u)^{-1}\frac{\delta}{\delta u}\mathcal{F}(u) \\
&= -\nabla \cdot (V_1(u)\nabla\frac{\delta}{\delta u}\mathcal{F}(u)) + V_2(u)\frac{\delta}{\delta u}\mathcal{F}(u),
\end{aligned}$$

which finishes the proof. Thus the gradient flow in $(\mathcal{M}(\Omega), g)$ satisfies

$$\partial_t u(t, x) = -\text{grad}\mathcal{F}(u)(t, x) = \nabla \cdot (V_1(u)\nabla\frac{\delta}{\delta u}\mathcal{F}(u)) - V_2(u)\frac{\delta}{\delta u}\mathcal{F}(u).$$

In particular, if $\mathcal{F}(u) = \int_\Omega G(u)dx$, we have

$$\begin{aligned}
\partial_t u &= \nabla \cdot (V_1(u)\nabla\frac{\delta}{\delta u}\mathcal{F}(u)) - V_2(u)\frac{\delta}{\delta u}\mathcal{F}(u) \\
&= \nabla \cdot (\frac{F'}{G''}\nabla G') + G' \cdot \frac{R}{G'} \\
&= \nabla \cdot (\frac{F'}{G''}G''\nabla u) + G' \cdot \frac{R}{G'} \\
&= \Delta F(u) + R(u).
\end{aligned}$$

$\square$

We next present the decay property of the Lyapunov functional along gradient flow equation (6).

Proposition 3 (Mean-Field Information De-Bruijn Identity) *Suppose $u(t, x)$ satisfies (6), then*

$$\frac{d}{dt}\mathcal{G}(u) = -\mathcal{I}(u),$$

where $\mathcal{I}: \mathcal{M}(\Omega) \to \mathbb{R}$ is a functional defined by

$$\mathcal{I}(u) = \int_\Omega \|\nabla G'(u)\|^2 V_1(u)dx + \int_\Omega |G'(u)|^2 V_2(u)dx. \tag{8}$$

Proof The proof follows from the definition of gradient flow. Notice that along the gradient flow (6),

$$\begin{aligned}
\frac{d}{dt}\mathcal{G}(u) &= \int_\Omega (\frac{\delta}{\delta u}\mathcal{G}(u), \partial_t u)dx \\
&= \int_\Omega (G'(u), \nabla \cdot (V_1(u)\nabla G'(u)) - V_2(u)G'(u))dx \\
&= -\int_\Omega \|\nabla G'(u)\|^2 V_1(u)dx - \int_\Omega |G'(u)|^2 V_2(u)dx \\
&= -\mathcal{I}(u),
\end{aligned}$$

where the third equality holds by the integration by parts formula. □

Remark 2 We remark that if $G = u\log u$, $V_1 = u$, $V_2 = 0$, then the gradient flow satisfies the heat equation. And the decay of Lyapunov functional along the heat flow satisfies

$$\mathcal{I}(u) = \int_\Omega \|\nabla \log u\|^2 u\, dx.$$

In literature, the relation $\frac{d}{dt}\mathcal{G}(u) = -\mathcal{I}(u)$ is often named the De-Bruijn identity. And $\mathcal{I}(u)$ is called the Fisher information functional. Following this spirit, we name the generalized dissipation property "mean-field information De-Bruijn identity". And we call $\mathcal{I}$ the "mean-field information functional".

Remark 3 The gradient flow not only works for a scalar function u. One can define a similar metric operator for a vector valued function u; see examples in [26]. It is worth mentioning that there are more general choices of V_1, which includes kernel functions; see examples in [6, 20].

Remark 4 In information geometry [1] and its applications in machine learning, the Fisher-Rao gradient flow is known as the natural gradient flow. The Fisher-Rao metric refers to $V_1(u) = 0$, $V_2(u) = u$. This metric and gradient flow has been widely used in machine learning. In addition, the mean-field information gradient flow is the generalization of the "natural gradient" flow. The terminology "natural" corresponds to the "projection" operation. In other words, one projects the infinite dimensional metric space into finite dimensional parameterized models, e.g. neural networks. In this paper, we focus on the infinite dimensional gradient flows, and design classical finite volume methods to solve the related dynamics. See an example in the time-implicit scheme in the next section.

2.2.2 Mean-Field Information Hamiltonian Flows

In this subsection, we state the main formulation of this paper, towards which we will design fast numerical methods. We first define Hamiltonian flows associated with F, G, R, which is the critical point of the following variational problem (9) in the positive density space. We call the derived Hamiltonian flows as *mean field information dynamics*.

Definition 4 (Mean-Field Information Variational Problems) Denote a given energy functional by $\mathcal{F}\colon \mathcal{M}(\Omega) \to \mathbb{R}$. Consider a variational problem:

$$\inf_{v_1,v_2,u} \int_0^1 \Big[\int_\Omega \frac{1}{2}\|v_1(t,x)\|^2 V_1(u(t,x)) + \frac{1}{2}|v_2(t,x)|^2 V_2(u(t,x))dx - \mathcal{F}(u)\Big]dt: \tag{9}$$

where the infimum is taken among all density functions $u\colon [0,1]\times\Omega \to \mathbb{R}$, vector fields $v_1\colon [0,1]\times\Omega \to \mathbb{R}^d$, and reaction rate functions $v_2\colon [0,1]\times\Omega \to \mathbb{R}$, such that

$$\partial_t u(t,x) + \nabla \cdot (V_1(u(t,x))v_1(t,x)) = v_2(t,x)V_2(u(t,x)),$$

with fixed initial and terminal density functions $u_0, u_1 \in \mathcal{M}(\Omega)$.

We next derive critical point systems of variational problem (9), which are proposed mean-field information Hamiltonian flows.

Proposition 5 (Mean-Field Information Hamiltonian Flows) *Assume $u(t,x) > 0$ for $t \in [0,1]$, then the critical point system of variational problem (9) satisfies*

$$v_1(t,x) = \nabla\Phi(t,x), \quad v_2(t,x) = \Phi(t,x),$$

with

$$\begin{cases} \partial_t u(t,x) + \nabla \cdot (V_1(u(t,x))\nabla\Phi(t,x)) = V_2(u(t,x))\Phi(t,x), \\ \partial_t \Phi(t,x) + \frac{1}{2}\|\nabla\Phi(t,x)\|^2 V_1'(u(t,x)) + \frac{1}{2}|\Phi(t,x)|^2 V_2'(u(t,x)) + \frac{\delta}{\delta u}\mathcal{F}(u)(t,x) \\ = 0. \end{cases} \tag{10}$$

Proof We first rewrite the variables in variational formula (9) as

$$m_1(t,x) = V_1(u)v(t,x), \quad m_2(t,x) = V_2(u)v_2(t,x),$$

Then variational problem (9) is

$$\inf_{m_1,m_2,u} \left\{ \int_0^1 \int_\Omega \frac{\|m_1(t,x)\|^2}{2V_1(u(t,x))} + \frac{|m_2(t,x)|^2}{2V_2(u(t,x))} - \mathcal{F}(u) dx dt : \right. \tag{11}$$
$$\left. \partial_t u(t,x) + \nabla \cdot m_1(t,x) = m_2(t,x), \quad \text{fixed } u_0, u_1 \right\}.$$

In this case, denote the Lagrangian multiplier of problem (11) by Φ. We then consider the following saddle point problem

$$\inf_{m_1,m_2,u} \sup_{\Phi} \mathcal{L}(m_1, m_2, u, \Phi),$$

with

$$\mathcal{L}(m_1, m_2, u, \Phi) = \int_0^1 \int_\Omega \left\{ \frac{\|m_1(t,x)\|^2}{2V_1(u(t,x))} + \frac{|m_2(t,x)|^2}{2V_2(u(t,x))} \right.$$
$$\left. + \Phi(t,x) \Big(\partial_t u(t,x) + \nabla \cdot m_1(t,x) - m_2(t,x) \Big) \right\} dx dt.$$

By finding the saddle point of $\mathcal{L}$, we have

$$\begin{cases} \frac{\delta}{\delta m_1} \mathcal{L} = 0 \\ \frac{\delta}{\delta m_2} \mathcal{L} = 0 \\ \frac{\delta}{\delta u} \mathcal{L} = 0 \\ \frac{\delta}{\delta \Phi} \mathcal{L} = 0 \end{cases} \Rightarrow \begin{cases} \frac{m_1}{V_1} = \nabla \Phi \\ \frac{m_2}{V_2} = \Phi \\ -\frac{1}{2} \frac{\|m_1\|^2}{V_1^2} V_1' - \frac{1}{2} \frac{|m_2|^2}{V_2^2} V_2' - \frac{\delta}{\delta u} \mathcal{F} - \partial_t \Phi = 0 \\ \partial_t u + \nabla \cdot m_1 - m_2 = 0, \end{cases}$$

where $\frac{\delta}{\delta m_1}, \frac{\delta}{\delta m_2}, \frac{\delta}{\delta u}, \frac{\delta}{\delta \Phi}$ are L^2 first variation derivatives w.r.t. functions m_1, m_2, u, Φ, respectively. Substituting the above two equations into the last two equations, we derive the PDE pair (10) in $\mathcal{M}(\Omega)$. □

Remark 5 If $V_1 = u$, $V_2 = 0$, the formulation corresponds to the well-known Benamou-Brenier formula [4] in optimal transport.

Remark 6 If V_1, V_2 are positive functions and are concave w.r.t. u, and the functional $\mathcal{F}$ is convex w.r.t. u, then the objective functional of problem (11) is convex. In this case, the derived Hamiltonian flow is a minimizer of variational problem (11).

2.2.3 Mean-Field Information Hamiltonian Gradient Flows and Variational Time Discretization

In this section, we demonstrate that the proposed gradient flow has a Hamiltonian formalism.

Denote the Hamiltonian functional $\mathcal{H}\colon L^2(\Omega) \times L^2(\Omega) \to \mathbb{R}$ as

$$\mathcal{H}(u, \Phi) = \int_\Omega \Big(\frac{1}{2}\|\nabla\Phi\|^2 V_1(u) + \frac{1}{2}|\Phi|^2 V_2(u)\Big) dx + \mathcal{F}(u). \tag{12}$$

Proposition 6 *Denote the mean-field information Hamiltonian by* (12) *with* $\mathcal{F} = 0$. *Then the mean-field information gradient flow satisfies*

$$\begin{cases} \partial_t u = \dfrac{\delta}{\delta \Phi}\mathcal{H}(u, \Phi), \\ \Phi = -\dfrac{\delta}{\delta u}\mathcal{G}(u). \end{cases}$$

Proof The proof is followed from a direct calculation. □

Remark 7 It is worth mentioning that the above Hamiltonian gradient flow formulation also holds for non-quadratic Hamiltonian functionals. See examples in [2, 27].

We notice that the Hamiltonian gradient formulation is useful in designing a variational backward Euler time discretization. This is well known in the literature, e.g. the Jordan-Kinderlehrer-Otto (JKO) scheme [16]. We present it here for the completeness of this paper. The goal of JKO scheme is to construct an iterative sequence of variational problems, which approximates Eq. (6) in a time domain.

Definition 7 (Iterative Variational Formulations for Reaction Diffusion Equations)
Denote a time stepsize by $\Delta t > 0$, and $t_k = k\Delta t$, $k = 0, 1, 2, \cdots$, and $u_0(x) = u(0, x)$. Consider the following iterative variational problem:

$$\inf_{v_1, v_2, u(t,\cdot), u_{k+1}} \int_{t_k}^{t_{k+1}} \int_\Omega \frac{1}{2}\|v_1(t,x)\|^2 V_1(u(t,x)) + \frac{1}{2}|v_2(t,x)|^2 V_2(u(t,x)) dx dt$$
$$+ \mathcal{G}(u(t_{k+1}, x)), \tag{13}$$

where the infimum is taken among all density functions $u\colon [t_k, t_{k+1}] \times \Omega \to \mathbb{R}$, vector fields $v_1\colon [t_k, t_{k+1}] \times \Omega \to \mathbb{R}^d$, and reaction functions $v_2\colon [t_k, t_{k+1}] \times \Omega \to \mathbb{R}$, such that

$$\partial_t u(t, x) + \nabla \cdot (V_1(u(t,x)) v_1(t,x)) = v_2(t,x) V_2(u(t,x)),$$

with a fixed initial value function $u(t_k, x) = u_k(x)$ and a given terminal energy functional $\mathcal{G}(u_{k+1})$. Denote the update as

$$u_{k+1}(x) = u(t_{k+1}, x), \quad k = 1, 2, 3 \cdots$$

where $u(t_{k+1}, x)$ is the minimizer for variational problem (13).

Proposition 8 *Assume that u, Φ are smooth functions w.r.t. t and x, then $\{u_k\}_{k=1}^{\infty}$ approximates the reaction diffusion equation (6) with the first order accuracy in time.*

Proof Similar as the proof in Proposition 5, we derive the minimizer system for variational problem (13). Again, denote the Lagrangian multiplier of problem (13) by Φ. Then we consider the following saddle point problem

$$\inf_{m_1, m_2, u(t, \cdot), u(h, \cdot)} \sup_{\Phi} \mathcal{L}_1(m_1, m_2, u, \Phi),$$

with

$$\mathcal{L}_1(m_1, m_2, u, \Phi) = \int_{t_k}^{t_{k+1}} \int_{\Omega} \left\{ \frac{\|m_1(t, x)\|^2}{2V_1(u(t, x))} + \frac{|m_2(t, x)|^2}{2V_2(u(t, x))} \right.$$

$$\left. + \Phi(t, x) \Big(\partial_t u(t, x) + \nabla \cdot m_1(t, x) - m_2(t, x) \Big) \right\} dx dt$$

$$+ \mathcal{G}(u(h, \cdot)).$$

Similarly, by finding the saddle point of $\mathcal{L}_1$, we have

$$\begin{cases} \frac{\delta}{\delta m_1} \mathcal{L}_1 = 0 \\ \frac{\delta}{\delta m_2} \mathcal{L}_1 = 0 \\ \frac{\delta}{\delta u} \mathcal{L}_1 = 0 \\ \frac{\delta}{\delta \Phi} \mathcal{L}_1 = 0 \\ \frac{\delta}{\delta u_{k+1}} \mathcal{G}(u_{k+1}) = 0, \end{cases} \Rightarrow \begin{cases} \frac{m_1}{V_1} = \nabla \Phi \\ \frac{m_2}{V_2} = \Phi \\ -\frac{1}{2} \frac{\|m_1\|^2}{V_1^2} V_1' - \frac{1}{2} \frac{|m_2|^2}{V_2^2} V_2' - \partial_t \Phi = 0 \\ \partial_t u + \nabla \cdot m_1 - m_2 = 0 \\ \Phi(t_{k+1}, x) + \frac{\delta}{\delta u_{k+1}} \mathcal{G}(u_{k+1}) = 0. \end{cases}$$

In other words, we have

$$\partial_t u = \frac{\delta}{\delta \Phi}\mathcal{H}(u, \Phi), \quad \partial_t \Phi = -\frac{\delta}{\delta u}\mathcal{H}(u, \Phi),$$

$$u_0 = u_k, \quad \Phi(t_{k+1}, x) = -\frac{\delta}{\delta u_{k+1}}\mathcal{G}(u_{k+1}).$$

Hence

$$\begin{aligned}
u_{k+1}(x) &= u_0(x) + \int_{t_k}^{t_{k+1}} \partial_t u(t, x) ds \\
&= u_k(x) + \int_{t_k}^{t_{k+1}} \frac{\delta}{\delta \Phi(t, x)}\mathcal{H}(u(t, x), \Phi(t, x)) dt \\
&= u_k(x) + (t_{k+1} - t_k)\frac{\delta}{\delta \Phi}\mathcal{H}(u, \Phi)|_{\Phi = \Phi(t_{k+1}, x)} + o(\Delta t) \\
&= u_k(x) + \Delta t \frac{\delta}{\delta \Phi}\mathcal{H}(u, \Phi)|_{\Phi = -\frac{\delta}{\delta u_{k+1}}\mathcal{G}(u_{k+1})} + o(\Delta t).
\end{aligned}$$

The above update is a time discretization for Eq. (6), which is true for a small order perturbation in term of Δt. A known fact is that the backward Euler method is a first order method w.r.t. the time stepsize Δt. Hence we finish the derivation. □

2.2.4 Examples

In this section, we list several examples of mean-field information metric, gradient and Hamiltonian flows. They are designed by both Lyapunov functionals and reaction diffusion equations. From now on, we also denote an additional energy $\mathcal{F}$ for the Hamiltonian flow associated with mean field control/variation problem (11). Toward the Hamiltonian flows in these examples, we shall design numerical schemes based on primal-dual variational methods.

Example 1 (Wasserstein Metric, Heat Flow and Density Hamilton Flows) Let $G(u) = u \log u$, $F = u$, $R = 0$. Then

$$V_1(u) = u, \quad V_2(u) = 0.$$

Then the mean-field information metric is

$$g(u)(\sigma_1, \sigma_2) = \int_\Omega (\nabla \Phi_1(x), \nabla \Phi_2(x)) u(x) dx,$$

with $\sigma_i = -\nabla \cdot (u \nabla \Phi_i)$, $i = 1, 2$. In this case, the mean-field information metric coincides with the Wasserstein-2 metric [2, 31, 35]. In this case, the gradient flow of $\mathcal{G}(u)$, named negative Boltzmann-Shannon entropy, in $(\mathcal{M}(\Omega), g)$ is the heat equation, i.e.

$$\partial_t u = \nabla \cdot (u \nabla G'(u)) = \nabla \cdot (u G''(u) \nabla u) = \Delta u.$$

In this case, the dissipation of $\mathcal{G}(u)$ is

$$\mathcal{I}(u) = \int_\Omega \|\nabla \log u(x)\|^2 u(x) dx.$$

Here the Hamiltonian flow in $(\mathcal{M}(\Omega), g)$ satisfies

$$\begin{cases} \partial_t u + \nabla \cdot (u \nabla \Phi) = 0 \\ \partial_t \Phi + \frac{1}{2}\|\nabla \Phi\|^2 + \frac{\delta}{\delta u}\mathcal{F}(u) = 0. \end{cases}$$

Example 2 (Generalized Wasserstein Metric, Nonlinear Heat Flow and Density Hamiltonian Flows) Choose functions F, G, R, such that

$$\frac{F'(u)}{G''(u)} = u^m, \quad R(u) = 0.$$

In this case, the metric is

$$g(u)(\sigma_1, \sigma_2) = \int_\Omega (\nabla \Phi_1(x), \nabla \Phi_2(x)) u^m(x) dx,$$

with $\sigma_i = -\nabla \cdot (u^m \nabla \Phi_i)$, $i = 1, 2$. And the gradient flow of $\mathcal{G}$ in $(\mathcal{M}(\Omega), g)$ is

$$\partial_t u = \nabla \cdot (V_1(u) \nabla G'(u)) = \nabla \cdot (\frac{F'(u)}{G''(u)} G''(u) \nabla u) = \nabla \cdot (F'(u) \nabla u) = \Delta F(u),$$

and the dissipation of $\mathcal{G}(u)$ satisfies

$$\mathcal{I}(u) = \int_\Omega \|\nabla G'(u)\|^2 u^m dx.$$

Here the Hamiltonian flow in $(\mathcal{M}(\Omega), g)$ satisfies

$$\begin{cases} \partial_t u + \nabla \cdot (u^m \nabla \Phi) = 0 \\ \partial_t \Phi + \frac{m}{2}\|\nabla \Phi\|^2 u^{m-1} + \frac{\delta}{\delta u}\mathcal{F}(u) = 0. \end{cases}$$

Example 3 (H^{-1} **Metric, Nonlinear Heat Flow and Density Hamiltonian Flow**)
Consider $m = 0$ in the above example. We choose functions F, G, R, such that

$$\frac{F'(u)}{G''(u)} = 1, \quad R(u) = 0.$$

In this case,

$$V_1(u) = 1, \quad V_2(u) = 0.$$

Hence the metric is

$$g(u)(\sigma_1, \sigma_2) = \int_\Omega (\nabla \Phi_1(x), \nabla \Phi_2(x)) dx,$$

with $\sigma_i = -\nabla \cdot (\nabla \Phi_i)$, $i = 1, 2$. And the gradient flow of $\mathcal{G}$ in $(\mathcal{M}(\Omega), g)$ is

$$\partial_t u = \Delta F(u),$$

and the dissipation of $\mathcal{G}(u)$ satisfies

$$\mathcal{I}(u) = \int_\Omega \|\nabla G'(u)\|^2 dx.$$

Here the Hamiltonian flow in $(\mathcal{M}(\Omega), g)$ satisfies

$$\begin{cases} \partial_t u + \nabla \cdot (\nabla \Phi) = 0 \\ \partial_t \Phi + \dfrac{\delta}{\delta u} \mathcal{F}(u) = 0. \end{cases}$$

Example 4 (Fisher-Rao Metric, Birth-Death Equation and Density Hamiltonian Flows) Consider

$$F(u) = 0, \quad G(u) = u \log u - u, \quad R(u) = -u \log u,$$

then

$$V_1(u) = \frac{F'(u)}{G''(u)} = 0, \quad V_2(u) = -\frac{R(u)}{G'(u)} = u.$$

Then the mean-field information metric satisfies

$$g(u)(\sigma_1, \sigma_2) = \int_\Omega \Phi_1(x) \Phi_2(x) u(x) dx,$$

with $\sigma_i(x) = \Phi_i u$, $i = 1, 2$. In this case, the mean-field information metric is the Fisher-Rao metric in positive density space; see information geometry [1]. Here the gradient flow of $\mathcal{G}$ in $(\mathcal{M}(\Omega), g)$ satisfies the birth-death dynamics

$$\partial_t u = -V_2(u)G'(u) = -V_2(u)\log u = -u \log u.$$

And the dissipation of $\mathcal{G}(u)$ is

$$\mathcal{I}(u) = \int_\Omega |\log u(x)|^2 u(x) dx.$$

Here the Hamiltonian flow in $(\mathcal{M}(\Omega), g)$ satisfies

$$\begin{cases} \partial_t u - u\Phi = 0 \\ \partial_t \Phi + \frac{1}{2}|\Phi|^2 + \frac{\delta}{\delta u}\mathcal{F}(u) = 0. \end{cases}$$

Example 5 Consider

$$F(u) = u, \quad G(u) = u \log u - u, \quad R(u) = -u^\alpha \log u,$$

where $\alpha \in \mathbb{R}$ is a given value. In this case,

$$V_1(u) = \frac{F'(u)}{G''(u)} = u, \quad V_2(u) = -\frac{R(u)}{G'(u)} = u^\alpha.$$

And the metric is

$$g(u)(\sigma_1, \sigma_2) = \int_\Omega (\nabla \Phi_1(x), \nabla \Phi_2(x)) u(x) dx + \int_\Omega \Phi_1(x) \Phi_2(x) u(x)^\alpha dx,$$

with $\sigma_i = -\nabla \cdot (u \nabla \Phi_i) + \Phi_i u^\alpha$, $i = 1, 2$. And the gradient flow of $\mathcal{G}$ in $(\mathcal{M}(\Omega), g)$ is

$$\partial_t u = \Delta u + u^\alpha \log u,$$

and the dissipation of $\mathcal{G}(u)$ satisfies

$$\mathcal{I}(u) = \int_\Omega \|\nabla \log u\|^2 u \, dx + \int_\Omega |\log u|^2 u^\alpha dx.$$

Here the Hamiltonian flow in $(\mathcal{M}(\Omega), g)$ satisfies

$$\begin{cases} \partial_t u + \nabla \cdot (u \nabla \Phi) = \Phi u^\alpha \\ \partial_t \Phi + \frac{1}{2}\|\nabla \Phi\|^2 + \alpha \Phi u^{\alpha-1} + \frac{\delta}{\delta u}\mathcal{F}(u) = 0. \end{cases}$$

Example 6 (Constant Regularized Optimal Transport Metric) Consider

$$F(u) = u, \quad G(u) = (u+1)\log(u+1), \quad R(u) = 0.$$

Thus

$$V_1(u) = \frac{F'(u)}{G''(u)} = u + 1, \quad V_2(u) = 0.$$

Hence the mean-field information metric is

$$g(u)(\sigma_1, \sigma_2) = \int_\Omega (\nabla \Phi_1, \nabla \Phi_2)(u+1)dx,$$

with $\sigma_i = -\nabla \cdot ((u+1)\nabla \Phi_i)$, $i = 1, 2$. In this case, the gradient flow of $\mathcal{G}$ in $(\mathcal{M}(\Omega), g)$ satisfies

$$\partial_t u(t, x) = \Delta u(t, x).$$

And the dissipation of $\mathcal{G}(u)$ satisfies

$$\mathcal{I}(u) = \int_\Omega \|\nabla \log(u+1)\|^2 (u+1) dx.$$

Here the Hamiltonian flow in $(\mathcal{M}(\Omega), g)$ satisfies

$$\begin{cases} \partial_t u + \nabla \cdot ((u+1)\nabla \Phi) = 0 \\ \partial_t \Phi + \frac{1}{2}\|\nabla \Phi\|^2 + \frac{\delta}{\delta u}\mathcal{F}(u) = 0. \end{cases}$$

Example 7 (Fisher-KPP Metric, Fisher-KPP Equation and Density Hamiltonian Flows) Consider the Fisher-KPP equation by

$$\partial_t u = u(1-u) + \Delta u.$$

Consider

$$F(u) = u, \quad G(u) = u \log u - u, \quad R(u) = u(1-u).$$

Thus
$$V_1(u) = \frac{F'(u)}{G''(u)} = u, \quad V_2(u) = -\frac{R(u)}{G'(u)} = \frac{u(u-1)}{\log u}.$$

Hence the mean-field information metric is
$$g(u)(\sigma_1, \sigma_2) = \int_\Omega (\nabla \Phi_1, \nabla \Phi_2) u \, dx + \int_\Omega \Phi_1 \Phi_2 \frac{u(u-1)}{\log u} dx,$$

with $\sigma_i = -\nabla \cdot (u \nabla \Phi_i) + \frac{u(u-1)}{\log u} \Phi_i$, $i = 1, 2$. In this case, the gradient flow of $\mathcal{G}$ in $(\mathcal{M}(\Omega), g)$ satisfies the Fisher-KPP equation

$$\partial_t u(t, x) = u(1 - u) + \Delta u.$$

And the dissipation of $\mathcal{G}(u)$ satisfies
$$\mathcal{I}(u) = \int_\Omega \|\nabla \log u\|^2 u \, dx + \int_\Omega |\log u|^2 \frac{u(u-1)}{\log u} dx.$$

Here the Hamiltonian flow in $(\mathcal{M}(\Omega), g)$ satisfies
$$\begin{cases} \partial_t u + \nabla \cdot (u \nabla \Phi) - \dfrac{u(u-1)}{\log u} \Phi = 0 \\ \partial_t \Phi + \dfrac{1}{2}\|\nabla \Phi\|^2 + |\Phi|^2 \dfrac{(2u-1)\log u + 1 - u}{(\log u)^2} + \dfrac{\delta}{\delta u}\mathcal{F}(u) = 0. \end{cases}$$

Example 8 (Allen-Cahn Metric, Allen-Cahn Equation and Density Hamiltonian Flows) Let $f \in C^2(\mathbb{R})$ be a given function. Consider

$$F(u) = u, \quad G(u) = f(u), \quad R(u) = -f'(u).$$

Thus
$$V_1(u) = \frac{F'(u)}{G''(u)} = f''(u)^{-1}, \quad V_2(u) = -\frac{R(u)}{G'(u)} = 1.$$

Hence the mean-field information metric is
$$g(u)(\sigma_1, \sigma_2) = \int_\Omega (\nabla \Phi_1, \nabla \Phi_2) f''(u)^{-1} dx + \int_\Omega \Phi_1 \Phi_2 \, dx,$$

with $\sigma_i = -\nabla \cdot (f''(u)^{-1}\nabla \Phi_i) + \Phi_i$, $i = 1, 2$. In this case, the gradient flow of $\mathcal{G}$ in $(\mathcal{M}(\Omega), g)$ satisfies

$$\partial_t u(t, x) = \Delta u(t, x) - f'(u(t, x)).$$

And the dissipation of $\mathcal{G}(u)$ satisfies

$$\mathcal{I}(u) = \int_\Omega \|\nabla f'(u)\|^2 f''(u)^{-1} dx + \int_\Omega |f'(u)|^2 dx.$$

Here the Hamiltonian flow in $(\mathcal{M}(\Omega), g)$ satisfies

$$\begin{cases} \partial_t u + \nabla \cdot (f''(u)^{-1}\nabla \Phi) - \Phi = 0 \\ \partial_t \Phi - \frac{1}{2}\|\nabla \Phi\|^2 \frac{f'''(u)}{f''(u)^2} + \frac{\delta}{\delta u}\mathcal{F}(u) = 0. \end{cases}$$

3 Deep Neural Network Based Minimization Movement Schemes

Deep learning has revolutionized many areas, fundamentally reshaping fields such as natural language processing, visual recognition, and beyond [17]. It has also emerged as the cornerstone of contemporary scientific computing. By harnessing the power of deep neural networks and their ability to approximate complex functions, it has become an invaluable tool for approximating solutions to partial differential equations (PDEs), particularly those of high dimensions. These equations often defy conventional computational techniques due to their complexity and dimensionality.

To be more specific, we use the following notations.

$$u = \rho, \quad V_1(\rho) = M(\rho), \quad V_2(\rho) = 0.$$

Consider the continuity equation of the form:

$$\partial_t \rho = \nabla_x \cdot [\rho \nabla_x (U'(\rho) + V + W * \rho)], \quad \rho(0, \cdot) = \rho_0. \tag{14}$$

Here, $\rho(t, x)$, with $x \in \mathbb{R}^d$, represents the particle density function, ρ_0 is the initial density function, $U(\rho) \in \mathbb{R}^1$ represents the internal potential, $V(x) \in \mathbb{R}^1$ is a drift potential, $W(x, y) = W(y, x) \in \mathbb{R}^1$ is an interaction potential involving x and y in $\mathbb{R}^d$, and $*$ denotes the convolution operator. This equation can be viewed as the gradient flow of the energy functional

$$\mathcal{E}(\rho) = \int_{\mathbb{R}^d} [U(\rho(x)) + V(x)\rho(x)] \, dx + \frac{1}{2} \int_{\mathbb{R}^d \times \mathbb{R}^d} W(x - y)\rho(x)\rho(y) dx dy \tag{15}$$

in Wasserstein-2 space, which is the space of probability measures equipped with Wasserstein-2 distance, denoted as $\mathcal{W}_2$.

Among all the recent endeavors in devising structure-preserving methodologies for (14), our focus in this paper will be exclusively on the Jordan-Kindelenr-Otto (JKO) scheme [16]. This scheme centers on the task of determining $\rho^{k+1}(\cdot) \approx \rho(t^{k+1}, \cdot)$ with $t^{k+1} = (k+1)\Delta t$, given the approximation $\rho^{k+1}(\cdot) \approx \rho(t^{k+1}, \cdot)$:

$$\rho^{k+1} \in \arg\min_{\rho}\{\mathcal{W}_2(\rho, \rho^k)^2 + 2\Delta t \mathcal{E}(\rho)\}. \tag{16}$$

Here, $\mathcal{W}_2(\rho, \rho^k)$ quantifies the Wasserstein-2 distance between ρ and ρ^k. The scheme in (16) can be conceptualized as a time-implicit variant of the fundamental gradient descent, with the expression on the right-hand side of (16) being recognized as the Wasserstein proximal operator associated with energy functional $\mathcal{E}$.

In general terms, there exist two methods for expressing the $\mathcal{W}_2$ distance, each leading to distinct implementations of (16). The first method is the Eulerian representation, which establishes a connection between two densities through a continuity equation involving a velocity field that minimizes kinetic energy during the transport. In contrast, the second method is grounded in the Lagrangian formulation [8]. This approach explores the diffeomorphism that maps one density onto another, a technique that has also gained prominence in modern machine learning applications.

This section adopts the Lagrangian viewpoint by employing the particle method, coupled with principles drawn from neural ordinary differential equations (neural ODEs) [9, 36]. To elaborate, starting from a set of particles $\{x_j^0\}$ that are i.i.d. samples from ρ^0, we seek a sequence of transformations governing the update of all particles' positions as follows:

$$\{x_j^0\} \xrightarrow{T^1} \{x_j^1\} \xrightarrow{T^2} \{x_j^2\} \cdots.$$

Here, $x_j^k = T^k(x_j^{k-1})$. As a consequence, the density evolves in accordance with $\rho^k = T^k \sharp \rho^{k-1}$ for $k = 1, 2, \ldots$, where $\sharp$ denotes the pushforward operator. We then approximate each map T^k by parameterizing the corresponding vector field using a deep neural network. Thanks to the Jacobi identity, this results in a simple update in estimating the density ρ^k; see (17c) below.

More precisely, our primary formulation is as follows. To transit from time t^k to t^{k+1}, we solve the following minimization problem:

$$\theta^* = \arg\min_{\theta \in \Theta} \mathbb{E}_{x \sim \rho}\Big[\int_0^1 |v_\theta(\tau, T(\tau, x))|^2 d\tau + 2\Delta t\Big(\frac{U(\rho(T(1, x)))}{\rho(T(1, x))} + V(T(1, x))\Big)\Big]$$
$$+ \Delta t \mathbb{E}_{(x,y) \sim \rho \otimes \rho}\Big[W(T(1, x), T(1, y))\Big],$$
(17a)

such that $T(\tau, x)$ satisfies the dynamical constraint:

$$\frac{\mathrm{d}}{\mathrm{d}\tau} T(\tau, x) = v_\theta(\tau, T(\tau, x)), \qquad T(0, x) = x, \tag{17b}$$

and the density $\rho(T(1, x))$ is computed as first evolving:

$$\frac{\partial}{\partial \tau} \log \det |\nabla_x T(\tau, x)| = \operatorname{div}(v_\theta)(\tau, T(\tau, x)), \quad \rho(T(0, x)) = \rho^k(x), \tag{17c}$$

followed by the pushforward relation:

$$\rho(T(1, x)) = \rho^k(x)/\det|\nabla_x T(1, x)|. \tag{17d}$$

In this problem, we employ Benamou-Brenier's dynamic formulation [4] to calculate the Wasserstein distance and reformulate the continuity equation using the flow map $T(\tau, x)$, with τ serving as an artificial time. The associated velocity field is denoted as v_θ, where $\theta \in \Theta$ represents the parameters of a neural network. The notation $\mathbb{E}_{x \sim \rho}$ denotes the expectation value for x distributed according to the probability density ρ, and $\mathbb{E}(x, y) \sim \rho \otimes \rho$ represents the expectation over pairs of independent variables (x, y), where x and y are independent and satisfy the joint density $\rho \otimes \rho$. Equation (17c) is a significant outcome from the perspective of neural ODEs, providing a straightforward means to compute $\det |\nabla_x T(\tau, x)|$. Consequently, this facilitates an efficient computation of density as outlined in (17d).

It's crucial to highlight that the proposed method goes beyond the current scope of merely approximating Wasserstein gradient flows. Similar approaches, as outlined in (17), can be employed for general gradient flows, including nonlinear mobility Wasserstein and Kalman-Wasserstein gradient flows. In such cases, the identity (17c), utilized in neural ODE density estimations, takes on even greater significance. This is due to the fact that the objective function in (17) becomes dependent on the density, introducing additional nonlinearities into the optimization objective (loss function), see (37) in the subsequent context.

3.1 Dynamic JKO Schemes in Lagrangian Coordinates

Given $\rho^n(x)$, then one can obtain $\rho^{n+1}(x)$ as $\rho(1, x)$ with $\rho(t, x)$ solving

$$\begin{cases} (\rho, v) = \arg\inf_{\rho, v} \int_0^1 \int_{\mathbb{R}^d} \rho|v|^2 \mathrm{d}x \mathrm{d}\tau + 2\Delta t \mathcal{E}(\rho(1, \cdot)) \\ \text{s.t.} \quad \partial_\tau \rho + \nabla \cdot (\rho v) = 0, \; \rho(0, x) = \rho^n(x). \end{cases} \tag{18}$$

In the above formulation, the minimization is over the probability density function $\rho(t, x)$, $t \in (0, 1)$, terminal time density $\rho(1, x)$, and vector fields $v(t, x)$, such that the continuity equation holds.

Now we translate the variational problem (18) into the Lagrangian formulation. Assume the velocity field is sufficiently regular, then the solution $\rho(\tau, x)$ to the constrained continuity equation can be written as

$$\rho(\tau, \cdot) = T(\tau, \cdot)\sharp \rho^n , \tag{19}$$

where T is the flow map that solves the following ODE:

$$\frac{d}{d\tau} T(\tau, x) = v(\tau, T(\tau, x)), \qquad T(0, x) = x . \tag{20}$$

Note that (19) implies that for any integrable test function ϕ, we have that

$$\int_{\mathbb{R}^d} \phi(x) \rho(\tau, x) dx = \int_{\mathbb{R}^d} \phi(T(\tau, x)) \rho^n(x) dx .$$

Then variational problem (18) rewrites as

$$\begin{cases} \min_v \int_0^1 \int_{\mathbb{R}^d} \rho^n(x) |v(\tau, T(t, x))|^2 d\tau dx + 2\Delta t \mathcal{E}(T(1, \cdot)\sharp \rho^n) \\ \text{s.t. } \frac{d}{d\tau} T(\tau, x) = v(\tau, T(\tau, x)), \qquad T(0, x) = x . \end{cases} \tag{21}$$

To proceed, it is necessary to derive an explicit formulation of $\mathcal{E}(T(1, \cdot)\sharp \rho^n)$ that can be interpreted as an expectation of a functional, allowing for its representation using particles in the subsequent discussion. In the following, we present a comprehensive formulation for each component of Eq. (15). Specifically, the external potential and interaction potential have straightforward expectation formulations through the push forward relations, whereas the internal energy necessitates further reformulation.

3.1.1 External Potential Energy

$$\int_{\mathbb{R}^d} V(x) \rho(1, x) dx = \int_{\mathbb{R}^d} V(T(1, x)) \rho^n(x) dx . \tag{22}$$

3.1.2 Interaction Energy

$$\int_{\mathbb{R}^d \times \mathbb{R}^d} W(x, y) \rho(1, x) \rho(1, y) dx dy$$
$$= \int_{\mathbb{R}^d \times \mathbb{R}^d} W(T(1, x), T(1, y)) \rho^n(x) \rho^n(y) dx dy . \tag{23}$$

3.1.3 Internal Energy/Entropy

$$\int_{\mathbb{R}^d} U(\rho(1,x))\mathrm{d}x$$
$$= \int_{\mathbb{R}^d} U(\rho(1, T(1,x))) \det|\nabla_x T(1,x)|\mathrm{d}x$$
$$= \int_{\mathbb{R}^d} U(\rho(1, T(1,x))) \frac{\det|\nabla_x T(1,x)|}{\rho^n(x)} \rho^n(x)\mathrm{d}x$$
$$= \int_{\mathbb{R}^d} U(\rho(1, T(1,x))) \frac{1}{\rho(1, T(1,x))} \rho^n(x)\mathrm{d}x , \qquad (24)$$

where we have used the Monge-Ampere equation

$$\rho(1, T(1,x)) \det|\nabla_x T(1,x)| = \rho^n(x) \qquad (25)$$

thanks to (19). In practice, computing the determinant of the Jacobian $\nabla_x T(1,x))$ poses a significant bottleneck due to its cubic cost with respect to the dimension of x. To address this issue, we draw inspiration from the concept of continuous normalizing flow and employ an efficient approach based on the instantaneous change of variable formula.

Proposition 9

$$\frac{\mathrm{d}}{\mathrm{d}\tau} \log \det|\nabla_x T(\tau, x)| = div(v)(\tau, T(\tau, x)). \qquad (26)$$

Proof Equation (26) is derived through a calculus that involves both ODE (20) and the Monge-Ampere equation (25). Assume that $T(\tau, x)$ is invertible, i.e. $\det|\nabla_x T(\tau, x)| \neq 0$, then

$$\frac{\mathrm{d}}{\mathrm{d}\tau} \log \det|\nabla_x T(\tau, x)| = \mathrm{tr}\Big((\nabla_x T(\tau, x))^{-1} \frac{\mathrm{d}}{\mathrm{d}\tau} \nabla_x T(\tau, x)\Big)$$
$$= \mathrm{tr}\Big((\nabla_x T(\tau, x))^{-1} \nabla_x \frac{\mathrm{d}}{\mathrm{d}\tau} T(\tau, x)\Big)$$
$$= \mathrm{tr}\Big((\nabla_x T(\tau, x))^{-1} \nabla_x v(\tau, T(\tau, x))\Big)$$
$$= \mathrm{div}(v)(\tau, T(\tau, x)),$$

where the first equality uses the Jacobi identity, and the last equality uses the chain rule. □

As a result, the computational burden shifts towards calculating $\text{div}(\boldsymbol{v})(\tau, \boldsymbol{T}(\tau, x))$, which only involves a trace calculation and can be accomplished much more efficiently. In sum, (25) and (26) play a crucial role in enabling efficient density estimation, as elaborated in the next section and Sect. 3.5.

3.2 A Particle Method

By interpreting the integral against $\rho(x)\mathrm{d}x$ as an expectation, (21) reveals a direct particle representation. More precisely, let $\{x_j^n\}_{j=1}^N$ be particles sampled from $\rho^n(x)$, we discretize (21) as follows. Note here we omit the superscript n in x_j^n.

Case 1: (21) with (22) and/or (23)

$$\begin{cases} \min_{\boldsymbol{v}} \frac{1}{N} \sum_{j=1}^N \left[\int_0^1 |\boldsymbol{v}(\tau, \boldsymbol{T}(\tau, x_j))|^2 \mathrm{d}\tau + 2\Delta t V(\boldsymbol{T}(1, x_j)) \right. \\ \qquad\qquad \left. + 2\Delta t \sum_{l=1}^N W(\boldsymbol{T}(1, x_j), \boldsymbol{T}(1, x_l)) \right] \\ \text{s.t. } \frac{\mathrm{d}}{\mathrm{d}\tau} \boldsymbol{T}(\tau, x_j) = \boldsymbol{v}(\tau, \boldsymbol{T}(\tau, x_j)), \qquad \boldsymbol{T}(0, x_j) = x_j \,. \end{cases} \quad (27)$$

Case 2: (21) with (24)

$$\begin{cases} \min_{\boldsymbol{v}} \frac{1}{N} \sum_{j=1}^N \left[\int_0^1 |\boldsymbol{v}(\tau, \boldsymbol{T}(\tau, x_j))|^2 \mathrm{d}\tau + 2\Delta t U\left(\frac{\rho^n(x_j)}{\det |\nabla_x \boldsymbol{T}(1, x_j)|} \right) \frac{\det |\nabla_x \boldsymbol{T}(1, x_j)|}{\rho^n(x_j)} \right] \\ \text{s.t. } \frac{\mathrm{d}}{\mathrm{d}\tau} \boldsymbol{T}(\tau, x_j) = \boldsymbol{v}(\tau, \boldsymbol{T}(\tau, x_j)), \qquad \boldsymbol{T}(0, x_j) = x_j \\ \qquad \frac{\partial}{\partial \tau} \log \det |\nabla_x \boldsymbol{T}(\tau, x_j)| = \text{div}(\boldsymbol{v})(\tau, \boldsymbol{T}(\tau, x_j)), \quad \log \det |\nabla_x \boldsymbol{T}(0, x_j)| = 0\,, \end{cases} \quad (28)$$

In cases where the energy solely comprises potential and interaction components, it suffices to monitor the particle positions. However, when internal energy is introduced, it becomes necessary also to track the density ρ, which is accomplished by monitoring the logarithmic determinant of the transport map.

More precisely, denote the minimizer of either problem $\boldsymbol{T}^n$, then

$$\boldsymbol{T}^n(1, x_j^n) =: x_j^{n+1}\,,$$

which can be viewed as samples from $\rho^{n+1}(x)$. Therefore, starting from $\{x_j^0\}$, we have a sequence of update

$$\{x_j^0\} \xrightarrow{\boldsymbol{T}^1} \{x_j^1\} \xrightarrow{\boldsymbol{T}^2} \{x_j^2\} \xrightarrow{\boldsymbol{T}^3} \{x_j^3\} \cdots, \quad (29)$$

where $x_j^k = T^k(1, x_j^{k-1})$. Likewise, the density evolves as

$$\rho^0 \xrightarrow{T^1} \rho^1 \xrightarrow{T^2} \rho^2 \xrightarrow{T^3} \rho^3 \cdots, \qquad (30)$$

where $\rho^k = T^k(1, \cdot) \sharp \rho^{k-1}$.

In practice, since we only evolve particles, a major difficulty in (30) lies in the *density estimation* of ρ^n, which can be very expensive and inaccurate in high dimensions. To resolve this issue, a key observation is that we don't need the full information of the density, but only the density evaluated along the trajectory of particles!

Suppose we are given a set of samples x_j^0 drawn from the known analytical expression of ρ_0. In the first JKO step, we can compute the updated density as follows:

$$\rho^1(\underbrace{T^1(1, x_j^0)}_{x_j^1}) = \frac{\rho^0(x_j^0)}{\det |\nabla_x T^1(1, x_j^0)|}, \qquad (31)$$

where T^1 represents the map obtained in the first JKO step. Here, x_j^1 corresponds to the transformed sample after applying T^1 to x_j^0. Following a similar procedure, we obtain the density in the subsequent JKO steps:

$$\rho^2(\underbrace{T^2(1, x_j^1)}_{x_j^2}) = \frac{\rho^1(x_j^1)}{\det |\nabla_x T^2(1, x_j^1)|}, \qquad (32)$$

and this iterative process can be continued for any subsequent JKO step.

Remark 8 We emphasize that this observation holds true for general nonlinear internal potential functions, denoted as U. In the existing literature, U is commonly assumed to be the negative Boltzmann-Shannon entropy, given by $U(z) = z \log z$. In this particular case, the aforementioned difficulty can be circumvented. Indeed, since

$$U\left(\frac{\rho^n(x_j)}{\det |\nabla_x T(1, x_j)|}\right) \frac{\det |\nabla_x T(1, x_j)|}{\rho^n(x_j)} = \log \rho^n(x_j) - \log \det |\nabla_x T(1, x_j)|,$$

we can simplify the computation by avoiding the need to directly calculate $\rho^n(x_j)$. This is because $\log \rho^n(x_j)$ is separated from $\log \det(\nabla_x T(1, x_j))$ in Eq. (28), enabling us to exclude it from the calculation. However, for general potential functions, the computation of $\rho^n(x_j)$ cannot be circumvented.

3.3 Nonlinear Mobility

In this section, we extend the previous method to the generalized gradient flow by considering the following equation:

$$\partial_t \rho = \nabla_x \cdot \left(M(\rho) \nabla_x \frac{\delta \mathcal{E}(\rho)}{\delta \rho} \right), \tag{33}$$

Here, the function $M(\rho) \geq 0$ represents a nonlinear mobility function [6]. This type of equation appears in various contexts, such as thin films, phase separation in binary alloys described by the Cahn-Hilliard equation, and the transport phenomena of biological systems with overcrowding prevention, among others. Recently, this formulation has been applied in the study of controlling droplet dynamics [12].

For general mobility function $M(\rho)$, Eq. (33) can also be viewed as a gradient flow. The generalized metric also admits a dynamic formulation, leading to the following extension of the dynamic JKO scheme (18):

$$\begin{cases} (\rho, \tilde{v}) = \arg\inf_{\rho, \tilde{v}} \int_0^1 \int_{\mathbb{R}^d} M(\rho) |\tilde{v}|^2 \mathrm{d}x \mathrm{d}\tau + 2\Delta t \mathcal{E}(\rho(1, \cdot)) \\ \text{s.t.} \quad \partial_\tau \rho + \nabla \cdot (M(\rho) \tilde{v}) = 0, \ \rho(0, x) = \rho^n(x). \end{cases} \tag{34}$$

In the above formulation, the minimization is over density function $\rho(t, x)$, $t \in (0, 1)$ terminal time density $\rho(1, x)$, and vector fields $\tilde{v}(t, x)$, such that the nonlinear mobility induced continuity equation holds.

Let $v = \frac{M(\rho)}{\rho} \tilde{v}$, Eq. (34) can be rewritten into

$$\begin{cases} (\rho, v) = \arg\inf_{\rho, \tilde{v}} \int_0^1 \int_{\mathbb{R}^d} \frac{\rho^2}{M(\rho)} |v|^2 \mathrm{d}x \mathrm{d}\tau + 2\Delta t \mathcal{E}(\rho(1, \cdot)) \\ \text{s.t.} \quad \partial_\tau \rho + \nabla \cdot (\rho v) = 0, \ \rho(0, x) = \rho^n(x). \end{cases} \tag{35}$$

This reformulation states that the minimization is constrained by the classical continuity equation. It is worth noting that such reformulation does not change the optimizer of the problem. Indeed, let ϕ be the Lagrangian multiplier, then one can check that the critical point system of variational problems (34) and (35) lead to the same Hamilton-Jacobi equation for ϕ:

$$\partial_t \phi + \frac{1}{4} M'(\rho) |\nabla \phi|^2 = 0.$$

We note that variational problem (35) now admits a similar Lagrangian representation as in (21), with the only distinction being the first term, which now becomes:

$$\int_0^1 \int_{\mathbb{R}^d} \frac{\rho^2(\tau, x)}{M(\rho(\tau, x))} |v(\tau, x)|^2 \mathrm{d}\tau \mathrm{d}x.$$

By employing the same technique as utilized in deriving (24), the integral with respect to x in the given expression rewrites as

$$\int_{\mathbb{R}^d} \frac{\rho^2(\tau, x)}{M(\rho(\tau, x))} |v(\tau, x)|^2 dx$$

$$= \int_{\mathbb{R}^d} \frac{\rho^2(\tau, T(\tau, x))}{M(\rho(\tau, T(\tau, x)))} |v(\tau, T(\tau, x))|^2 \det |\nabla_x T(\tau, x)| dx$$

$$= \int_{\mathbb{R}^d} \frac{\rho^2(\tau, T(\tau, x))}{M(\rho(\tau, T(\tau, x)))} |v(\tau, T(\tau, x))|^2 \frac{\det |\nabla_x T(\tau, x)|}{\rho^n(x)} \rho^n(x) dx$$

$$= \int_{\mathbb{R}^d} \frac{\rho^2(\tau, T(\tau, x))}{M(\rho(\tau, T(\tau, x)))} \frac{|v(\tau, T(\tau, x))|^2}{\rho(\tau, T(\tau, x))} \rho^n(x) dx$$

$$= \int_{\mathbb{R}^d} \frac{\rho(\tau, T(\tau, x))}{M(\rho(\tau, T(\tau, x)))} |v(\tau, T(\tau, x))|^2 \rho^n(x) dx. \tag{36}$$

As in (31), the density $\rho(\tau, T(\tau, x))$ is determined by the Monge-Ampéré equation:

$$\rho(\tau, T(\tau, x)) = \frac{\rho^n(x)}{\det |\nabla_x T(\tau, x)|}.$$

Similar to the JKO scheme presented in (27) or (28), given $\{x_j^n\}_j$ and $\{\rho^n(x_j^n)\}_j$ (we omit the superscript n in x_j^n in the following formula), the updated formulation can now be expressed as follows:

$$\begin{cases} \min_T \frac{1}{N} \sum_{j=1}^N \left[\int_0^1 \frac{\rho(T(\tau, x_j))}{M(\rho(T(\tau, x_j)))} |v(\tau, T(\tau, x_j))|^2 d\tau + 2\Delta t \frac{U(\rho(T(1, x_j)))}{\rho(T(1, x_j))} \right] \\ \text{s.t. } \frac{d}{d\tau} T(\tau, x_j) = v(\tau, T(\tau, x_j)), \qquad T(0, x_j) = x_j \\ \frac{\partial}{\partial \tau} \log \det |\nabla_x T(\tau, x_j)| = \text{div}(v)(\tau, T(\tau, x_j)), \quad \log \det |\nabla_x T(0, x_j)| = 0 \\ \rho(\tau, T(\tau, x_j)) = \frac{\rho^n(x_j)}{\det |\nabla_x T(\tau, x_j)|}. \end{cases} \tag{37}$$

It is obvious that when $M(\rho) = \rho$, (37) reduces to (28).

3.4 Neural ODE Empowered JKO Schemes

In this section, we leverage neural network functions to approximate the vector field v as described in Eqs. (27), (28), or, more broadly, Eq. (37). As a result, the optimization procedure aims at refining the parameters of the neural network. Building upon the efficiency gains achieved through the use of continuous normalizing flow for density estimation, as previously discussed, we incorporate a recursive relation formula proposed in a prior work [29] to compute derivatives, significantly improving the efficiency of backpropagation. Lastly, we conclude this section by presenting a concise summary of the algorithm.

3.5 Learning the Potential Function

In view of (27) and (28), the primary objective is to determine the optimal transport map, denoted as T. However, this task can be computationally demanding, especially in high-dimensional scenarios. Fortunately, by applying the principles of optimal transport [35], we can express the velocity field $\mathbf{v}(t, x)$ as the gradient of a potential function $\phi(t, x)$ such that $\mathbf{v}(t, x) = \nabla_x \phi(t, x)$. Consequently, our focus shifts to finding the potential function ϕ. To achieve this, we utilize a neural network to approximate $\phi(t, x)$, denoted as $\phi_\theta(t, x)$. As a result, we obtain $\mathbf{v}_\theta(t, x) = \nabla_x \phi_\theta(t, x)$ as the corresponding approximation of the velocity field.

To be more specific, let us consider solving Eq. (27). By substituting $\mathbf{v}$ with $\mathbf{v}_\theta = \nabla_x \phi_\theta$, the constraint in (27) becomes:

$$\frac{\mathrm{d}}{\mathrm{d}\tau} T(\tau, x_j) = \nabla_x \phi_\theta(\tau, T(\tau, x_j)), \qquad T(0, x_j) = x_j.$$

Similarly, in the case of the more complex equation (28), computing $\rho(T(1, x_j))$ calls for the following update:

$$\frac{\partial}{\partial \tau} \log \left| \det(\nabla_x^2 \phi_\theta(\tau, T(\tau, x_j))) \right| = \operatorname{div}(\mathbf{v}_\theta)(\tau, T(\tau, x_j)) = -\operatorname{tr}(\nabla_x^2 \phi_\theta(\tau, T(\tau, x_j))).$$

Detailed computations of $\nabla^2 \phi_\theta$ will be given in the next subsection.

Remark 9 We remark that using $v = \nabla \Phi$ only works for $M(\rho) = \rho$. We need to carefully design the vector field v for general mobility functions. This is left for future work.

3.6 Deep JKO Algorithms

We hereby provide comprehensive details for computing the minimizer of (27), (28) or (37), when $\mathbf{v}$ is replaced by $\mathbf{v}_\theta$. Let N_τ denote the number of time discretization such that $[0, 1]$ is discretized into $0, \frac{1}{N_\tau}, \frac{2}{N_\tau}, \cdots, \frac{N_\tau - 1}{N_\tau}$, and θ is a vector of parameters for neural network functions. Given the density ρ^n at the n-th JKO step and sampled points $\{x_j\}_{j=1}^N$, the loss function to compute the density at the $n+1$-th JKO step is defined as follows:

$$\begin{aligned}
L(\theta, \{x_j\}_{j=1}^N) = \frac{1}{N} \sum_{j=1}^N \Bigg[& d\tau \sum_{k=0}^{N_\tau - 1} \left| \nabla_x \phi_\theta(k d\tau, \mathbf{z}_j^k) \right|^2 \\
& + 2\Delta t \left(\frac{U(\rho(\mathbf{z}_j^{N_\tau}))}{\rho(\mathbf{z}_j^{N_\tau})} + V(\mathbf{z}_j^{N_\tau}) + \frac{1}{N} \sum_{l=1}^N W(\mathbf{z}_j^{N_\tau}, \mathbf{z}_l^{N_\tau}) \right) \Bigg],
\end{aligned}$$
(38)

where the inner time step is denoted as $d\tau = 1/N_\tau$ and the update equations are given by:

$$\mathbf{z}_j^{k+1} = \mathbf{z}_j^k - d\tau \, \nabla_x \phi_\theta(kd\tau, \mathbf{z}_j^k), \quad \mathbf{z}_j^0 = x_j,$$

$$l_j^{k+1} = l_j^k - d\tau \, \text{tr}(\nabla_x^2 \phi_\theta(kd\tau, \mathbf{z}_j^k)), \quad l_j^0 = \log\left|\det\left(\nabla_x^2 \phi_\theta(kd\tau, \mathbf{z}_j^k)\right)\right|,$$

for $k = 0, \cdots, N_\tau - 1$ and the terminal density ρ at the location $z_j^{N_\tau}$ is defined as

$$\rho(z_j^{N_\tau}) = \frac{\rho^n(x_j)}{\exp(l_j^{N_\tau})}. \tag{39}$$

The above equations for $\mathbf{z}_j^{k+1}$ and $\log \rho(\mathbf{z}_j^{k+1})$ utilize the forward Euler method to solve the ordinary differential equations (ODEs) in the constraints of (28). However, alternative ODE solvers such as Runge-Kutta can also be employed. For our numerical experiment, we utilize the RK4 as the ODE solver.

The loss function, denoted as $L(\theta, \{x_j\}_{j=1}^N)$, takes two inputs: the neural network parameter θ and a set of sampled points $\{x_j\}_{j=1}^N$. The objective is to minimize this loss function with respect to θ. This θ plays a vital role in approximating the potential function $\phi_\theta(x, \tau)$, which relies on two variables: a state vector $x \in \mathbb{R}^d$ and a time value $t \in [0, 1]$. The velocity is computed as:

$$v_\theta(\tau, x) = -\nabla_x \phi_\theta(\tau, x),$$

and its divergence is obtained as the trace of the Hessian:

$$\text{div}(v_\theta)(\tau, x) = -\text{tr}(\nabla_x^2 \phi_\theta(\tau, x)).$$

To simplify the calculation of gradients and Hessians in the formulas above, we utilize the explicit expression introduced in [29]. This approach capitalizes on the smoothness of the activation function within the neural network. Consequently, our optimization process experiences significant acceleration, as we no longer rely on automatic differentiation (AD) for backpropagation to compute gradients and Hessians in each iteration.

More precisely, we adopt a multi-layer ResNet that takes the input $\mathbf{s} = (\tau, x) \in \mathbb{R}^{d+1}$, where $x \in \mathbb{R}^d$ represents the initial state, and $\tau \in [0, 1]$ denotes time, yielding an output in $\mathbb{R}$. The neural network function consists of L hidden layers with recursive relations defined as follows:

$$\mathbf{u}_1 = \sigma(W_0 \mathbf{s} + \mathbf{b}_0),$$

$$\mathbf{u}_{l+1} = \mathbf{u}_l + \sigma(W_l \mathbf{u}_l + \mathbf{b}_l), \quad l = 1, \cdots L - 1,$$

$$NN(\mathbf{s}; \theta) = \mathbf{w}^\top \mathbf{u}_L.$$

Here, $\theta = (W_0, W_l, b_l)$, with $W_0 \in \mathbb{R}^{m \times (d+1)}$ and $W_l \in \mathbb{R}^{m \times m}$ ($l = 1, \cdots, L$) are dense matrices. Additionally, $b_k \in \mathbb{R}^m$ ($l = 0, \cdots, L$) are biases, while m denotes the number of nodes in each hidden layer and $w \in \mathbb{R}^m$ is a vector we perform dot product at the end of the ResNet. The element-wise activation function $\sigma(x) = \log(\exp(x) + \exp(-x))$ is used, which is differentiable. The dual variable ϕ_θ is defined through

$$\phi_\theta(\tau, x) = NN((\tau, x); \theta). \tag{40}$$

Denote by ∇_x a gradient operator with respect to the space variable x and ∇_s be a gradient operator with respect to the state vector $\mathbf{s}$. Both the gradient $\nabla_x \phi_\theta(x, \tau)$ and the Hessian $\nabla_x^2 \phi_\theta(x, \tau)$ can be effortlessly computed using the explicit formulas presented in [29]. Specifically, for the case where $L = 2$, these formulas are as follows:

$$\nabla_\mathbf{s} \phi_\theta(x, \tau) = W_0^\top \mathrm{diag}(\sigma'(W_0 \mathbf{s} + b_0)) z_1, \tag{41a}$$

$$z_1 = w + W_1^\top \mathrm{diag}(\sigma'(W_1 u_1 + b_1)) w, \tag{41b}$$

and

$$\nabla_\mathbf{s}^2 \phi_\theta(x, \tau) = t_0 + t_1, \tag{41c}$$

where

$$t_0 = W_0^\top \mathrm{diag}(\sigma''(W_0 \mathbf{s} + b_0) \odot z_1) W_0,$$
$$t_1 = J_0^\top W_1^\top \mathrm{diag}(\sigma''(W_1 u_1 + b_1) \odot w) W_1 J_0,$$
$$J_0 = W_0^\top \mathrm{diag}(\sigma'(W_0 \mathbf{s} + b_0)).$$

Here $\odot$ represents the element-wise multiplication. Then, $\nabla_x \phi_\theta(x, \tau)$ is the first d elements of $\nabla_\mathbf{s} \phi_\theta(x, \tau)$, and $\nabla_x^2 \phi_\theta(x, \tau) = [\nabla_\mathbf{s}^2 \phi_\theta(x, \tau)]_{1:d, 1:d}$ where the subscript $1 : d, 1 : d$ denotes the submatrix of size $d \times d$.

We now provide a summary of the proposed algorithms. Algorithm 2 serves as the primary algorithm, guiding the movement of particles along the gradient flow. Whenever sampling/resampling is necessary, Algorithm 1 will be invoked. Please note that the l update and density computation in Algorithm 1 are only required when working with general nonlinear internal energy and mobility. Additionally, in the algorithm, we utilize forward Euler, but it is adaptable for replacement with explicit Runge-Kutta type ODE solvers.

Algorithm 1 Generate samples and densities at n-th JKO step

Input:

- Outer JKO time step Δt.
- Inner dynamic formulation time step $d\tau$ and total number of inner steps N_τ.
- Initial distribution ρ^0 and initial samples $\{x_j^0\}_{j=1}^N$ and densities $\{\rho(x_j^0)\}_{j=1}^N$
- Potential functions from previous steps ($\phi_\theta^0, \cdots, \phi_\theta^{n-1}$).

Output:

- Sampled points $\{x_j^n\}_{j=1}^N$ at n-th outer iteration.
- Density values $\{\rho(x_j^n)\}_{j=1}^N$ at n-th outer iteration.

for $i = 0, \cdots, n-1$ **do**
 for $j = 1, \cdots, N$ **do**
  $z_j^0 = x_j^i, l_j^0 = \log \left| \det \left(\nabla_x^2 \phi_\theta(k d\tau, z_j^k) \right) \right|$
  for $k = 0, \cdots, N_\tau - 1$ **do**
   $z_j^{k+1} = z_j^k - d\tau \, \nabla_x \phi_\theta(k d\tau, z_j^k)$
   $l_j^{k+1} = l_j^k - d\tau \, \text{tr}(\nabla_x^2 \phi_\theta(k d\tau, z_j^k))$
  end for
  $x_j^{i+1} = z_j^{N_\tau}, \rho(x_j^{i+1}) = \frac{\rho^n(x_j)}{\exp(l_j^{N_\tau})}$
 end for
end for

In Algorithm 2, the error in the stopping condition is defined as

$$\text{error} = \frac{|L(\theta^{(current)}, \{x_j^n\}_{j=1}^N) - L(\theta^{(previous)}, \{x_j^n\}_{j=1}^N)|}{|L(\theta^{(previous)}, \{x_j^n\}_{j=1}^N)|}$$

where $\theta^{(previous)}$ and $\theta^{(current)}$ are the parameters computed at the previous and current iterations.

3.7 Numerical Examples

This section presents several numerical examples of the proposed deep JKO schemes for computing gradient flows with various choices of energy functionals and mobility functions.

Algorithm 2 Deep JKO scheme

Input:

- Outer JKO time step Δt
- Inner dynamic formulation time step $d\tau$
- Initial distribution ρ^0
- Learning rate α for the Adam optimizer
- Total number of outer iterations K
- Error tolerance TOL for the optimization process
- Resampling frequency C

Output:

- Neural network parameter θ and corresponding potential function $\phi_\theta^n(t, x), n = 1, 2, \cdots, K$
- Particle locations: $\{x_j^n\}, n = 1, 2, \cdots, K, j = 1, 2, \cdots, N$.
- Density along particle trajectories: $\rho^n(x_j^n), n = 1, 2, \cdots, K, j = 1, 2, \cdots, N$.

Initialize θ from a standard normal distribution.

for $n = 1, \cdots, K$ **do**
    $c = 0$.
    while error $> TOL$ **do**
        if $mod(c, C) = 0$ **then do**
            Sample $\{x_j^0\}_{j=1}^N \overset{\text{i.i.d.}}{\sim} \rho^0$
            Compute $\{x_j^n\}_{j=1}^N$ using Algorithm 1 and update the density value through (39).
        end if
        Update $\theta \leftarrow \theta - \alpha \nabla_\theta L(\theta, \{x_j^n\}_{j=1}^N)$
        where L is from (38) and the gradient is computed using automatic differentiation (AD).
        $c \leftarrow c + 1$
    end while
end for

3.7.1 Wasserstein Gradient Flow of the KL Divergence

In this experiment, we present the computed solutions of (28) with the energy functional defined as

$$U(\rho) = D_{\text{KL}}(\rho \| q) = \int \rho(x) \log \frac{\rho(x)}{q(x)} dx, \quad (42)$$

where q represents a reference density. The initial density ρ^0 is characterized by two Gaussian distributions centered at $(\pm 1.2, 0)$ with a standard deviation of $\sigma = 0.5$. The

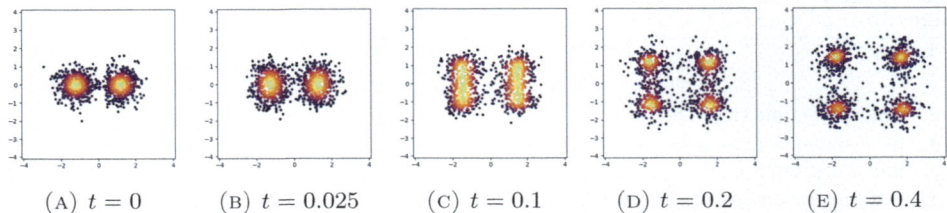

(A) $t = 0$ (B) $t = 0.025$ (C) $t = 0.1$ (D) $t = 0.2$ (E) $t = 0.4$

Fig. 1 Computed solutions of Fokker-Planck equation, i.e. the Wasserstein gradient flow of KL divergence (42). An initial density is a mixture of two Gaussians and the target density is a mixture of four Gaussians. The figures visually depict the evolution of these densities from $t = 0$ to $t = 0.4$. (a) $t = 0$. (b) $t = 0.025$. (c) $t = 0.1$. (d) $t = 0.2$. (e) $t = 0.4$

Fig. 2 3D plot (left) and the energy decay plot (right) from Fig. 1. The left plot shows the trajectories of each particles where z-axis represents time from $t = 0$ to $t = 0.35$. The right plot illustrates the decay of energy over 20 outer iterations (i.e., JKO steps) from $t = 0$ to $t = 0.5$

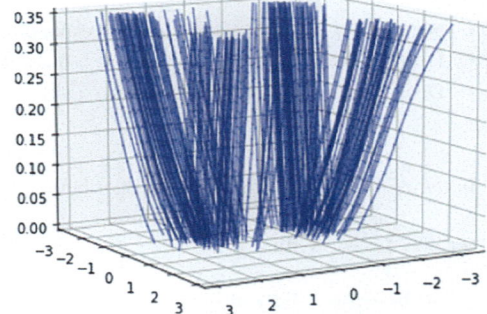

reference density q is defined as four Gaussian distributions centered at $(\pm 2, \pm 2)$ with a standard deviation of $\sigma = 0.5$.

For the numerical experiment, the inner timestep is set as $N_\tau = 1$ and the JKO time stepsize is set as $\Delta t = 0.025$. The model architecture includes 3 layers with $m = 64$ nodes in the hidden layers. The training is performed using a batch size of 1000, and the learning rate is chosen to be 10^{-5}. The total number of iterations is set to 10,000.

The two-dimensional results are depicted in Figs. 1 and 2. Figure 1 illustrates the evolution of the density from $t = 0$ to $t = 0.4$. In this figure, the color of the points represents the value of the density $\rho^n(x)$ at the corresponding location x. Brighter colors indicate higher density values. The computation involves a total of 16 JKO steps. Figure 2 displays the trajectories of each particle with respect to time t and the energy value computed from the algorithm with respect to the iterations. As the iterations progress, the energy decreases and eventually converges to the stationary value.

We further extended our experiments to a 10-dimensional space while retaining the same energy functional (42). The initial density ρ^0 was set as a Gaussian distribution centered at the origin, and the reference density q was constructed using a combination of four Gaussian distributions centered at $(2, 2.5)$, $(-2.5, 2)$, $(-2, -2.5)$, and $(2.5, -2)$. In this experimental setup, we utilized a time step size of $\tau = 0.2$, a learning rate of 10^{-5}, and executed a total of 10,000 iterations. The results are illustrated in Fig. 3, highlighting

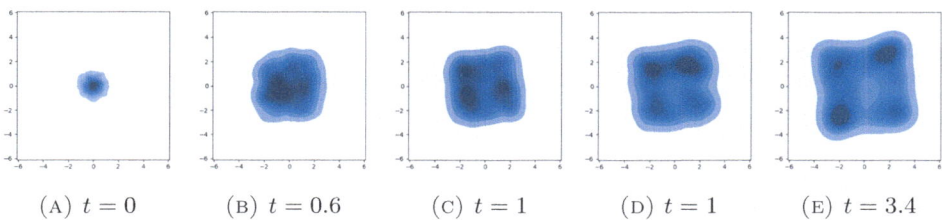

(A) $t = 0$ (B) $t = 0.6$ (C) $t = 1$ (D) $t = 1$ (E) $t = 3.4$

Fig. 3 Evolution of densities to the Fokker-Planck equation in $\mathbb{R}^{10}$ from $t = 0$ to $t = 3.4$. An initial density is a Gaussian distribution centered at the origin, and the target density is a mixture of four Gaussians. (**a**) $t = 0$. (**b**) $t = 0.6$. (**c**) $t = 1$. (**d**) $t = 1$. (**e**) $t = 3.4$

the initial 2D cross sections of the density's evolution from $t = 0$ to $t = 3.4$. These visual representations present kernel density plots derived from the point clouds generated by the algorithm at each time instance t.

4 Acceleration Optimization Methods in Probability Density Spaces

Optimization problems in probability space, arising from Bayesian inference and inverse problems, attract attention in machine learning communities. One typical example here is to draw samples from an intractable target distribution. Such a sampling problem is essential in exploring the distribution of interest and quantifying uncertainty among data. From an optimization viewpoint, this problem suffices to minimize an objective function, such as Kullback-Leibler (KL) divergence, which measures the closeness between the current density and the target distribution.

Gradient descent methods play essential roles in solving these optimization problems. Here, the gradient direction relies on the information metric in probability space. In literature, two important metrics, such as the Fisher-Rao metric and the Wasserstein-2 (in short, Wasserstein) metric, are of great interest. For classical optimization problems in Euclidean space, Nesterov's accelerated gradient method [28] is a widely applied optimization that accelerates gradient descent methods. The continuous-time limit of this method is known as the accelerated gradient flow [32]. Natural questions arise: *What is the accelerated gradient flow in probability space under general information metrics? What is the corresponding discrete-time sampling algorithm?* The probability space embedded with the information metric can be viewed as a Riemannian manifold, known as a density manifold. Several previous works explore accelerated methods in this manifold under the Wasserstein metric.

In this section, we present a unified framework of accelerated gradient flows in probability space embedded with information metrics, named Accelerated Information Gradient (AIG) flows. From a transport-information-geometry perspective, we derive AIG flows by damping Hamiltonian flows. Examples include Fisher-Rao metric, Wasserstein-2 metric, Kalman-Wasserstein metric and Stein metric.

This note is organized as follows. Section 4.1 briefly reviews gradient flows and accelerated gradient flows in Euclidean space. Then, the information metrics in probability space and their corresponding gradient and Hamiltonian flows are introduced. In Sect. 4.2, we formulate AIG flows under the Fisher-Rao metric, the Wasserstein metric, the Kalman-Wasserstein metric, and the Stein metric. We theoretically prove the convergence rate of AIG flows in Sect. 4.3. Section 4.4 presents the discrete-time algorithm for W-AIG flows, including the BM method and the adaptive restart technique. Section 4.5 provides numerical experiments.

4.1 Reviews

In this section, we review gradient flows and accelerated gradient flows in Euclidean space. Then, we introduce the optimization problems in probability spaces, and review several definitions of information metrics therein. Based on these metrics, we demonstrate gradient and Hamiltonian flows in probability space. These formulations serve as necessary preparations for us to derive accelerated gradient flows in probability space.

4.1.1 Accelerated Gradient Flows in Euclidean Space

Consider an optimization problem in Euclidean space:

$$\min_{x \in \mathbb{R}^d} f(x),$$

where $f(x)$ is a given convex function with L-Lipschitz continuous gradient. Here $(\cdot, \cdot)$ and $\|\cdot\|$ are the Euclidean inner product and norm in $\mathbb{R}^d$. The gradient descent method has the update rule

$$x_{k+1} = x_k - \tau_k \nabla f(x_k),$$

where $\tau_k > 0$ is a step size. With the limit $\tau_k \to 0$, the continuous-time limit of the gradient descent method is the gradient flow (GF)

$$\dot{x} = -\nabla f(x).$$

To accelerate the gradient descent method, Nesterov introduced an accelerated method [28]:

$$\begin{cases} x_k = y_{k-1} - \tau_k \nabla f(y_{k-1}), \\ y_k = x_k + \alpha_k (x_k - x_{k-1}). \end{cases}$$

Here α_k depends on the convexity of $f(x)$. If $f(x)$ is β-strongly convex, then $\alpha_k = \frac{\sqrt{L}-\sqrt{\beta}}{\sqrt{L}+\sqrt{\beta}}$; otherwise, $\alpha_k = \frac{k-1}{k+2}$. Su et al. [32] show that the continuous-time limit of Nesterov's accelerated method satisfies an ODE, which is known as the accelerated gradient flow (AGF):

$$\ddot{x} + \alpha_t \dot{x} + \nabla f(x) = 0. \tag{43}$$

Here $\alpha_t = 2\sqrt{\beta}$ if $f(x)$ is β-strongly convex; $\alpha_t = 3/t$ for general convex $f(x)$. An important observation is that the accelerated gradient flow (43) can be formulated as a damped Hamiltonian flow:

$$\begin{bmatrix} \dot{x} \\ \dot{p} \end{bmatrix} + \begin{bmatrix} 0 \\ \alpha_t p \end{bmatrix} - \begin{bmatrix} 0 & I \\ -I & 0 \end{bmatrix} \begin{bmatrix} \nabla_x H^E(x,p) \\ \nabla_p H^E(x,p) \end{bmatrix} = 0.$$

where x is the state variable and p is the momentum variable. The Hamiltonian function satisfies $H^E(x,p) = \frac{\|p\|^2}{2} + f(x)$, which consists of Euclidean kinetic function $\frac{\|p\|^2}{2}$ and potential function $f(x)$. In other words, one can formulate an accelerated gradient flow by adding a linear momentum term into the Hamiltonian flow. Later on, we follow this damped Hamiltonian perspective and derive related accelerated gradient flows in probability space.

4.1.2 Metrics in Probability Space

In practice, machine learning problems, especially Bayesian sampling problems, can be formulated as optimization problems in probability space. In other words, consider

$$\min_{\rho \in \mathcal{P}(\Omega)} \mathcal{F}(\rho),$$

where $\Omega \subset \mathbb{R}^d$ is a region and the set of probability density is denoted by $\mathcal{P}(\Omega) = \{\rho \in \mathcal{F}(\Omega) : \int_\Omega \rho dx = 1, \ \rho \geq 0\}$. Here $\mathcal{F}(\Omega)$ represents the set of smooth functions on Ω. In practice, $\mathcal{F}(\rho)$ is often chosen as a divergence or metric functional between ρ and a target density $\rho^* \in \mathcal{P}(\Omega)$.

In literature, it has been shown that various sampling algorithms correspond to gradient flows of $\mathcal{F}(\rho)$, depending on the metrics in probability space. We brief review the definition of metrics in probability space as follows.

Definition 10 (Metric in Probability Space) Denote the tangent space at $\rho \in \mathcal{P}(\Omega)$ by $T_\rho \mathcal{P}(\Omega) = \{\sigma \in \mathcal{F}(\Omega) : \int \sigma dx = 0.\}$. The cotangent space at ρ, $T_\rho^* \mathcal{P}(\Omega)$, can be treated as the quotient space $\mathcal{F}(\Omega)/\mathbb{R}$. A metric tensor $G(\rho) : T_\rho \mathcal{P}(\Omega) \to T_\rho^* \mathcal{P}(\Omega)$ is an invertible mapping from $T_\rho \mathcal{P}(\Omega)$ to $T_\rho^* \mathcal{P}(\Omega)$. This metric tensor defines the metric (inner product) on tangent space $T_\rho \mathcal{P}(\Omega)$:

$$g_\rho(\sigma_1, \sigma_2) = \int \sigma_1 G(\rho) \sigma_2 dx = \int \Phi_1 G(\rho)^{-1} \Phi_2 dx, \quad \sigma_1, \sigma_2 \in T_\rho \mathcal{P}(\Omega)$$

where Φ_i is the solution to $\sigma_i = G(\rho)^{-1} \Phi_i$, $i = 1, 2$.

Along with a given metric, the probability space $\mathcal{P}(\Omega)$ can be viewed as an infinite-dimensional Riemannian manifold known as the density manifold. We review four examples of metrics in $\mathcal{P}(\Omega)$: the Fisher-Rao metric from information geometry, the Wasserstein metric from optimal transport, the Kalman-Wasserstein metric from ensemble Kalman sampling and the Stein metric from Stein variational gradient method. For simplicity, we denote $\mathbb{E}_\rho[\Phi] = \int \Phi \rho dx$.

Example 9 (Fisher-Rao Metric) The inverse of the Fisher-Rao metric tensor is defined by

$$G^F(\rho)^{-1} \Phi = \rho(\Phi - \mathbb{E}_\rho[\Phi]), \quad \Phi \in T_\rho^* \mathcal{P}(\Omega).$$

Example 10 (Wasserstein Metric) The inverse of the Wasserstein metric tensor writes

$$G^W(\rho)^{-1} \Phi = -\nabla \cdot (\rho \nabla \Phi), \quad \Phi \in T_\rho^* \mathcal{P}(\Omega).$$

Example 11 (Kalman-Wasserstein Metric, [13]) The inverse of the metric tensor is defined by

$$G^{KW}(\rho)^{-1} \Phi = -\nabla \cdot (\rho C^\lambda(\rho) \nabla \Phi), \quad \Phi \in T_\rho^* \mathcal{P}(\Omega).$$

Here $\lambda \geq 0$ is a given regularization constant and $C^\lambda(\rho) \in \mathbb{R}^{n \times n}$ follows

$$C^\lambda(\rho) = \int (x - m(\rho))(x - m(\rho))^T \rho dx + \lambda I, \quad m(\rho) = \int x \rho dx.$$

Example 12 (Stein Metric, [23]) The inverse of Stein metric tensor is defined by

$$G^S(\rho)^{-1} \Phi(x) = -\nabla_x \cdot (\rho(x) \int k(x, y) \rho(y) \nabla_y \Phi(y) dy).$$

Here $k(x, y)$ is a given positive kernel function.

4.1.3 Gradient Flows and Hamiltonian Flows in Probability Space

The gradient flow for $\mathcal{F}(\rho)$ in $(\mathcal{P}(\Omega), g_\rho)$ takes the form

$$\partial_t \rho = -G(\rho)^{-1} \frac{\delta \mathcal{F}}{\delta \rho}.$$

Here $\frac{\delta \mathcal{F}}{\delta \rho}$ is the L^2 first variation w.r.t. ρ. For example, the Wasserstein gradient flow writes

$$\partial_t \rho = -G^W(\rho)^{-1} \frac{\delta \mathcal{F}}{\delta \rho} = \nabla \cdot (\rho \nabla \frac{\delta \mathcal{F}}{\delta \rho}).$$

We then briefly review Hamiltonian flows in probability space. Given a metric $\mathcal{G}(\rho)$, denote the density function ρ as a state variable while function Φ as a momentum variable. The Hamiltonian flow in probability space follows

$$\partial_t \begin{bmatrix} \rho \\ \Phi \end{bmatrix} - \begin{bmatrix} 0 & 1 \\ -1 & 0 \end{bmatrix} \begin{bmatrix} \frac{\delta}{\delta \rho} \mathcal{H}(\rho, \Phi) \\ \frac{\delta}{\delta \Phi} \mathcal{H}(\rho, \Phi) \end{bmatrix} = 0, \qquad (44)$$

with respect to the Hamiltonian in density space by

$$\mathcal{H}(\rho, \Phi) = \frac{1}{2} \int \Phi G(\rho)^{-1} \Phi dx + \mathcal{F}(\rho).$$

Similar to the Euclidean Hamiltonian function, the Hamiltonian functional in density space consists of a kinetic energy $\frac{1}{2} \int \Phi G(\rho)^{-1} \Phi dx$ and a potential energy $\mathcal{F}(\rho)$.

4.2 Accelerated Information Gradient Flow

We introduce the accelerated gradient flow in probability density space as follows. Let $\alpha_t \geq 0$ be a scalar function of t. We add a damping term $\alpha_t \Phi$ to the Hamiltonian flow (44):

$$\partial_t \begin{bmatrix} \rho \\ \Phi \end{bmatrix} + \begin{bmatrix} 0 \\ \alpha_t \Phi \end{bmatrix} - \begin{bmatrix} 0 & 1 \\ -1 & 0 \end{bmatrix} \begin{bmatrix} \frac{\delta}{\delta \rho} \mathcal{H}(\rho, \Phi) \\ \frac{\delta}{\delta \Phi} \mathcal{H}(\rho, \Phi) \end{bmatrix} = 0. \qquad (45)$$

We call dynamics (45) *Accelerated Information Gradient (AIG) flow*.

Proposition 11 *The accelerated information gradient flow satisfies*

$$\begin{cases} \partial_t \rho - G(\rho)^{-1}\Phi = 0, \\ \partial_t \Phi + \alpha_t \Phi + \dfrac{1}{2}\dfrac{\delta}{\delta \rho}(\int \Phi G(\rho)^{-1}\Phi dx) + \dfrac{\delta \mathcal{F}}{\delta \rho} = 0, \end{cases} \quad \text{(AIG)}$$

with initial values $\rho|_{t=0} = \rho_0$ and $\Phi|_{t=0} = 0$.

We give examples of AIG flows under several metrics, such as the Fisher-Rao metric, the Wasserstein metric, the Kalman-Wasserstein metric, and the Stein metric. See detailed derivations in the supplementary material.

Example 13 (Fisher-Rao AIG Flow)

$$\begin{cases} \partial_t \rho - (\Phi - \mathbb{E}_\rho[\Phi])\rho = 0, \\ \partial_t \Phi + \alpha_t \Phi + \dfrac{1}{2}\Phi^2 - \mathbb{E}_\rho[\Phi]\Phi + \dfrac{\delta \mathcal{F}}{\delta \rho} = 0. \end{cases} \quad \text{(F-AIG)}$$

Example 14 (Wasserstein AIG Flow, [7, 10, 33])

$$\begin{cases} \partial_t \rho + \nabla \cdot (\rho \nabla \Phi) = 0, \\ \partial_t \Phi + \alpha_t \Phi + \dfrac{1}{2}\|\nabla \Phi\|^2 + \dfrac{\delta \mathcal{F}}{\delta \rho} = 0. \end{cases} \quad \text{(W-AIG)}$$

Example 15 (Kalman-Wasserstein AIG Flow)

$$\begin{cases} \partial_t \rho + \nabla \cdot (\rho C^\lambda(\rho)\nabla \Phi) = 0, \\ \partial_t \Phi + \alpha_t \Phi + \dfrac{1}{2}\Big((x - m(\rho))^T B_\rho(\Phi)(x - m(\rho) + \nabla \Phi(x)^T C^\lambda(\rho)\nabla \Phi(x)\Big) + \dfrac{\delta \mathcal{F}}{\delta \rho} = 0. \end{cases}$$
$$\text{(KW-AIG)}$$

Here we denote $B_\rho(\Phi) = \int \nabla \Phi \nabla \Phi^T \rho dx$.

Example 16 (Stein AIG Flow)

$$\begin{cases} \partial_t \rho(x) + \nabla_x \cdot (\rho(x) \int k(x,y)\rho(y)\nabla_y \Phi(y) dy) = 0, \\ \partial_t \Phi(x) + \alpha_t \Phi(x) + \int \nabla \Phi(x)^T \nabla \Phi(y) k(x,y) \rho(y) dy + \dfrac{\delta \mathcal{F}}{\delta \rho}(x) = 0. \end{cases} \quad \text{(S-AIG)}$$

To design fast sampling algorithms, we need to reformulate the evolution of probability in terms of samples. In other words, PDEs in terms of (ρ, Φ) is the Eulerian formulation in fluid dynamics, while the particle formulation is the flow map equation, known as the Lagrangian formulation. We present examples of the W-AIG flow, the KW-AIG flow, and the S-AIG flow, which have particle formulations. We suppose that $x \sim \rho$ and $V = \nabla \Phi(x)$ are the position and velocity of a particle at time t.

Example 17 (Particle W-AIG Flow) The particle dynamical system for the flow (W-AIG) writes

$$\begin{cases} \dfrac{d}{dt} x = V, \\ \dfrac{d}{dt} V = -\alpha_t V - \nabla \left(\dfrac{\delta \mathcal{F}}{\delta \rho} \right)(x). \end{cases} \tag{46}$$

Example 18 (Particle KW-AIG Flow) The particle dynamical system for the flow (KW-AIG) writes

$$\begin{cases} \dfrac{dx}{dt} = C^{\lambda}(\rho) V, \\ \dfrac{dV}{dt} = -\alpha_t V - \mathbb{E}[V V^T](x - \mathbb{E}[x]) - \nabla \left(\dfrac{\delta \mathcal{F}}{\delta \rho} \right)(x). \end{cases} \tag{47}$$

Here, the expectation is taken over the particle system.

Example 19 (Particle S-AIG Flow) The particle dynamical system for the flow (S-AIG) writes

$$\begin{cases} \dfrac{dx}{dt} = \int k(x, y) \nabla \Phi(y) \rho(y) dy, \\ \dfrac{dV}{dt} = -\alpha_t V - \int V^T \nabla \Phi(y) \nabla_x k(x, y) \rho(y) dy - \nabla \left(\dfrac{\delta \mathcal{F}}{\delta \rho} \right)(x). \end{cases} \tag{48}$$

We notice that dynamics in Examples 17–19 are mean-field dynamics. Here the mean-field represents that the dynamics evolves its own probability density function in its path. In addition, they are also the mean field of the Markov process. Here, the Markov property holds that the update of dynamics only depends on the current time probability density. Shortly, we will design a finite-dimensional particle dynamical system to simulate these proposed dynamics.

In later on algorithm and convergence analysis, the choice of α_t is important. Similar as the ones in Euclidean space, α_t depends on the convexity of $\mathcal{F}(\rho)$ w.r.t. given metrics.

Definition 12 (Convexity in Probability Space) For a functional $\mathcal{F}(\rho)$ defined on the probability space, we say that $\mathcal{F}(\rho)$ is β-strongly convex w.r.t. metric g_ρ if there exists a constant $\beta \geq 0$ such that for any $\rho \in \mathcal{P}(\Omega)$ and any $\sigma \in T_\rho \mathcal{P}(\Omega)$, we have

$$g_\rho(\text{Hess}\mathcal{F}(\rho)\sigma, \sigma) \geq \beta g_\rho(\sigma, \sigma).$$

Here Hess is the Hessian operator w.r.t. g_ρ. If $\beta = 0$, we say that $\mathcal{F}(\rho)$ is convex w.r.t. metric g_ρ.

Again, if $\mathcal{F}(\rho)$ is β-strongly convex for $\beta > 0$, then $\alpha_t = 2\sqrt{\beta}$; if $\mathcal{F}(\rho)$ is convex, then $\alpha_t = 3/t$.

We can also formulate W-AIG flows in probability models. For instance, the W-AIG flow in Gaussian families becomes an ODE system, which corresponds to updates of covariance matrices.

Proposition 13 (W-AIG Flows in Gaussian Families) *Suppose that ρ_0, ρ^* are Gaussian distributions with zero means and their covariance matrices are Σ_0 and Σ^*. $E(\Sigma)$ evaluates the KL divergence from ρ to ρ^*:*

$$E(\Sigma) = \frac{1}{2}(tr(\Sigma(\Sigma^*)^{-1}) - \log\det(\Sigma(\Sigma^*)^{-1}) - d), \tag{49}$$

Let (Σ_t, S_t) be the solution to

$$\begin{cases} \dot{\Sigma}_t - 2(S_t\Sigma_t + \Sigma_t S_t) = 0, \\ \dot{S}_t + \alpha_t S_t + 2S_t^2 + \nabla_{\Sigma_t} E(\Sigma_t) = 0, \end{cases} \tag{W-AIG-G}$$

with initial values $\Sigma_t|_{t=0} = \Sigma_0$ and $S_t|_{t=0} = 0$. Here Σ_t and S_t are symmetric matrices. Then, for any $t \geq 0$, Σ_t is well-defined and stays positive definite. Furthermore, we denote

$$\rho(x) = \frac{(2\pi)^{-n/2}}{\sqrt{\det(\Sigma_t)}} \exp(-\frac{1}{2}x^T \Sigma_t^{-1} x), \quad \Phi(x) = x^T S_t x + C(t),$$

where $C(t) = -t + \frac{1}{2}\int_0^t \log\det(\Sigma_s(\Sigma^*)^{-1})ds$. Then, (ρ, Φ) is the solution to (W-AIG) with initial values $\rho|_{t=0} = \rho_0$ and $\Phi|_{t=0} = 0$.

4.3 Convergence Rate Analysis on AIG Flows

In this section, we prove the convergence rates of AIG flows under either the Wasserstein metric or the Fisher-Rao metric. This validates the acceleration effect. The proof is motivated by Lyapunov functions of Euclidean accelerated gradient flows in Sect. 4.1.1.

Theorem 14 *Suppose that $\mathcal{F}(\rho)$ is β-strongly convex for $\beta > 0$ and $\mathcal{F}(\rho^*) = 0$. The solution ρ to (F-AIG) or (W-AIG) with $\alpha_t = 2\sqrt{\beta}$ satisfies*

$$\mathcal{F}(\rho) \leq C_0 e^{-\sqrt{\beta}t} = C_0 e^{-\sqrt{\beta}t}.$$

If $\mathcal{F}(\rho)$ is convex, then the solution ρ to (F-AIG) or (W-AIG) with $\alpha_t = 3/t$ satisfies

$$\mathcal{F}(\rho) \leq C_0' t^{-2} = \mathcal{O}(t^{-2}).$$

Here the constants C_0, C_0' only depend on ρ_0.

In Euclidean case, the convergence rate of accelerated gradient flow is based on the construction of Lyapunov functions. Namely, for β-strongly convex $f(x)$, consider a Lyapunov function:

$$\mathcal{E}(t) = \frac{e^{\sqrt{\beta}t}}{2}\|\sqrt{\beta}(x-x^*) + \dot{x}\|^2 + e^{\sqrt{\beta}t}(f(x) - f(x^*)).$$

For general convex $f(x)$, consider a Lyapunov function

$$\mathcal{E}(t) = \frac{1}{2}\left\|(x-x^*) + \frac{t}{2}\dot{x}\right\|^2 + \frac{t^2}{4}(f(x) - f(x^*)).$$

Based on different assumptions on the convexity of $f(x)$, we can prove that these Lyapunov functions are not increasing w.r.t. t. Hence, the convergence rates are obtained.

Following Lyapunov functions in Euclidean space, we provide a sketch of the proof for Theorem 14. We first consider the case where $\mathcal{F}(\rho)$ is β-strongly convex for $\beta > 0$. Let T_t denote the optimal transport plan from ρ to ρ^*. Consider a Lyapunov function

$$\mathcal{E}(t) = \frac{e^{\sqrt{\beta}t}}{2}\int \left\|-\sqrt{\beta}(T_t(x) - x) + \nabla\Phi(x)\right\|^2 \rho(x)dx \qquad (50)$$
$$+ e^{\sqrt{\beta}t}(\mathcal{F}(\rho) - \mathcal{F}(\rho^*)).$$

Here, the $-(T_t(x) - x)$ term can be viewed as $x - x^*$ and $\nabla\Phi$ can be viewed as $\dot{x}_t$. Different from the Euclidean case, we introduce an important lemma in proving that $\mathcal{E}(t)$ is non-increasing.

Lemma 15 *Denote $u_t = \partial_t(T_t)^{-1} \circ T_t$. Then, u_t satisfies*

$$\nabla \cdot (\rho(u_t - \nabla\Phi)) = 0.$$

We also have

$$\partial_t T_t(x) = -\nabla T_t(x) u_t(x).$$

More importantly, we have

$$\int (\nabla \Phi - u_t, \nabla T_t \nabla \Phi) \rho dx \geq 0,$$

$$\int (\nabla \Phi - u_t, \nabla T_t(x)(T_t(x) - x)) \rho = 0.$$

We then show that $\mathcal{E}(t)$ is not increasing w.r.t. t.

Proposition 16 *Suppose that $\mathcal{F}(\rho)$ satisfies Hess(β) for $\beta > 0$ and $\mathcal{F}(\rho^*) = 0$. ρ is the solution to (W-AIG) with $\alpha_t = 2\sqrt{\beta}$. Then, $\mathcal{E}(t)$ defined in (50) satisfies $\dot{\mathcal{E}}(t) \leq 0$. As a result,*

$$\mathcal{F}(\rho) \leq e^{-\sqrt{\beta}t} \mathcal{E}(t) \leq e^{-\sqrt{\beta}t} \mathcal{E}(0) = \mathcal{O}(e^{-\sqrt{\beta}t}).$$

Note that $\mathcal{E}(0)$ only depends on ρ_0. This proves the first part of Theorem 1. We now focus on the case where $\mathcal{F}(\rho)$ is convex. Similarly, we construct the following Lyapunov function.

$$\begin{aligned}\mathcal{E}(t) = &\frac{1}{2} \int \left\| -(T_t(x) - x) + \frac{t}{2} \nabla \Phi(x) \right\|^2 \rho(x) dx \\ &+ \frac{t^2}{4}(\mathcal{F}(\rho) - \mathcal{F}(\rho^*)).\end{aligned} \quad (51)$$

Proposition 17 *Suppose that $\mathcal{F}(\rho)$ satisfies Hess(0) and $\mathcal{F}(\rho^*) = 0$. ρ is the solution to (W-AIG) with $\alpha_t = 3/t$. Then, $\mathcal{E}(t)$ defined in (51) satisfies $\dot{\mathcal{E}}(t) \leq 0$. As a result,*

$$\mathcal{F}(\rho) \leq \frac{4}{t^2} \mathcal{E}(t) \leq \frac{4}{t^2} \mathcal{E}(0) = \mathcal{O}(t^{-2}).$$

Because $\mathcal{E}(0)$ only depends on ρ_0, we complete the proof.

4.4 Discrete-Time Algorithms for AIG Flows

In this section, we present the discrete-time particle implementation of the flow (W-AIG) based on the particle W-AIG flow (46). Similar discrete-time algorithms of (KW-AIG) and (S-AIG) are provided in the supplementary material. Here we mainly introduce a kernel bandwidth selection method and an adaptive restart technique to deal with difficulties in numerical implementations.

A typical choice of $\mathcal{F}(\rho)$ for sampling is the KL divergence

$$D_{KL}(\rho\|\rho^*) = \int \rho \log \frac{\rho}{e^{-f}} dx - \log Z,$$

where the target density $\rho^*(x) \propto \exp(-f(x))$ and $Z = \int \exp(-f(x))dx$. Then, (46) is equivalent to

$$\begin{cases} dx = V dt, \\ dV = -\alpha_t V dt - \nabla f(x) dt - \nabla \log \rho(x) dt. \end{cases} \quad (52)$$

Consider a particle system $\{X_0^i\}_{i=1}^N$ and let $V_0^i = 0$. In k-th iteration, the update rule follows

$$\begin{cases} X_{k+1}^i = X_k^i + \sqrt{\tau_k} V_{k+1}^i, \\ V_{k+1}^i = \alpha_k V_k^i - \sqrt{\tau_k}(\nabla f(X_k^i) + \xi_k(X_k^i)), \end{cases} \quad (53)$$

for $i = 1, 2 \ldots N$. If $\mathcal{F}(\rho)$ is β-strongly convex, then $\alpha_k = \frac{1-\sqrt{\beta\tau_k}}{1+\sqrt{\beta\tau_k}}$; if $\mathcal{F}(\rho)$ is convex or β is unknown, then $\alpha_k = \frac{k-1}{k+2}$. Here $\xi_k(x)$ is an approximation of $\nabla \log \rho_k(x)$. For a general distribution, we use the kernel density estimation (KDE), $\tilde{\rho}_k(x) = \frac{1}{N} \sum_{i=1}^N K(x, X_k^i)$ to approximate $\rho_k(x)$. Here $K(x, y)$ is a positive kernel function. Then, ξ_k writes

$$\xi_k(x) = \nabla \log \tilde{\rho}_k(x) = \frac{\sum_{i=1}^N \nabla_x K(x, X_k^i)}{\sum_{i=1}^N K(x, X_k^i)}. \quad (54)$$

A common choice of $K(x, y)$ is a Gaussian kernel with the bandwidth h, $K(x, y) = (2\pi h)^{-n/2} \exp(-\|x-y\|^2/(2h))$.

There are two difficulties in the time discretization. For one thing, the bandwidth h strongly affects the estimation of $\nabla \log \rho$, so we propose the BM method to learn the bandwidth from Brownian motion samples. For another, the second equation in (W-AIG) is the Hamilton-Jacobi equation, which usually has strong stiffness. In numerical trials, we observe that the densities from the particles may collapse in certain dimensions following W-AIG flows, even for Gaussian target density. Therefore, we propose an adaptive restart technique to deal with this problem.

4.4.1 Learn the Bandwidth via Brownian Motion

SVGD uses a median (MED) method to choose the bandwidth, i.e.,

$$h_k = \frac{1}{2\log(N+1)} \text{median}(\{\|X_k^i - X_k^j\|^2\}_{i,j=1}^N). \quad (55)$$

Given the bandwidth h, $\{X_k^i\}_{i=1}^N$ and a step size s, we can compute two particle systems:

$$Y_k^i(h) = X_k^i - s\xi_k(x;h), \quad Z_k^i = X_k^i + \sqrt{2s}B^i, \quad i = 1,\ldots N$$

where B^i is the standard Brownian motion. Denote the empirical distributions of $\{X_k^i\}_{i=1}^N$, $\{Y_k^i\}_{i=1}^N$ and $\{Z_k^i\}_{i=1}^N$ by $\hat\rho_X$, $\hat\rho_Y$ and $\hat\rho_Z$. With $n \to \infty$, we shall have $\hat\rho_Y = \hat\rho_Z = \rho|_{t=s}$, where $\hat\rho$ satisfies $\partial_t\hat\rho = \Delta\hat\rho = \nabla \cdot (\hat\rho\nabla \log \hat\rho)$ with initial value $\hat\rho|_{t=0} = \hat\rho_X$. With an appropriate bandwidth h, we shall also have $\hat\rho_Y = \rho|_{t=s}$. Hence, we consider the following optimization problem

$$\min_h \mathrm{MMD}(\hat\rho_Y, \hat\rho_Z) = \int\int (\hat\rho_Y(y) - \hat\rho_Z(y))k(y,z)(\hat\rho_Y(z) - \hat\rho_Z(z))dydz. \tag{56}$$

where MMD (maximum mean discrepancy) evaluates the similarity between $\{Y_k^i\}_{i=1}^N$ and $\{Z_k^i\}_{i=1}^N$. Here, the kernel $k(y,z)$ in MMD is chosen as a Gaussian kernel with bandwidth 1. So we optimize (56) using the bandwidth h_{k-1} from the last iteration as the initialization. For simplicity, we denote

$$\mathrm{BM}(h_{k-1}, \{X_k^i\}_{i=1}^N, s)$$

as the minimizer of problem (56). It is the output of the BM method.

4.4.2 Adaptive Restart

To enhance the practical performance, we introduce an adaptive restart technique, which shares the same idea of gradient restart under the Euclidean case. Consider

$$\varphi_k = -\sum_{i=1}^N (V_{k+1}^i, \nabla f(X_k^i) + \xi_k(X_k^i)), \tag{57}$$

which can be viewed as discrete-time approximation of

$$-g_\rho^W(\partial_t\rho, G^W(\rho)^{-1}\frac{\delta\mathcal{F}}{\delta\rho}) = -\partial_t\mathcal{F}(\rho).$$

If $\varphi_k < 0$, then we restart the algorithm with initial values $X_0^i = X_k^i$ and $V_0^i = 0$. This essentially keeps $\partial_t\mathcal{F}(\rho)$ negative along the trajectory.

4.5 Numerical Experiments

In this section, we present several numerical experiments to demonstrate the effectiveness of the BM method, the acceleration effect of AIG flows, and the strength of the adaptive restart technique. Implementation details are provided in the supplementary material.

4.5.1 Toy Examples

We first generate samples from a toy bi-modal distribution in viwnf. We compare sampling algorithms based on gradient flows and accelerated gradient flows under the Wasserstein metric, the Kalman-Wasserstein metric, and the Stein metric. The number of particles follows $N = 200$. The initial distribution of the particle system follows $\mathcal{N}([0, 10]', I)$.

For the approximation of $\nabla \log \rho_k$, we use a Gaussian kernel and the kernel bandwidth is selected by the BM method. We apply the restart technique for discrete-time algorithms of AIG flows. For W-GF, W-AIG, SVGD and S-AIG, we take the step size $\tau_k = 0.1$. For KW-GF and KW-AIG, we set the regularization parameter $\lambda = 1$ and the step size $\tau_k = 0.02$. We choose a smaller step size for the Kalman-Wasserstein metric because the particle system may blow up for a larger step size. For SVGD and S-AIG, we use a Gaussian kernel with fixed bandwidth 1. The step size of SVGD is adjusted by Adagrad.

From Fig. 4, the convergence rate of the particle system depends on the metric. For a fixed metric, samples generated by accelerated gradient flows always converge faster than the ones generated by gradient flows.

4.5.2 Effect of BM Method

We first investigate the validity of the BM method in selecting the bandwidth. The target density ρ^* is a toy bi-modal distribution. We compare two types of particle implementations of the Wasserstein gradient flow over KL divergence:

$$X^i_{k+1} = X^i_k - \tau \nabla f(X^i_k) + \sqrt{2\tau} B^i_k,$$
$$X^i_{k+1} = X^i_k - \tau (\nabla f(X^i_k) + \xi_k(X^i_k)).$$

Here $B^i_k \sim \mathcal{N}(0, 1)$ is the standard Brownian motion and ξ_k is estimated via KDE. The first method is known as the Langevin MCMC method and the second method is called the ParVI method. For ParVI methods, the bandwidth h is selected by MED/HE/BM, respectively. The initial distribution of the particle system follows the standard Gaussian $\mathcal{N}(0, I)$. The objective density function follows

$$\rho^*(x) \propto \exp(-2(\|x\| - 3)^2)$$
$$\times (\exp(-2(x_1 - 3)^2) + \exp(-2(x_1 + 3)^2)).$$

All methods run for 200 iterations using the same fixed step size $\tau = 0.1$.

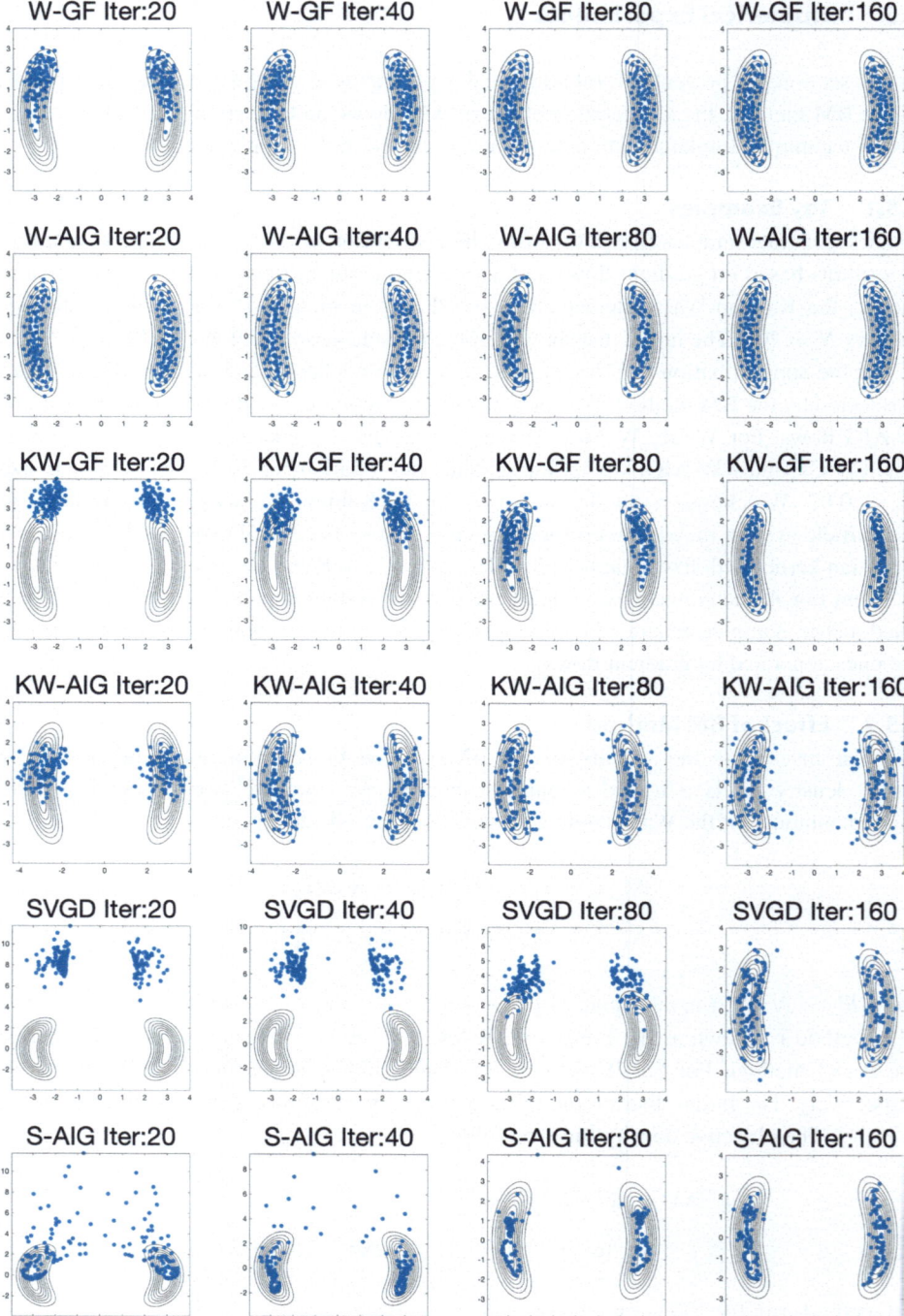

Fig. 4 Comparison of different AIG flows on a toy example

Figure 4 shows the distribution of 200 samples based on different methods. Samples from MCMC match the target distribution in a stochastic way; samples from MED collapse; samples from HE align tidily around contour lines; samples from BM arrange neatly and are closer to samples from MCMC. This indicates that the BM method makes the particle system behave similar to MCMC, though in a deterministic way (Fig. 5).

4.5.3 Bayesian Logistic Regression

We perform the standard Bayesian logistic regression experiment on the Covertype dataset. We select the kernel bandwidth using either the MED method or the proposed BM method. Figure 6 indicates that the BM method accelerates and stabilizes the performance of GFs and AIGs. The performance of MCMC and WGF are similar, and they achieve the best log-likelihood. For a given metric, AIG flows have better test accuracy and test log-likelihood in the first 2000 iterations. W-AIG and KW-AIG achieve 75% test accuracy in less than 500 iterations.

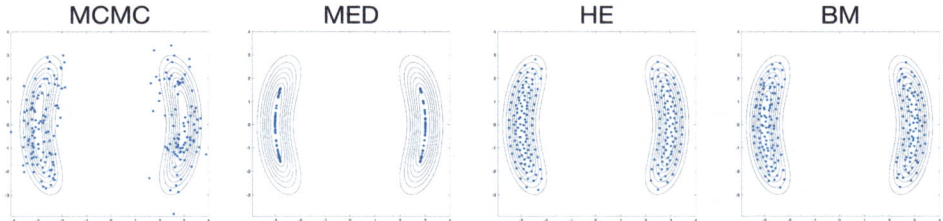

Fig. 5 The effect of the BM method. Samples are plotted as blue dots. Left to right: MCMC, MED, HE, and BM. All methods are run for 200 iterations with the same initialization

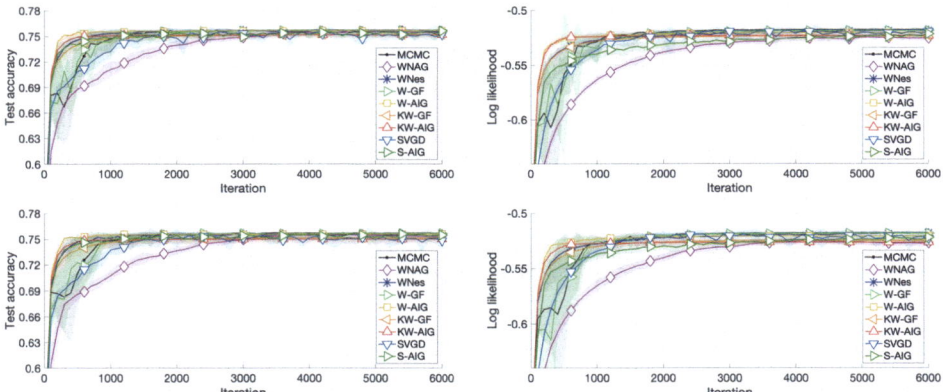

Fig. 6 Results on Bayesian logistic regression, averaged over 10 independent trials. The shaded areas show the variance. Top: BM; Bottom: MED. Left: Test accuracy; Right: Test log-likelihood

Table 1 Test root-mean-square-error (RMSE)

Dataset	AIG	WGF	SVGD
Boston	$2.871_{\pm 3.41e-3}$	$3.077_{\pm 5.52e-3}$	$\mathbf{2.775_{\pm 3.78e-3}}$
Combined	$\mathbf{4.067_{\pm 9.27e-1}}$	$4.077_{\pm 3.85e-4}$	$4.070_{\pm 2.02e-4}$
Concrete	$\mathbf{4.440_{\pm 1.34e-1}}$	$4.883_{\pm 1.93e-1}$	$4.888_{\pm 1.39e-1}$
Kin8nm	$\mathbf{0.094_{\pm 5.56e-6}}$	$0.096_{\pm 3.36e-5}$	$0.095_{\pm 1.32e-5}$
Wine	$0.606_{\pm 1.40e-5}$	$0.614_{\pm 3.48e-4}$	$\mathbf{0.604_{\pm 9.89e-5}}$
Year	$8.876_{\pm 3.71e-4}$	$\mathbf{8.872_{\pm 2.81e-4}}$	$8.873_{\pm 7.19e-4}$

The bold numbers indicate that the AIG method performs better than other methods in terms of RMSE and log-likelihood values

Table 2 Test log-likelihood

Dataset	AIG	WGF	SVGD
Boston	$\mathbf{-2.609_{\pm 1.34e-4}}$	$-2.694_{\pm 2.83e-4}$	$-2.611_{\pm 1.36e-4}$
Combined	$\mathbf{-2.822_{\pm 5.72e-3}}$	$-2.825_{\pm 2.36e-5}$	$-2.823_{\pm 1.24e-5}$
Concrete	$\mathbf{-2.884_{\pm 8.84e-3}}$	$-2.971_{\pm 8.93e-3}$	$-2.978_{\pm 6.05e-3}$
Kin8nm	$\mathbf{0.951_{\pm 6.43e-4}}$	$0.923_{\pm 3.37e-3}$	$0.932_{\pm 1.43e-3}$
Wine	$-0.961_{\pm 1.28e-4}$	$-0.961_{\pm 3.17e-4}$	$\mathbf{-0.952_{\pm 9.89e-5}}$
Year	$-3.654_{\pm 1.00e-5}$	$-3.655_{\pm 7.82e-6}$	$\mathbf{-3.652_{\pm 1.28e-5}}$

The bold numbers indicate that the AIG method performs better than other methods in terms of RMSE and log-likelihood values

4.6 Bayesian Neural Network

We apply our proposed method to the Bayesian neural network over the UCI dataset. We compare W-AIG, W-GF and SVGD. For all methods, we use $N = 10$ particles. The averaged results over 20 independent trials are collected in Tables 1 and 2. We observe that on most datasets, W-AIG has better test root-mean-square-error and test log-likelihood than W-GF and SVGD. This indicates that W-AIG may have better generalization than W-GF and SVGD.

5 Conclusions and Future Directions

In this note, we embark on a journey to explore the fascinating formulations of generalized Wasserstein dynamics, viewing them through the unique lenses of their gradient flows and Hamiltonian flows.

In scientific computing, we introduce neural network-based implicit particle methods for computing high-dimensional Wasserstein-type gradient flows with linear and nonlinear mobility functions. Our approach centers around the Lagrangian formulation within the Jordan–Kinderlehrer–Otto (JKO) framework, where neural networks approximate the velocity field. Neural ODE techniques are harnessed to efficiently compute the time-implicit updates of both particle locations and the densities they carry. Additionally, we leverage an explicit recurrence relation to compute derivatives, significantly streamlining the backpropagation process. Our methodology exhibits versatility and can handle a broad spectrum of gradient flows while accommodating diverse potential functions and nonlinear mobility scenarios. There are several promising directions for future work. One important avenue for exploration is the convexity of the problem, investigating how it depends on parameters like the step size Δt and network architectures, including the choices of neural network activation functions. One important property is the accuracy of the computation of neural JKO schemes. Some accurate high order accuracy numerical methods in solving Wasserstein dynamics are also important here. A coherent study between high-order numerical schemes and machine learning methods with applications in Wasserstein dynamics are important research directions [11]. Understanding the convexity properties can contribute to faster training and provide insights into the convergence properties of the algorithm. Additionally, we aim to extend this approach to accommodate more complex energy functionals and scenarios where a gradient flow structure is not fully presented or is only part of the physical dynamics. This extension would broaden the applicability of our method to a broader range of problems in simulating physics-oriented partial differential equations beyond gradient flows.

In optimization with large dimensional data samples, we propose the framework of AIG flows by damping Hamiltonian flows concerning specific information metrics in probability space. In theory, we establish the convergence rate of F-AIG and W-AIG flows. We propose particle formulations for W-AIG flow, KW-AIG, and S-AIG flow in the algorithm. Numerically, we propose discrete-time algorithms and an adaptive restart technique to overcome numerical stiffness of AIG flows. To efficiently approximate $\nabla \log \rho_k(x)$, we introduce a novel kernel selection method by learning from Brownian motion samples. Numerical experiments verify the acceleration effect of AIG flows and the strength of adaptive restart. In future works, we will systematically explain the stiffness of AIG flows and the effects of the adaptive restart. We shall apply our results to general information metrics, especially for generalized Wasserstein metrics. We expect to study the related sampling efficient optimization methods and discrete-time algorithms. We also plan to incorporate Hessian operators in probability space to design higher-order accelerated algorithms. We shall compare these information metrics-induced methods regarding computational complexity and sampling efficiency.

A central problem in mathematical machine learning lies in the optimization procedure and scientific computation problems. What type of energy functional, mobility functions V_1, V_2, can help us design fast, efficient, and hopefully simple particle-level algorithms? Again, a new class of loss functional [18, 19] and its convergence analysis combining optimal transport theory, such as displacement convexities or Ricci curvatures on sample spaces [3, 35], with machine learning models, such as neural ODEs [9, 36] and normalization flows [15, 24], is a long and fruitful research direction.

References

1. S. Amari, *Information Geometry and Its Applications*, 1st edn. (Springer Publishing Company, Incorporated, New York, 2016)
2. L. Ambrosio, N. Gigli, G. Savaré, *Gradient Flows in Metric Spaces and in the Space of Probability Measures*, 2nd edn. Lectures in Mathematics ETH Zurich (Birkhauser Verlag, Basel 2008)
3. E. Bayraktar, Q. Feng, W. Li, Exponential entropy dissipation for weakly self-consistent Vlasov–Fokker–Planck equations. J. Nonlinear Sci. **34**, 7 (2024)
4. J.D. Benamou, Y. Brenier, A computational fluid mechanics solution to the Monge-Kantorovich mass transfer problem. Numer. Math. **84**(3), 375–393 (2000)
5. P. Cardaliaguet, F. Delarue, J. Lasry, P. Lions, The master equation and the convergence problem in mean-field games. arXiv:1509.02505 (2015)
6. J.A. Carrillo, S. Lisini, G. Savaré, D. Slepcev, Nonlinear mobility continuity equations and generalized displacement convexity. J. Funct. Anal. **258**(4), 1273–1309 (2009)
7. J.A. Carrillo, Y.-P. Choi, O. Tse, Convergence to equilibrium in Wasserstein distance for damped Euler equations with interaction forces. Commun. Math. Phys. **365**(1), 329–361 (2019)
8. J.A. Carrillo, D. Matthes, M.-T. Wolfram, Lagrangian schemes for Wasserstein gradient flows, in *Handbook of Numerical Analysis*, vol. 22 (Elsevier, Amsterdam, 2021), pp. 271–311
9. R.T. Chen, Y. Rubanova, J. Bettencourt, D.K. Duvenaud, Neural ordinary differential equations. Adv. Neural Inf. Process. Syst. **31**, 6571–6583 (2018)
10. S. Chen, Q. Li, O. Tse, S.J. Wright, Accelerating optimization over the space of probability measures. arXiv:2310.04006 (2024)
11. G. Fu, S. Osher, W. Li, High order spatial discretization for variational time implicit schemes: Wasserstein gradient flows and reaction-diffusion systems. arXiv:2303.08950 (2023)
12. G. Fu, H. Ji, W. Pazner, W. Li, Mean field control of droplet dynamics with high order finite element computations. arXiv:2402.05923 (2024)
13. A. Garbuno-Inigo, F. Hoffmann, W. Li, A.M. Stuart, Interacting Langevin diffusions: gradient structure and ensemble Kalman sampler. SIAM J. Appl. Dyn. Syst. **19**, 412–441 (2020)
14. I. Goodfellow et al., Generative adversarial nets, in *Advances in Neural Information Processing Systems*, vol. 27 (2014)
15. Z. Hu, C. Liu, Y. Wang, Z. Xu, Energetic variational neural network discretizations to gradient flows. SIAM J. Sci. Computing **46**(4) (2024)
16. R. Jordan, D. Kinderlehrer, F. Otto, The variational formulation of the Fokker–Planck equation. SIAM J. Math. Anal. **29**, 1–17 (1998)
17. Y. LeCun, Y. Bengio, G. Hinton, Deep learning. Nature **521**, 436–444 (2015)
18. W. Li, Transport information geometry: Riemannian calculus on probability simplex. Inf. Geom. **5**, 161–207 (2022)
19. W. Li, Transport information Bregman divergences. Inf. Geom. **4**, 435–470 (2021)
20. W. Li, Hessian metric via transport information geometry. J. Math. Phys. **62**, 033301 (2021)
21. W. Li, W. Lee, S. Osher, Computational mean-field information dynamics associated with reaction-diffusion equations. J. Comput. Phys. **466**, 111409 (2022)
22. A.T. Lin, W. Li, S. Osher, G. Montúfar, Wasserstein proximal of GANs, in *Geometric Science of Information*, ed. by F. Nielsen, F. Barbaresco (Springer International Publishing, Cham, 2021), pp. 524–533
23. Q. Liu, Stein variational gradient descent as gradient flow, in *Advances in Neural Information Processing Systems* (Curran Associates, Inc., 2017), pp. 3115–3123
24. S. Liu, W. Li, H. Zha, H. Zhou, Neural parametric Fokker–Planck equation. SIAM J. Numer. Anal. **60**, 1385–1449 (2022)

25. A. Mielke, A gradient structure for reaction-diffusion systems and for energy-drift-diffusion systems. Nonlinearity **4**(4), 1329 (2011)
26. A. Mielke, Free energy, free entropy, and a gradient structure for thermoplasticity, in *Innovative Numerical Approaches for Multi-Field and Multi-Scale Problems*, ed. by K. Weinberg, A. Pandolfi. Lecture Notes in Appl. Comp. Mechanics, vol. 81 (Springer, Berlin, 2016), pp. 135–160
27. A. Mielke, D.R.M. Renger, M.A. Peletier, A generalization of Onsagers reciprocity relations to gradient flows with nonlinear mobility. J. Non-Equil. Thermodyn. **41**(2), 141–149 (2016)
28. Y.E. Nesterov, A method of solving a convex programming problem with convergence rate $O(1/k^2)$, in *Doklady Akademii Nauk*, vol. 269 (Russian Academy of Sciences, 1983), pp. 543–547
29. D. Onken, S.W. Fung, X. Li, L. Ruthotto, OT-flow: fast and accurate continuous normalizing flows via optimal transport, in *Proceedings of the AAAI Conference on Artificial Intelligence*, vol. 35 (2021), pp. 9223–9232
30. L. Onsager, Reciprocal relations in irreversible processes, I+II. Phys. Rev. **37**, 405–426 (1931)
31. F. Otto, The geometry of dissipative evolution equations: the porous medium equation. Commun. Partial Differ. Equ. **26**(1–2), 101–174 (2001)
32. W. Su, S. Boyd, E.J. Candes, A differential equation for modeling Nesterov's accelerated gradient method: theory and insights. J. Mach. Learn. Res. **17**, 1–43 (2016)
33. A. Taghvaei, P. Mehta, Accelerated flow for probability distributions, in *International Conference on Machine Learning*. PMLR (2019), pp. 6076–6085
34. C. Villani, Hypocoercivity. Mem. Am. Math. Soc. **202**(950), iv+141 (2009)
35. C. Villani, *Optimal Transport: Old and New*, vol. 338 (Springer, Berlin, 2009)
36. C. Xu, X. Cheng, Y. Xie, Invertible normalizing flow neural networks by JKO scheme, in *Proceedings of the 37th International Conference on Neural Information Processing Systems*, 2023

Flow Matching: Markov Kernels, Stochastic Processes and Transport Plans

Christian Wald and Gabriele Steidl

Abstract

Among generative neural models, flow matching techniques stand out for their simple applicability and good scaling properties. Here, velocity fields of curves connecting a simple latent and a target distribution are learned. Then the corresponding ordinary differential equation can be used to sample from a target distribution, starting in samples from the latent one. This paper reviews from a mathematical point of view different techniques to learn the velocity fields of absolutely continuous curves in the Wasserstein geometry. We show how the velocity fields can be characterized and learned via transport plans (couplings) between latent and target distributions, Markov kernels and stochastic processes, where the latter two include the coupling approach, but are in general broader.

Besides this main goal, we show how flow matching can be used for solving Bayesian inverse problems, where the definition of conditional Wasserstein distances plays a central role.

Finally, we briefly address continuous normalizing flows and score matching techniques, which approach the learning of velocity fields of curves from other directions.

C. Wald · G. Steidl (✉)
Institute of Mathematics, Technische Universität Berlin, Berlin, Germany
http://tu.berlin/imageanalysis
e-mail: wald@math.tu-berlin.de; steidl@math.tu-berlin.de

1 Introduction

Generative models with neural network approximations have shown impressive results in many applications, in particular in inverse problems and data assimilation. First developments in this direction were generative adversarial networks by Goodfellow et al. [15] in 2014 and variational autoencoders by Kingma and Welling [25] in 2013. In this paper, we are interested in recent flow matching techniques pioneered by Liu et al. [29, 30] in 2022/2023 from the point of view of stochastic processes, and by Lipman et al. [28] in 2023 via conditional probability paths. Flow matching is closely related to continuous normalizing flows (CNFs), also called neural ordinary differential equations by Chen et al. [9] in 2018, and to score-based diffusion models introduced by Sohl-Dickstein et al. [38] and Ermon and Song [39] in 2015 and 2019, respectively. The later ones can be considered as an instance of CNFs, but add the perspective of stochastic differential equations. To provide a more complete picture, we will briefly consider these approaches in Sects. 10 and 11. Flow matching techniques stand out for their simple applicability and good scaling properties. They can be easily incorporated into other optimization techniques like plug-and-play approaches [31] and can be generalized for solving Bayesian inverse problems as we will see in Sect. 9. Recently, flow matching was generalized towards generator matching in [20, 22].

We will approach the topic of flow matching from a mathematical point of view, where we intend to make the paper to a certain degree self-contained. However, for the practical implementation of flow matching, we assume that the reader is familiar with training neural networks once a tractable loss function is given.

The basic task of generative modeling consists in generating new samples from a probability distribution P_{data}, where we have only access to a set of samples, e.g. from the distribution of face images. This is a quite classical task in stochastics, since it is in general hard to sample from a high-dimensional distribution even if its density is known. Few examples, where it is easy to sample from in multiple dimensions, are the Gaussian distribution and their relatives. In this paper, we focus on the standard Gaussian distribution as so-called "latent" distribution P_{latent}. Then, the idea is to approximate a "nice" curve $t \mapsto \mu_t$, mapping a time $t \in [0, 1]$ to a probability measure μ_t, where

$$\mu_0 = P_{\text{latent}} \quad \text{and} \quad \mu_1 = P_{\text{data}}.$$

Our curves of interest are absolutely continuous curves in the Wasserstein space, which we introduce in Sect. 3. Due to the special metric of the Wasserstein space, such a curve possesses a vector field $v : [0, 1] \times \mathbb{R}^d \to \mathbb{R}^d$ such that the curve-velocity pair (μ_t, v_t) fulfills the continuity equation

$$\partial_t \mu_t + \nabla_x \cdot (\mu_t v_t) = 0.$$

For the associated vector field v_t, under mild additional conditions, there exists a solution $\phi : [0, 1] \times \mathbb{R}^d \to \mathbb{R}^d$ of the ordinary differential equation (ODE)

$$\partial_t \phi(t, x) = v_t(\phi(t, x)), \quad \phi(0, x) = x,$$

and the curve μ_t is just the push-forward of the latent distribution by this solution

$$\mu_t = \phi(t, \cdot)_\sharp \mu_0 := \mu_0 \circ \phi^{-1}(t, \cdot).$$

Thus, if the vector field v_t is known, it remains to solve the ODE by some standard solver starting at $t = 0$ with a sample of the latent distribution to produce a desired data sample at time $t = 1$. An illustration of the flow of samples from the one-dimensional standard Gaussian distribution to a Gaussian mixture is given in Fig. 1. For a higher-dimensional example, see Fig. 8.

Before learning the velocity fields via a neural network approximation, we have to determine appropriate curve-velocity pairs (μ_t, v_t) fulfilling in particular the continuity equation. In Sects. 4–6, we provide three approaches. We will see that the first one follows as a special case both from the second and third one. Ultimately, we develop the second approach with Markov kernels to explain the setting of Lipman et al. in [28] also for measures without densities.

1. **Curves induced by couplings**: Given a probability measure α on $\mathbb{R}^d \times \mathbb{R}^d$ with marginals μ_0 and μ_1, also called "coupling" of μ_0 and μ_1, the curve-velocity pair induced by α is given by the push-forwards

$$\mu_t := e_{t,\sharp} \alpha \quad \text{and} \quad v_t \mu_t := e_{t,\sharp}[(y - x)\alpha]$$

with

$$e_t(x, y) := (1 - t)x + ty, \quad t \in [0, 1].$$

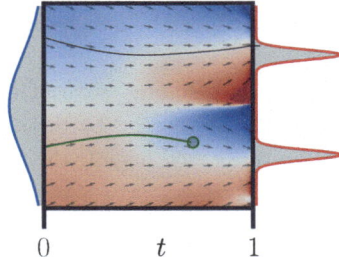

Fig. 1 Illustration of a curve from the standard Gaussian distribution to a Gaussian mixture in one dimension. The plot shows $t \mapsto \phi(t, x^i)$, $i = 1, 2$ for two different samples x^i (black, green), the vectors $(1, \partial_t \phi)$ and the red-blue color-coded velocity field $\partial_t \phi$. Courtesy: Blogpost [5]

Here $v_t \mu_t = e_{t,\sharp}[(y-x)\alpha]$ is a vector valued measures, where you may note that for $\gamma \in \mathcal{P}_2(\mathbb{R}^d)$, $f \in L^1(\mathbb{R}^d, \gamma)$ the vector valued measure $f\gamma$ is defined by

$$\int_{\mathbb{R}^d} \varphi(x) \mathrm{d} f\gamma(x) = \int_{\mathbb{R}^d} \varphi(x) f(x) \mathrm{d}\gamma(x)$$

for every bounded measurable function $\varphi : \mathbb{R}^d \to \mathbb{R}$. Most common couplings are independent couplings $\alpha = \mu_0 \times \mu_1$ and so-called "optimal" transport couplings with respect to the Wasserstein distance. The optimal couplings stand out by the fact that the induced curves are geodesics in the Wasserstein space, which in particular are minimal length curves connecting μ_0 and μ_1. In an alternative formulation, we may start with random variables with joint law $\alpha = P_{X_0, X_1}$ so that $\mu_0 = P_{X_0}$ and $\mu_1 = P_{X_1}$. Then the induced curve by α becomes $\mu_t = P_{X_t}$, where

$$X_t := e_t(X_0, X_1) = (1-t)X_0 + tX_1.$$

2. **Curves via Markov kernels**: We start with a family of Markov kernels $(\mathcal{K}_t)_{t \in [0,1]}$ with the property that "conditioned to $y \in \mathbb{R}^d$", the curve $t \mapsto \mathcal{K}_t(y, \cdot)$ is absolutely continuous with associated vector field v_t^y. Then we define a time-dependent sequence of measures on $\mathbb{R}^d \times \mathbb{R}^d$ by gluing the Markov kernel with μ_1 as second marginal, i.e.

$$\alpha_t := \mathcal{K}_t(y, \cdot) \times_y \mu_1.$$

Denoting by v_t its first marginal and using the disintegration $\alpha_t = \alpha_t^x \times_x v_t$, we will show that the curve $t \mapsto v_t$ is also absolutely continuous with associated velocity field

$$v_t(x) := \int_{\mathbb{R}^d} v_t^y \, \mathrm{d}\alpha_t^x(y).$$

Note that the curve v_t does not necessarily fulfill $v_1 = \mu_1$ or admit a v_0 that is easy to sample from. But for reasonable choices it is possible to obtain that v_0 is a standard Gaussian distribution and $v_1 \cong \mu_1$. An tractable pair (v_t, v_t) which cannot be deduced from a curve-velocity pair induced by a coupling between μ_0 and μ_1 is given in Example 5.4.

However, for the following special setting, we obtain our induced curves from couplings α of μ_0 and μ_1: Let $\alpha = \alpha^y \times_y \mu_1$ be the disintegration with respect to the second marginal. Then, $\alpha_t := (e_t, \pi^2)_\sharp \alpha$ fits into the above setting with

$$\mathcal{K}_t(y, \cdot) = e_{t,\sharp}(\alpha^y \times \delta_y) \quad \text{and} \quad v_t^y = \frac{y-x}{1-t}, \quad t \in (0, 1)$$

and α_t has the first marginal $v_t = e_{t,\sharp}\alpha$ which is exactly μ_t from item 1.

3. **Curves via stochastic processes**: Starting with a time differentiable stochastic process $(X_t)_{t \in [0,1]}$, Liu et al. [29, 30] determined (μ_t, v_t) using the conditional expectation

$$\mu_t := P_{X_t} \quad \text{and} \quad v_t(x) := \mathbb{E}[\partial_t X_t | X_t = x].$$

For the special process $X_t := (1-t)X_0 + tX_1$, $t \in [0,1]$, this results again in a curve-velocity pair induced by the coupling from the joint distribution $\alpha = P_{X_0, X_1}$.

Section 7 deals with the actual flow matching, i.e., how we can learn the above vector field v_t by a neural network u_t^θ. Clearly, e.g. in the Markov kernel approach, the function

$$F(\theta) := \int_0^1 \int_{\mathbb{R}^d \times \mathbb{R}^d} \|u_t^\theta(x) - v_t(x)\|^2 \, dv_t(x) dt$$

would be a nice loss, but unfortunately we do not have access to v_t. On the other hand, we know the conditioned velocities v_t^y from (1). Fortunately, we will see that the vector field v of the curve-velocity pair (v_t, v_t) appears as minimizer of the functional

$$v = \underset{u}{\operatorname{argmin}} \int_0^1 \int_{\mathbb{R}^d \times \mathbb{R}^d} \|u_t(x) - v_t^y(x)\|^2 \, d\alpha_t(x, y) dt \tag{1}$$

or in other words the above loss F coincides up to a constant with the loss function

$$\operatorname{FM}_{\alpha_t}(\theta) := \int_0^1 \int_{\mathbb{R}^d \times \mathbb{R}^d} \|u_t^\theta(x) - v_t^y(x)\|^2 \, d\alpha_t(x, y) dt.$$

In the original paper [28], the notation "conditional flow matching" (CFM) was used instead of FM which refers to a "conditional flow path". We prefer FM, since we will later deal with Bayesian inverse problems, where the notation of being "conditional" is occupied in a different way. For the numerical simulation, we have on the one hand to sample from $\alpha_\ell = \mathcal{K}_\ell(y, \cdot) \times_y \mu_1$ and on the other hand to compute $v_\ell^y(x)$. Samples y_i, $i = 1, \ldots, n$ from the data distribution μ_1 are usually available and it should be easy to sample from the distributions $\mathcal{K}_t(y_i, \cdot)$, $i = 1, \ldots, n$. In the special case, where $(v_t, v_t) = (\mu_t, v_t)$ is induced by a coupling α, we have a simple formula for v_t^y for which the loss function becomes

$$\operatorname{FM}_\alpha(\theta) = \int_0^1 \int_{\mathbb{R}^d \times \mathbb{R}^d} \|u_t^\theta(e_t(x, y)) - (y - x)\|^2 \, d\alpha(x, y) dt. \tag{FM}$$

Similarly, for the approach via a stochastic process, we get

$$v = \underset{u}{\operatorname{argmin}} \int_0^1 \mathbb{E}\left[\|\partial_t X_t - u_t(X_t)\|^2\right] dt.$$

Again, for the special stochastic process $X_t = (1-t)X_0 + tX_1$ leading to a curve-velocity pair induced by the plan $\alpha = P_{X_0, X_1}$ this results in (FM). For the independent coupling $\alpha = \mu_0 \times \mu_1$ and the "optimal" one, this can be minimized numerically.

Let us start with the necessary notation agreements in the next section.

2 Preliminaries and Notation

Throughout this paper, we equip $\mathbb{R}^d$ with the σ-algebra of Borel sets $\mathcal{B}(\mathbb{R}^d)$, and by "measurable" sets/functions we always refer to "Borel measurable". Speaking about absolutely continuous measures on $\mathbb{R}^d$, we mean absolutely continuous with respect to the Lebesgue measure $\mathcal{L}$ on $\mathbb{R}^d$. By $\mathcal{L}_A$ we denote the Lebesgue measure restricted to the measurable set $A \subset \mathbb{R}^d$. By e_i, $i = 1, \ldots, d$ we denote the canonical basis elements of $\mathbb{R}^d$.

2.1 Special Metric Spaces

Let $C_b(\mathbb{R}^d)$ be the Banach space of bounded, continuous, real-valued functions with norm

$$\|\varphi\|_\infty := \sup_{x \in \mathbb{R}^d} |\varphi(x)|,$$

$C_0(\mathbb{R}^d)$ its closed subspace of continuous functions vanishing at infinity, and $C_c(\mathbb{R}^d)$ the space of continuous functions with compact support. Further, $C^l(\mathbb{R}^d)$, $l \geq 1$, denotes the space of l times continuously differentiable functions and $C^\infty(\mathbb{R}^d)$ the space of infinity often differentiable functions likewise combined with the above compact support property $C_c^\infty(\mathbb{R}^d)$. Instead of $\mathbb{R}^d$, functions may be also defined on time intervals $I \subset \mathbb{R}$ or direct products $I \times \mathbb{R}^d$ with the corresponding adaptation of the notation. We will also consider the vector-valued functions mapping into $\mathbb{R}^m$ and use the notation $C^l(\mathbb{R}^d, \mathbb{R}^m)$ here.

By $\mathcal{M}(\mathbb{R}^d)$, we denote the Banach space of finite Borel measures on $\mathbb{R}^d$ equipped with the *total variation norm*

$$\|\mu\|_{\mathrm{TV}} := \sup \left\{ \sum_{k=1}^\infty |\mu(B_k)| : \bigcup_{k=1}^\infty B_k = \mathbb{R}^d \right\}$$

for any pairwise disjoint Borel sets B_k, $k \in \mathbb{N}$. It is the dual space of $C_0(\mathbb{R}^d)$, in particular, every measure $\mu \in \mathcal{M}(\mathbb{R}^d)$ defines a continuous, linear functional on $C_0(\mathbb{R}^d)$ via

$$\varphi \mapsto \langle \mu, \varphi \rangle = \int_{\mathbb{R}^d} \varphi(x) \, d\mu(x)$$

and

$$\|\mu\|_{\mathcal{M}(\mathbb{R}^d)} = \sup_{\|\varphi\|_{C_0(\mathbb{R}^d)} \leq 1} |\langle \mu, \varphi \rangle|.$$

A sequence $(\mu_n)_n \subset \mathcal{M}(\mathbb{R}^d)$ *converges weak-** to a measure $\mu \in \mathcal{M}(\mathbb{R}^d)$, if

$$\lim_{n\to\infty} \int_{\mathbb{R}^d} \varphi \, d\mu_n = \int_{\mathbb{R}^d} \varphi \, d\mu \quad \text{for all} \quad \varphi \in C_0(\mathbb{R}^d).$$

We are mainly interested in the subset of probability measures

$$\mathcal{P}(\mathbb{R}^d) := \{\mu \in \mathcal{M}(\mathbb{R}^d) : \mu \geq 0, \; \mu(\mathbb{R}^d) = 1\}.$$

Unfortunately, a sequence of probability measures $(\mu_n)_n$ may not converge in the weak-* sense to a probability measure. For example, $\mu_n := \delta_n$, $n \in \mathbb{N}$, converges weak-* towards $\mu \equiv 0$ which is not a probability measure. Therefore, the concept of narrow convergence[1] was introduced: a sequence $(\mu_n)_n \subset \mathcal{M}(\mathbb{R}^d)$ *converges narrowly* to a measure $\mu \in \mathcal{M}(\mathbb{R}^d)$, written $\mu_n \rightharpoonup \mu$ as $n \to \infty$, if

$$\lim_{n\to\infty} \int_{\mathbb{R}^d} \varphi \, d\mu_n = \int_{\mathbb{R}^d} \varphi \, d\mu \quad \text{for all} \quad \varphi \in C_b(\mathbb{R}^d).$$

If $\mu_n \in \mathcal{P}(\mathbb{R}^d)$ with $\mu_n \rightharpoonup \mu$ as $n \to \infty$, then also $\mu \in \mathcal{P}(\mathbb{R}^d)$. However, note that $\mathcal{M}(\mathbb{R}^d)$ is not the dual space of $C_b(\mathbb{R}^d)$, which is indeed the space $\mathrm{rba}(\mathbb{R}^d)$ of finitely additive set functions. For more information, see [34, Section 4.4].

Remark 2.1 (Test Functions) The interpretation of $\mathcal{M}(\mathbb{R}^d)$ as the dual of $C_0(\mathbb{R}^d)$ implies that $\mu = \nu$ in $\mathcal{M}(\mathbb{R}^d)$ if and only if

$$\int_{\mathbb{R}^d} \varphi \, d\mu = \int_{\mathbb{R}^d} \varphi \, d\nu \quad \text{for all} \quad \varphi \in C_0(\mathbb{R}^d).$$

Since $C_c^\infty(\mathbb{R}^d)$ is dense in $C_0(\mathbb{R}^d)$ with respect to the $\|\cdot\|_\infty$ norm, we could also just test against all $\varphi \in C_c^\infty(\mathbb{R}^d)$. ◇

For $\mu \in \mathcal{P}(\mathbb{R}^d)$, let $L^p(\mathbb{R}^m, \mu)$, $p \in [1, \infty)$, denote the Banach space of (equivalence classes of) μ-measurable functions $f : \mathbb{R}^d \to \mathbb{R}^m$ with

$$\|f\|_{L^p(\mathbb{R}^m, \mu)} := \left(\int_{\mathbb{R}^d} \|f\|^p \, d\mu \right)^{\frac{1}{p}} < \infty,$$

where $\|\cdot\|$ is the Euclidean norm on $\mathbb{R}^m$. In case of the Lebesgue measure $\mu = \mathcal{L}$, we will skip the μ in the notation. Furthermore, we write L^1 for $L^1(\mathbb{R}, \mathcal{L})$.

---

[1] In measure theory, narrow convergence of measures is also called weak convergence. We do not use the later notation to avoid confusion with the notation of weak convergence in functional analysis.

2.2 Push-Forward Operator

The *push-forward measure* of $\mu \in \mathcal{M}(\mathbb{R}^d)$ by a measurable map $T : \mathbb{R}^d \to \mathbb{R}^n$ is defined by

$$T_\#\mu = \mu \circ T^{-1}$$

with the preimage $T^{-1}(B)$ of $B \in \mathcal{B}(\mathbb{R}^n)$.

Then, a function $f : \mathbb{R}^n \to \mathbb{R}$ is integrable with respect to $T_\#\mu$ if $f \circ T$ is integrable with respect to μ and

$$\int_{\mathbb{R}^n} f(y) \, dT_\#\mu(y) = \int_{T^{-1}(\mathbb{R}^n)} f(T(x)) \, d\mu(x).$$

If $T : \mathbb{R}^d \to \mathbb{R}^d$ is a C^1 diffeomorphism and μ is absolutely continuous with density p_μ, then we have by the *transformation theorem*, also known as *change-of-variable formula* for all measurable, bounded functions f that

$$\int_{\mathbb{R}^d} f(y) \, dT_\#\mu(y) = \int_{\mathbb{R}^d} f(T(x)) \, p_\mu(x) \, dx = \int_{\mathbb{R}^d} f(y) \, p_\mu(T^{-1}(y)) |\det(\nabla T^{-1}(y))| \, dy, \tag{2}$$

where ∇T denotes the Jacobian of T. In particular, $T_\#\mu$ has the density

$$p_{T_\#\mu}(y) = p(T^{-1}(y)) |\det(\nabla T^{-1}(y))| = p(T^{-1}(y)) / |\det(\nabla T(y))|. \tag{3}$$

Recall that the convolution of measures $\mu, \nu \in \mathcal{P}(\mathbb{R}^d)$ is defined to be the measure $\mu * \nu \in \mathcal{P}(\mathbb{R}^d)$ which fulfills

$$\int_{\mathbb{R}^d} \varphi(x) \, d(\mu * \nu)(x) = \int_{\mathbb{R}^d \times \mathbb{R}^d} \varphi(x+y) \, d\mu(x) d\nu(y)$$

for all $\varphi \in C_b(\mathbb{R}^d)$. If μ, ν have densities $p_\mu, p_\nu \in L^1$, then $\mu * \nu$ has the density $p_{\mu*\nu} \in L^1$ given by

$$p_{\mu*\nu} = \int_{\mathbb{R}^d} p_\mu(x) p_\nu(\cdot - x) \, dx.$$

Measures can be characterized via laws of random variables.

Remark 2.2 (Random Variables) Let $(\Omega, \Sigma, \mathbb{P})$ be a probability space. Recall that a random variable $X: \Omega \to \mathbb{R}^d$ is a measurable map $X: \Omega \to \mathbb{R}^d$ and the push-forward measure

$$P_X := X_\# \mathbb{P} = \mathbb{P} \circ X^{-1}$$

is known as law of X. Note that different random variables can have the same law. For every measure $\mu \in \mathcal{P}_2(\mathbb{R}^d)$ there exists a random variable with $P_X = \mu$. If $X_0, X_1 : \Omega \to \mathbb{R}^d$ are independent random variables, then

(i) $(X_0, X_1) : \Omega \to \mathbb{R}^d \times \mathbb{R}^d$ has law $P_{X_0, X_1} = P_{X_0} \times P_{X_1}$, and
(ii) $Z := a_0 X_0 + a_1 X_1 : \Omega \to \mathbb{R}^d$, $a_0, a_1 \neq 0$ has law $P_Z = a_{0,\sharp} P_{X_0} * a_{1,\sharp} P_{X_1}$, where we abbreviated a_i Id by a_i in the last formula. If P_{X_i} has a density p_{X_i}, then $a_{i,\sharp} P_{X_i}$ has the density $p_{X_i}(a_i^{-1} \cdot)$, $i = 0, 1$, and Z has the density

$$p_Z = \int_{\mathbb{R}^d} p_{X_0}(a_0^{-1} x) p_{X_1}\left(a_2^{-1}(\cdot - x)\right) \, dx. \qquad \diamond$$

For comparing measures, in particular $T_\sharp P_{\text{latent}}$ and P_{data}, divergences between measures can be used. A divergence has the main property that it achieves a minimum exactly if both measures coincide. One of the most frequently used divergence is the Kullback-Leibler one.

Remark 2.3 (Kullback-Leibler Divergence) The *Kullback-Leibler (KL) divergence* $\mathrm{KL} : \mathcal{P}(\mathbb{R}^d) \times \mathcal{P}(\mathbb{R}^d) \to [0, +\infty]$ of two measures $\mu, \nu \in \mathcal{P}(\mathbb{R}^d)$ with existing Radon-Nikodym derivative $\frac{d\mu}{d\nu}$ of μ with respect to ν is defined by

$$\mathrm{KL}(\mu, \nu) := \int_{\mathbb{R}^d} \log\left(\frac{d\mu}{d\nu}\right) d\mu.$$

If the Radon-Nikodym derivative does not exist, we set $\mathrm{KL}(\mu, \nu) := +\infty$. If μ, ν have densities p_μ, p_ν, where $p_\nu(x) = 0$ implies $p_\mu(x) = 0$, then

$$\mathrm{KL}(\mu, \nu) := \int_{\mathbb{R}^d} \log\left(\frac{p_\mu}{p_\nu}\right) p_\mu \, dx$$
$$= \mathbb{E}_{x \sim p_\mu}[\log p_\mu] - \mathbb{E}_{x \sim p_\mu}[\log p_\nu].$$

The Kullback-Leibler divergence is a so-called Bregman distance related to the Shannon entropy. It is not a distance, since it is neither symmetric nor fulfills a triangular inequality. However, it holds $\mathrm{KL}(\mu, \nu) \geq 0$ for all $\mu, \nu \in \mathcal{P}(\mathbb{R}^d)$ with equality if and only if the measures $\nu = \mu$ coincide.

If the latent distribution P_latent has a density p_latent, and $T: \mathbb{R}^d \to \mathbb{R}^d$ is a C^1 diffeomorphism, then by (2) also $T_\sharp P_\text{latent}$ has a density which we denote by $T_\sharp p_\text{latent}$. Thus, if the data distribution admits a density p_data, then we get

$$\mathrm{KL}(P_\text{data}, T_\sharp P_\text{latent}) = \underbrace{\mathbb{E}_{x \sim p_\text{data}}[\log p_\text{data}]}_{\text{constant}} - \mathbb{E}_{x \sim p_\text{data}}[\log T_\sharp p_\text{latent}].$$

◇

2.3 Disintegration and Markov Kernels

In this paper, we consider projections $\pi^i : \mathbb{R}^d \times \mathbb{R}^d \to \mathbb{R}^d$, $i = 1, 2$, defined by

$$\pi^1(x, y) := x, \quad \pi^2(x, y) := y.$$

Then, for a measure $\alpha \in \mathcal{P}(\mathbb{R}^d \times \mathbb{R}^d)$, we have that $\pi^1_\sharp \alpha$ and $\pi^2_\sharp \alpha$ are the left and right marginals of α, respectively.

For a measure $\alpha \in \mathcal{P}(\mathbb{R}^d \times \mathbb{R}^d)$ with marginal $\pi^1_\sharp \alpha = \mu$, there exists a μ-a.e. uniquely defined family of probability measures $\{\alpha^x\}_x$, called *disintegration of α with respect to π^1*, such that the map $x \mapsto \alpha^x(B)$ is measurable for every $B \in \mathcal{B}(\mathbb{R}^d)$, and

$$\alpha = \alpha^x \times_x \mu$$

meaning that

$$\int_{\mathbb{R}^d \times \mathbb{R}^d} f(x, y) \, d\alpha(x, y) = \int_{\mathbb{R}^d} \int_{\mathbb{R}^d} f(x, y) \, d\alpha^x(y) d\mu(x)$$

for every measurable, bounded function $f : \mathbb{R}^d \times \mathbb{R}^d \to \mathbb{R}$. For an illustration, see Fig. 2. Similarly, we define for a measure $\alpha \in \mathcal{P}(\mathbb{R}^d \times \mathbb{R}^d)$ with marginal $\pi^2_\sharp \alpha = \nu$ the *disintegration of α with respect to π^2* as

$$\alpha = \alpha^y \times_y \nu.$$

The notation of disintegration is directly related to Markov kernels. A *Markov kernel* is a map $\mathcal{K} : \mathbb{R}^d \times \mathcal{B}(\mathbb{R}^d) \to \mathbb{R}$ such that

(i) $\mathcal{K}(x, \cdot)$ is a probability measure on $\mathbb{R}^d$ for every $x \in \mathbb{R}^d$, and
(ii) $\mathcal{K}(\cdot, B)$ is a Borel measurable map for every $B \in \mathcal{B}(\mathbb{R}^d)$.

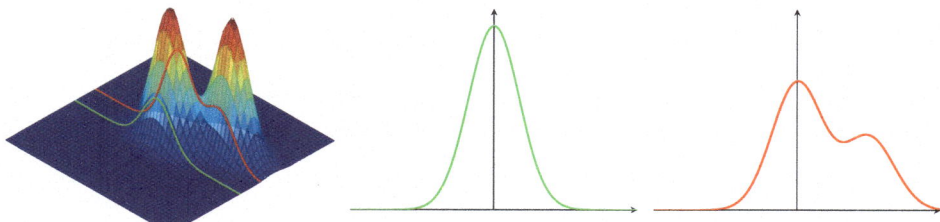

Fig. 2 Disintegration of the measure $\alpha \in \mathcal{P}(\mathbb{R} \times \mathbb{R})$ (left). Measures $\alpha^{-0.3} \in \mathcal{P}(\mathbb{R})$ (middle, green) and $\alpha^{0.2} \in \mathcal{P}(\mathbb{R})$ (right, red)

Hence, given a probability measure $\mu \in \mathcal{P}(\mathbb{R}^d)$, we can define a new measure $\alpha := \alpha^x \times_x \mu \in \mathcal{P}(\mathbb{R}^d \times \mathbb{R}^d)$ by

$$\int_{\mathbb{R}^d \times \mathbb{R}^d} f(x, y) \, d\alpha(x, y) := \int_{\mathbb{R}^d} \int_{\mathbb{R}^d} f(x, y) \, d\mathcal{K}(x, \cdot)(y) d\mu(x)$$

for all measurable, bounded functions f. Identifying $\alpha^x(B)$ with $\mathcal{K}(x, B)$, we see that conversely, $\{\alpha^x\}_x$ is the disintegration of α with respect to π^1.

Remark 2.4 (Random Variables) Let $X_0, X_1 : \Omega \to \mathbb{R}^d$ be random variables with joint distribution P_{X_0, X_1}. Then the conditional distribution $\{P_{X_1|X_0=x}\}_x$ provides the disintegration of P_{X_0, X_1}, i.e.

$$\int_{\mathbb{R}^d \times \mathbb{R}^d} f(x, y) \, dP_{X_0, X_1}(x, y) = \int_{\mathbb{R}^d} \int_{\mathbb{R}^d} f(x, y) \, dP_{X_1|X_0=x}(y) dP_{X_0}(x).$$

In other words, $\mathcal{K} = P_{X_1|X_0}$ is a Markov kernel. If P_{X_0, X_1} admits a density p_{X_0, X_1} and P_{X_0} a density $p_{X_0} > 0$, then $P_{X_1|X_0=x}$ has the density $p_{X_1|X_0=x} = p_{X_0, X_1}(x, \cdot)/p_{X_0}(x)$. ◇

Example 2.5 For $\mathcal{X} := \{x_i \in \mathbb{R}^d : i = 1, \ldots, n\}$ and $\mathcal{Y} := \{y_j \in \mathbb{R}^d, j = 1, \ldots, m\}$, the discrete probability measure

$$\alpha := \sum_{i,j=1}^{n,m} \alpha_{i,j} \delta_{(x_i, y_j)},$$

has first and second marginals

$$\mu = \pi^1_\sharp \alpha = \sum_{i=1}^{n} \Big(\sum_{j=1}^{m} \alpha_{i,j} \Big) \delta_{x_i} = \sum_{i=1}^{n} \mu_i \delta_{x_i},$$

	ν_1	...	ν_j	...	ν_m		ν_1/μ		ν_j/μ		ν_m/μ
μ_1	$\alpha_{1,1}$	...	$\alpha_{1,j}$	...	$\alpha_{1,m}$	1	$\frac{\alpha_{1,1}}{\mu_1}$	...	$\frac{\alpha_{1,j}}{\mu_1}$	...	$\frac{\alpha_{1,m}}{\mu_1}$
$\vdots$			$\vdots$		$\vdots$	$\vdots$					
μ_i	$\alpha_{i,1}$	...	$\alpha_{i,j}$	...	$\alpha_{i,m}$	1	$\frac{\alpha_{i,1}}{\mu_i}$	...	$\frac{\alpha_{i,j}}{\mu_i}$	...	$\frac{\alpha_{i,m}}{\mu_i}$
$\vdots$			$\vdots$		$\vdots$	$\vdots$			$\vdots$		$\vdots$
μ_n	$\alpha_{n,1}$	...	$\alpha_{n,j}$	...	$\alpha_{n,m}$	1	$\frac{\alpha_{n,1}}{\mu_n}$	...	$\frac{\alpha_{n,j}}{\mu_n}$	...	$\frac{\alpha_{n,m}}{\mu_n}$

Fig. 3 Plan/Coupling of two discrete measures μ and ν (left) and Markov kernel/disintegration (right) with row and column sums

$$\nu = \pi_\sharp^2 \alpha = \sum_{j=1}^m \Big(\sum_{i=1}^n \alpha_{i,j}\Big) \delta_{y_j} = \sum_{j=1}^m \nu_j \delta_{y_j},$$

see Fig. 3. The disintegration of α with respect to π^1 is given by the family of measures

$$\alpha^{x_i} = \mathcal{K}(x_i, \cdot) = \sum_{j=1}^m \frac{\alpha_{i,j}}{\mu_i} \delta_{y_j} = \sum_{j=1}^m k_{i,j} \delta_{y_j},$$

$$\alpha^{x_i}(B) = \mathcal{K}(x_i, B) = \sum_{y_j \in B} k_{i,j}$$

for all $B \in \mathcal{B}(\mathbb{R}^d)$. The Markov kernel fulfills

$$\sum_{j=1}^m k_{i,j} = 1, \quad \sum_{i=1}^n k_{i,j} \mu_i = \nu_j.$$

In other words, with the vector $\mathbf{1}_m$ consisting of m entries one, and $\boldsymbol{\mu} := (\mu_i)_{i=1}^n$, $\boldsymbol{\nu} := (\nu_j)_{j=1}^m$, as well as $\boldsymbol{K} := (k_{i,j})_{i,j=1}^{n,m}$, we have

$$\boldsymbol{K} \mathbf{1}_n = \mathbf{1}_m, \quad \boldsymbol{K}^{\mathrm{T}} \boldsymbol{\mu} = \boldsymbol{\nu}$$

meaning that $\boldsymbol{K}^{\mathrm{T}}$ is a stochastic matrix and a "transfer matrix" from μ to ν. ◇

2.4 Couplings and Wasserstein Distance

Let

$$\mathcal{P}_2(\mathbb{R}^d) := \{\mu \in \mathcal{P}(\mathbb{R}^d) : \int_{\mathbb{R}^d} \|x\|^2 \, d\mu < \infty\}$$

be the probability measures with finite second moments. For $\mu, \nu \in \mathcal{P}_2(\mathbb{R}^d)$, we define the set of *plans* or *couplings* with marginals μ and ν by

$$\Gamma(\mu, \nu) := \left\{ \alpha \in \mathcal{P}(\mathbb{R}^d \times \mathbb{R}^d) : \pi^1_\sharp \alpha = \mu, \ \pi^2_\sharp \alpha = \nu \right\}.$$

The set $\Gamma(\mu, \nu)$ is a compact subset of $\mathcal{P}_2(\mathbb{R}^d \times \mathbb{R}^d)$ with respect to the narrow topology. The *Wasserstein(-2) distance* is defined by

$$W_2(\mu, \nu) := \min_{\alpha \in \Gamma(\mu, \nu)} \|x - y\|_{L^2(\mathbb{R}^d \times \mathbb{R}^d, \alpha)} = \min_{\alpha \in \Gamma(\mu, \nu)} \left(\int_{\mathbb{R}^d \times \mathbb{R}^d} \|x - y\|^2 \, d\alpha(x, y) \right)^{\frac{1}{2}}. \tag{4}$$

Indeed, the above minimum is attained, see, e.g. [36, Theorem 1.7], and the *set of optimal plans* is denoted by $\Gamma_o(\mu, \nu)$. The space $(\mathcal{P}_2(\mathbb{R}^d), W_2)$ is a complete metric space. Speaking about $\mathcal{P}_2(\mathbb{R}^d)$, we will always equip it with the W_2 metric. Then we have for a sequence $(\mu_n)_n$ of measures in $\mathcal{P}_2(\mathbb{R}^d)$ that

$$\mu_n \to \mu \quad \text{if and only if} \quad \mu_n \rightharpoonup \mu \quad \text{and} \quad \lim_{n \to \infty} \int_{\mathbb{R}^d} \|x\|^2 \, d\mu_n = \int_{\mathbb{R}^d} \|x\|^2 \, d\mu.$$

In general, the minimizer in (4) is not unique. However, if μ is absolutely continuous, then uniqueness is ensured by the following theorem.

Theorem 2.6 (Brenier) *Let $\mu \in \mathcal{P}_2(\mathbb{R}^d)$ be absolutely continuous and $\nu \in \mathcal{P}_2(\mathbb{R}^d)$. Then there is a unique plan $\alpha \in \Gamma_o(\mu, \nu)$, which is induced by a unique measurable optimal transport map, also called Monge map, $T : \mathbb{R}^d \to \mathbb{R}^d$, i.e.,*

$$\alpha = (\mathrm{Id}, T)_\sharp \mu$$

and

$$W_2^2(\mu, \nu) = \min_{T: \mathbb{R}^d \to \mathbb{R}^d} \int_{\mathbb{R}^d} \|x - T(x)\|_2^2 \, d\mu(x) \quad \text{subject to} \quad T_\sharp \mu = \nu.$$

Further, $T = \nabla \psi$, where $\psi : \mathbb{R}^d \to (-\infty, +\infty]$ is convex, lower semi-continuous (lsc) and μ-a.e. differentiable. Conversely, if ψ is convex, lsc and μ-a.e. differentiable with $\nabla \psi \in L_2(\mu, \mathbb{R}^d)$, then $T := \nabla \psi$ is an optimal map from μ to $\nu := T_\sharp \mu \in \mathcal{P}_2(\mathbb{R}^d)$.

The transport map between Gaussian distributions can be given analytically.

Example 2.7 For two Gaussians $\mu = \mathcal{N}(m_\mu, \Sigma_\mu)$ and $\nu = \mathcal{N}(m_\nu, \Sigma_\nu)$, where $\mathcal{N}(m, \Sigma)$ has the density

$$p(x) := (2\pi)^{-\frac{d}{2}} (\det \Sigma)^{-\frac{1}{2}} e^{-\frac{1}{2}(x-m)^{\mathsf{T}} \Sigma^{-1}(x-m)},$$

the transport map is given by

$$T(x) = m_\nu + A(x - m_\mu), \quad A := \Sigma_\mu^{-\frac{1}{2}} \left(\Sigma_\mu^{\frac{1}{2}} \Sigma_\nu \Sigma_\mu^{\frac{1}{2}} \right) \Sigma_\mu^{-\frac{1}{2}}, \tag{5}$$

see e.g. [33, Remark 2.31]. ◇

3 Absolutely Continuous Curves in $(\mathcal{P}_2(\mathbb{R}^d), W_2)$

In the rest of this paper, let $I := [a, b]$, $a < b$. The main player in this paper are curves

$$\mu_t : I \to \mathcal{P}_2(\mathbb{R}^d), \quad t \mapsto \mu_t.$$

Unfortunately, there is an ambiguity here which is usual in the literature, namely that the curve itself as well as the value of the curve at time $t \in I$ is denoted by μ_t. Usually it becomes clear from the context what is meant. We are only interested in *narrowly continuous curves*, meaning that for $t \to t'$ we have $\mu_t \rightharpoonup \mu_{t'}$. Indeed, to make the definition of integrals

$$\int_I \int_{\mathbb{R}^d} f(t, x) \, d\mu_t \, dt$$

meaningful for any measurable, bounded function f, we have to consider $\mu_t : I \times \mathcal{B}(\mathbb{R}^d) \to \mathbb{R}$ as a Markov kernel, which is possible by the following theorem whose proof is given in Appendix A.

Theorem 3.1 *Let $\mu_t : I \to \mathcal{P}_2(\mathbb{R}^d)$ be a narrowly continuous curve. Then, for every Borel set $B \subseteq \mathbb{R}^d$, we have that $t \mapsto \mu_t(B)$ is measurable, i.e., $\mu_t : I \times \mathcal{B}(\mathbb{R}^d) \to \mathbb{R}$ is a Markov kernel.*

Recall that in a complete metric space (X, d), a curve $\gamma : I \to X$ is called *absolutely continuous*, if there exists as $m \in L^1([a, b])$ such that

$$d(\gamma(s), \gamma(t)) \leq \int_s^t m(r) \, dr \quad \text{for all} \quad a \leq s \leq t \leq b. \tag{6}$$

Here $L^1([a, b])$ denotes the space of (equivalence classes of) absolutely integrable real-valued function defined on an interval I. Then the *metric derivative* of γ defined by

$$|\gamma'(t)| := \lim_{h\to 0} \frac{d(\gamma(t), \gamma(t+h))}{|h|}$$

exists a.e. and is the smallest function m in (6). The space of absolutely continuous curves is denoted by $AC^1(X, d)$. If we replace in the above definition $m \in L^p(I)$, $1 \le p < \infty$, we obtain the spaces $AC^p(X, d)$.

For our special space $\mathcal{P}_2(\mathbb{R}^d)$ with $d = W_2$, absolute continuity of a curve can be characterized by a continuity equation. To this end, let $\mu_t : I \to \mathcal{P}_2(\mathbb{R}^d)$ be a narrowly continuous curve and $v : I \times \mathbb{R}^d \to \mathbb{R}^d$ be a measurable vector field. For $v(t, \cdot) : \mathbb{R}^d \to \mathbb{R}^d$ we alternatively write v_t. We say that the curve-velocity pair (μ_t, v_t) satisfies the *continuity equation*

$$\partial_t \mu_t + \nabla_x \cdot (\mu_t v_t) = 0 \tag{CE}$$

in the sense of distributions, if

$$\int_I \int_{\mathbb{R}^d} \partial_t \varphi + \langle \nabla_x \varphi, v_t \rangle \, d\mu_t \, dt = 0$$

for all $\varphi \in C_c^\infty((a, b) \times \mathbb{R}^d)$. Here ∇_x is the gradient with respect to x for $\varphi = \varphi(t, x)$.

Remark 3.2 The term "in the sense of distributions" for the (CE) has the following meaning: Consider the measure $\mu = \mu_t \times_t dt$. Then $\partial_t \mu$ is the distribution

$$\partial_t \mu(\varphi) = -\mu(\partial_t \varphi) = -\int_I \int_{\mathbb{R}^d} \partial_t \varphi \, d\mu, \quad \varphi \in C_c^\infty((a, b) \times \mathbb{R}^d).$$

Further $\mu v(\varphi) = \mu(v\varphi)$ and

$$\mathrm{div}(\mu v)(\varphi) = \sum_{i=1}^d \partial_i (\mu v_i)(\varphi) = -\sum_{i=1}^d (\mu v_i)(\partial_i \varphi) = -\sum_{i=1}^d \mu(v_i \partial_i \varphi)$$

$$= -\mu(\langle \nabla_x \varphi, v \rangle) = -\int_I \int_{\mathbb{R}^d} \langle \nabla_x \varphi, v \rangle \, d\mu.$$

Thus, the continuity equation can be seen as an equation of distributions. ◇

Now we have the following fundamental theorem.

Theorem 3.3 ([1, Theorem 8.3.1]) *Let $\mu_t : I \to \mathcal{P}_2(\mathbb{R}^d)$ be a narrowly continuous curve. Then μ_t is absolutely continuous if and only if there exists a Borel measurable vector field $v : I \times \mathbb{R}^d \to \mathbb{R}^d$ such that the following two conditions are fulfilled:*

(i) $\|v_t\|_{L^2(\mathbb{R}^d,\mu_t)} \in L^1(I)$,
(ii) (μ_t, v_t) *fulfills* (CE).

In this case, the metric derivative $|\mu'_t|$ exists and we have $|\mu'_t| \leq \|v_t\|_{L^2(\mathbb{R}^d,\mu_t)}$ for a.e. $t \in I$.

We call a measurable vector field $v : I \times \mathbb{R}^d \to \mathbb{R}^d$ *associated to an absolutely continuous curve* $\mu_t : I \to \mathcal{P}_2(\mathbb{R}^d)$, if (μ_t, v_t) fulfill (CE) and $\|v_t\|_{L^2(\mathbb{R}^d,\mu_t)} \in L^1(I)$.

Indeed, an absolutely continuous curve may admit different associated velocity fields. For an illustration, see Example 4.9 and Fig. 5. By the following remark, a unique associated vector field is characterized by having minimal $L^2(\mathbb{R}^d, \mu_t)$-norm for a.e. $t \in I$, or equivalently, by being in the tangent space of μ_t.

Remark 3.4 (Minimal Velocity Fields of Absolutely Continuous Curves) Let $v_t, \tilde{v}_t$ be associated vector fields of an absolutely continuous curve μ_t with minimal norm $\|v_t\|_{L^2(\mathbb{R}^d,\mu_t)} = \|\tilde{v}_t\|_{L^2(\mathbb{R}^d,\mu_t)} =: z_t$ for a.e. $t \in I$ among all associated vector fields of μ_t. By the linear structure of (CE), also $\frac{v_t + \tilde{v}_t}{2}$ is associated to μ_t. Assume that $v_t \neq \tilde{v}_t$. By the strict convexity of $L^2(\mathbb{R}^d, \mu_t)$, it follows the contradiction $\|\frac{v_t+\tilde{v}_t}{2}\|_{L^2(\mathbb{R}^d,\mu_t)} < z_t$. Hence, we have the uniqueness $v_t = \tilde{v}_t$.

For an equivalent description, consider the *regular tangent space* $\mathcal{T}_\mu = T_\mu \mathcal{P}_2(\mathbb{R}^d)$ at $\mu \in \mathcal{P}_2(\mathbb{R}^d)$ which is defined by

$$\mathcal{T}_\mu := \overline{\{\nabla \phi : \phi \in C_c^\infty(\mathbb{R}^d)\}}^{L^2(\mathbb{R}^d,\mu)} \qquad (7)$$

$$= \overline{\{\lambda(T - \mathrm{Id}) : (\mathrm{Id}, T)_\# \mu \in \Gamma_o(\mu, T_\# \mu), \lambda > 0\}}^{L^2(\mathbb{R}^d,\mu)},$$

see [1, § 8]. The second description can be interpreted as locally moving mass in an "optimal way". Note that $\mathcal{T}_\mu$ is an infinite-dimensional subspace of $L^2(\mathbb{R}^d, \mu)$ if μ is absolutely continuous, and it is just $\mathbb{R}^d$ if $\mu = \delta_x$, $x \in \mathbb{R}^d$. There is another description of $\mathcal{T}_\mu$, namely, for a vector field $v \in L^2(\mathbb{R}^d, \mu)$, we have

$$v \in \mathcal{T}_\mu \iff \int_{\mathbb{R}^d} \langle w, v \rangle \, \mathrm{d}\mu = 0 \text{ for all } w \in L^2(\mathbb{R}^d, \mu) \text{ with } \nabla \cdot (w\mu) = 0, \qquad (8)$$

$$\iff \|v + w\|_{L^2(\mathbb{R}^d,\mu)} \geq \|v\|_{L^2(\mathbb{R}^d,\mu)} \text{ for all } w \in L^2(\mathbb{R}^d, \mu) \text{ with } \nabla \cdot (w\mu) = 0,$$

where the last equation is meant again in the distributional sense $\int_{\mathbb{R}^d} \langle w, \nabla\varphi \rangle \, \mathrm{d}\mu = 0$ for all $\varphi \in C_c^\infty(\mathbb{R}^d)$, see [1, Lemma 8.4.2].

Now let $v_t \in \mathcal{T}_{\mu_t}$ be a vector field associated to μ_t. For any other vector field $\tilde{v}_t$ associated to μ_t, it holds $\nabla \cdot ((v_t - \tilde{v}_t)\mu) = 0$, and hence by (8), $\|\tilde{v}_t\|_{L^2(\mathbb{R}^d,\mu)} \geq \|v_t\|_{L^2(\mathbb{R}^d,\mu)}$, meaning that v_t has minimal $L^2(\mathbb{R}^d, \mu_t)$-norm. On the other hand, let v_t be associated to μ_t having minimal norm. By [1, Theorem 8.3.1], there *exists* a vector field $\tilde{v}_t \in \mathcal{T}_{\mu_t}$ being associated to μ_t and having minimal norm. By the uniqueness shown above, it follows $v_t = \tilde{v}_t \in \mathcal{T}_{\mu_t}$.

Therefore, we have proven that the associated velocity field v_t of an absolutely continuous curve μ_t is unique if we require that it has minimal $L^2(\mathbb{R}^d, \mu_t)$-norm, or equivalently, lies in the tangent space $\mathcal{T}_{\mu_t}$. ◇

The following theorem connects absolutely continuous curves with flow ODEs.

Theorem 3.5 ([1, Theorem 8.1.8]) *Let $\mu_t : I \to \mathcal{P}_2(\mathbb{R}^d)$ be an absolutely continuous curve with associated vector field v_t such that for every compact Borel set $B \subset \mathbb{R}^d$ it holds*

$$\int_I \sup_{x \in B} \|v_t(x)\| + \mathrm{Lip}(v_t, B) \mathrm{d}t < \infty.$$

Then there exists a solution $\phi : I \times \mathbb{R}^d \to \mathbb{R}^d$ of the ODE

$$\partial_t \phi(t, x) = v_t(\phi(t, x)), \quad \phi(0, x) = x \tag{9}$$

and $\phi(t, \cdot)_\sharp \mu_0 = \mu_t$.

Theorem 3.5 indicates why absolutely continuous curves can be used as a tool for sampling. A common practice considers such curves starting in $\mu_0 := \mathcal{N}(0, 1)$ towards a target measure $\mu_1 = P_{\mathrm{data}}$ and try to approximate the vector field v_t by a neural network v_t^θ for trainable parameters θ. To sample from μ_1, we then can just sample from $\mu_0 = \mathcal{N}(0, 1)$ and solve the ODE (9) for these samples.

Remark 3.6 Also the other direction in Theorem 3.5 is true under some conditions. To see this, assume that ϕ is measurable and satisfies (9) for some locally bounded vector field v_t. For $\varphi \in C_c^\infty((a, b) \times \mathbb{R}^d)$, we can compute

$$0 = \varphi(b, \phi(b, x)) - \varphi(a, \phi(a, x)) = \int_{\mathbb{R}^d} \varphi(b, \phi(b, x)) - \varphi(a, \phi(a, x)) \mathrm{d}\mu_0$$

$$= \int_{\mathbb{R}^d} \int_a^b \frac{\mathrm{d}}{\mathrm{d}t}(\varphi(t, \phi(t, x))) \, \mathrm{d}t \mathrm{d}\mu_0$$

$$= \int_{\mathbb{R}^d} \int_a^b \langle \nabla_x \varphi(t, \phi(t, x)), v_t(t, \phi(t, x)) \rangle + (\partial_t \varphi)(t, \phi(t, x)) \, \mathrm{d}t \mathrm{d}\mu_0$$

$$= \int_a^b \int_{\mathbb{R}^d} \langle \nabla_x \varphi, v_t \rangle + \partial_t \varphi \, \mathrm{d}[\phi(t, \cdot)_\sharp \mu_0] \mathrm{d}t$$

and thus $\mu_t := \phi(t, \cdot)_\sharp \mu_0$ satisfies the continuity equation (in a weak sense). ◇

In the following Sects. 4–6 we show different methods for the construction of curve-velocity pairs fulfilling the conditions of Theorem 3.3, i.e. providing absolutely continuous curves in the Wasserstein geometry. Having (9) in mind, this will lead to a sampling procedure.

4 Curves Induced by Couplings

We start with curves induced by couplings $\alpha \in \Gamma(\mu_0, \mu_1)$. There are three important kinds of couplings:

- optimal couplings $\alpha \in \Gamma_o(\mu_0, \mu_1)$, since they correspond to geodesics in $\mathcal{P}_2(\mathbb{R}^d)$.
- couplings arising from maps, since their induced curves and vector fields are particularly simple, as we will see later.
- product couplings $\mu_0 \times \mu_1$, since we can easily sample from them by sampling from μ_0 and μ_1 and no extra computation is needed.

Using couplings it is also easy to see, that the space $\mathcal{P}_2(\mathbb{R}^d)$ is *path connected*, i.e., any two measures can be connected by a continuous curve (with respect to W_2). Even more, $\mathcal{P}_2(\mathbb{R}^d)$ is a *geodesic space*, meaning that any two measures can be connected by a geodesic. Recall that a curve $\mu_t : [0, 1] \to \mathcal{P}_2(\mathbb{R}^d)$ is called a (constant speed) *geodesic* if

$$W_2(\mu_s, \mu_t) = (t - s) W_2(\mu_0, \mu_1) \quad \text{for all} \quad 0 \leq s \leq t \leq 1.$$

Let $e_t : \mathbb{R}^d \times \mathbb{R}^d \to \mathbb{R}^d$ be defined by

$$e_t(x, y) := (1 - t)x + ty, \quad t \in [0, 1].$$

Let $\mu_0, \mu_1 \in \mathcal{P}_2(\mathbb{R}^d)$ and $\alpha \in \Gamma(\mu_0, \mu_1)$. We call a *curve* $\mu_t : [0, 1] \to \mathcal{P}_2(\mathbb{R}^d)$ *induced by the coupling* or *plan* α, if

$$\mu_t := e_{t,\sharp}\alpha, \tag{10}$$

i.e., by definition of the push-forward measure,

$$\int_{\mathbb{R}^d} f \, \mathrm{d}\mu_t = \int_{\mathbb{R}^d \times \mathbb{R}^d} f\bigl(e_t(x, y)\bigr) \, \mathrm{d}\alpha = \int_{\mathbb{R}^d \times \mathbb{R}^d} f\bigl((1 - t)x + ty\bigr) \, \mathrm{d}\alpha$$

for every measurable, bounded function $f : \mathbb{R}^d \to \mathbb{R}$.

Flow Matching: Markov Kernels, Stochastic Processes and Transport Plans

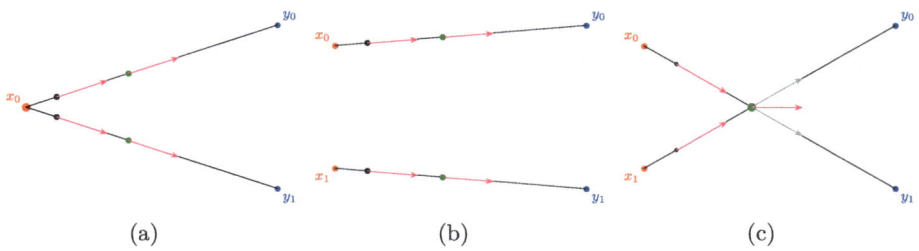

(a)   (b)   (c)

Fig. 4 Curve induced by $\alpha = (\mathrm{Id}, T)_\sharp \mu_0$ from $\mu_0 = \delta_{x_0}$, resp. $\mu_0 = \frac{1}{2}(\delta_{x_0} + \delta_{x_1})$ to $\mu_1 = \frac{1}{2}(\delta_{y_0} + \delta_{y_1})$. In (**c**), at the crossing time s of the path, there does not exist a map T_s that induces an element in $\Gamma_o(\mu_s, \mu_1)$. Red arrows: vector fields computed via (23). (**a**) μ_t for an optimal plan without map. (**b**) μ_t for an optimal plan with Monge map T. (**c**) μ_t for a non-optimal plan with map T

Remark 4.1 (Couplings Induced by Maps) For couplings induced by maps, i.e., $\alpha = (\mathrm{Id}, T)_\sharp \mu_0$, the induced curves admit a simple form, namely

$$\mu_t = (T_t)_\sharp \mu_0 \quad \text{with} \quad T_t(x) = (1-t)x + t T(x). \tag{11}$$

In general, also optimal plans $\alpha \in \Gamma_o(\mu_0, \mu_1)$ may not be induced by a Monge map. However, if α is optimal and $\mu_t = e_{t,\sharp}\alpha$, then, for $s \in (0, 1)$, there exists an optimal map T^s such that $\Gamma_o(\mu_s, \mu_1) = \{(\mathrm{Id}, T^s)_\sharp \mu_t\}$, see [1, Lemma 7.2.1]. Moreover, by [1, Theorem 7.2.2], the curve induced by T^s coincides, up to time parametrizations, with μ_t for $t \in [s, 1]$, $s > 0$. Figure 4a shows a curve from an optimal plan which does not arise from a map and Fig. 4b a curve induced by a plan with Monge map. However, in each case, the plan $\Gamma_o(\mu_s, \mu_1)$ is induced by a map for $s \in (0, 1)$.

This behavior is not necessarily the case for non-optimal plans even if they are induced by a map. As an example, see Fig. 4c, where the paths of two particles cross at a certain time $s \in (0, 1)$, and $\Gamma_o(\mu_s, \mu_1)$ cannot be obtained from a map. ◇

Remark 4.2 (Random Variables) Curves induced by plans can also be interpreted via random variables: Let $X_0, X_1 : \Omega \to \mathbb{R}^d$ be random variables with laws $P_{X_0} = \mu_0$ resp. $P_{X_1} = \mu_1$. Then $\alpha = P_{X_0, X_1} \in \Gamma(\mu_0, \mu_1)$ and the induced curve in (10) reads as

$$\mu_t = P_{X_t}, \quad \text{where} \quad X_t := (1-t)X_0 + tX_1.$$

In particular, if X_0 and X_1 are independent, then $P_{X_0, X_1} = \mu_0 \times \mu_1$ and

$$\mu_t = P_{X_t} = \left((1-t)_\sharp P_{X_0}\right) * \left(t_\sharp P_{X_1}\right).$$

Conversely, for $\alpha \in \Gamma(\mu_0, \mu_1)$, there is a random variable $Z : \Omega \to \mathbb{R}^d \times \mathbb{R}^d$ with law $P_Z = \alpha$, we can choose $X_0 := \pi^1 \circ Z$, $X_1 = \pi^2 \circ Z$ and obtain $\mu_0 = P_{X_0}$, $\mu_1 = P_{X_1}$ and $P_{X_0, X_1} = \alpha$. ◇

First of all, curves induced by plans are narrowly continuous, as the following lemma shows.

Lemma 4.3 Let $\mu_0, \mu_1 \in \mathcal{P}_2(\mathbb{R}^d)$ and $\alpha \in \Gamma(\mu_0, \mu_1)$. Then $\mu_t := e_{t, \sharp} \alpha$ is narrowly continuous.

Proof Let $f \in C_b(\mathbb{R}^d)$. Then $f \circ e_t \in C_b(\mathbb{R}^d \times \mathbb{R}^d)$ is measurable and bounded by $|(f \circ e_t)(x, y)| \leq \|f\|_\infty$ for all $(x, y) \in \mathbb{R}^d \times \mathbb{R}^d$ and $f \circ e_{t'} \to f \circ e_t$ pointwise for $t' \to t$. Thus, by the dominated convergence theorem, we have

$$\lim_{t' \to t} \int_{\mathbb{R}^d} f \, d\mu_{t'} = \lim_{t' \to t} \int_{\mathbb{R}^d} f \circ e_{t'} \, d\alpha = \int_{\mathbb{R}^d} \lim_{t' \to t} f \circ e_{t'} \, d\alpha$$
$$= \int_{\mathbb{R}^d} f \circ e_t \, d\alpha = \int_{\mathbb{R}^d} f \, d\mu_t$$

and hence the claim. □

Moreover, curves induced by optimal plans are geodesics, see [1, Chapter 7.2].

Proposition 4.4 Let $\mu_0, \mu_1 \in \mathcal{P}_2(\mathbb{R}^d)$ and $\alpha \in \Gamma_o(\mu_0, \mu_1)$. Then $\mu_t := e_{t, \sharp} \alpha$ is a geodesic between μ_0 and μ_1 and every geodesic connecting μ_0 and μ_1 is of this form.

Next, we want to show that a curve induced by a plan is also absolutely continuous. For this, we have to find a velocity field fulfilling (i) and (ii) in Theorem 3.3 so that (μ_t, v_t) satisfy (CE). To this end, we associate to a plan α the velocity field $v : [0, 1] \times \mathbb{R}^d \to \mathbb{R}^d$ as follows: Consider $\boldsymbol{\alpha} := \mathcal{L}_{[0,1]} \times \alpha$ and

$$e : [0, 1] \times \mathbb{R}^d \times \mathbb{R}^d \to [0, 1] \times \mathbb{R}^d, \quad (t, x, y) \mapsto (t, e_t(x, y)).$$

Then, by [2, Section 17], there exists a *unique vector field* $v \in L^2(e_\sharp \boldsymbol{\alpha}, \mathbb{R}^d)$ such that

$$v(e_\sharp \boldsymbol{\alpha}) = e_\sharp[(y - x)\boldsymbol{\alpha}]. \tag{12}$$

Since every $e_\sharp \boldsymbol{\alpha}$-measurable function has a Borel measurable representative, see [6, Proposition 2.1.11], we can choose v Borel measurable. We call a measurable *vector field* $v : [0, 1] \times \mathbb{R}^d \to \mathbb{R}^d$ *induced by a plan* α, if it fulfills (12).

Lemma 4.5 *Let $\mu_t = e_{t,\sharp}\alpha$ be the curve induced by the plan α. Then (12) is equivalent to*

$$v_t \mu_t = e_{t,\sharp}[(y-x)\alpha] \quad \text{for a.e. } t \in [0,1]. \tag{13}$$

Proof For $f \in C_b([0,1] \times \mathbb{R}^d)$, the left-hand side of (12) means

$$\int_0^1 \int_{\mathbb{R}^d} fv \, \mathrm{d}e_{\sharp}\underline{\alpha} = \int_0^1 \int_{\mathbb{R}^d \times \mathbb{R}^d} f(t, e_t(x,y)) v(t, e_t(x,y)) \, \mathrm{d}\underline{\alpha}$$

$$= \int_0^1 \int_{\mathbb{R}^d} fv \, \mathrm{d}(e_{t,\sharp}\alpha) \mathrm{d}t$$

$$= \int_0^1 \int_{\mathbb{R}^d} f(t,x) v(t,x) \, \mathrm{d}\mu_t \mathrm{d}t$$

and the right-hand side

$$\int_0^1 \int_{\mathbb{R}^d} f \, \mathrm{d}e_{\sharp}[(y-x)\underline{\alpha}] = \int_0^1 \int_{\mathbb{R}^d \times \mathbb{R}^d} f(t, e_t(x,y))(y-x) \, \mathrm{d}\underline{\alpha}$$

$$= \int_0^1 \int_{\mathbb{R}^d} f \, \mathrm{d}e_{t,\sharp}[(y-x)\mathrm{d}\alpha] \mathrm{d}t.$$

Thus, (13) implies (12).

Conversely, assume that (12) holds true. Then we have

$$\int_0^1 \int_{\mathbb{R}^d} f(t,x) v(t,x) \, \mathrm{d}\mu_t \mathrm{d}t = \int_0^1 \int_{\mathbb{R}^d} f \, \mathrm{d}e_{t,\sharp}[(y-x)\mathrm{d}\alpha] \mathrm{d}t.$$

Let $\{g_n\}_{n \in \mathbb{N}} \subset C_b(\mathbb{R}^d)$ be a dense subset. Then we obtain for any $h \in C_b([0,1])$ and all $n \in \mathbb{N}$ that

$$\int_0^1 h \left(\int_{\mathbb{R}^d} g_n v_t \, \mathrm{d}\mu_t - \int_{\mathbb{R}^d} g_n \, \mathrm{d}e_{t,\sharp}[(y-x)\alpha] \right) \mathrm{d}t = 0$$

and consequently

$$\int_{\mathbb{R}^d} g_n v_t \, \mathrm{d}\mu_t = \int_{\mathbb{R}^d} g_n \, \mathrm{d}e_{t,\sharp}[(y-x)\alpha] \quad \text{a.e. } t \in [0,1]. \tag{14}$$

Since the number of test functions g_n is countable, there exists a zero set $\mathcal{I} \subset [0,1]$ such that (14) holds true on $[0,1] \setminus \mathcal{I}$ for all $n \in \mathbb{N}$. Since $\{g_n\}_{n \in \mathbb{N}}$ is dense in $C_b(\mathbb{R}^d)$ we obtain that $v_t \mu_t = e_{t,\sharp}[(y-x)\alpha]$ for a.e. $t \in [0,1]$. More precisely, the last equation is an equation of vector-valued finite signed measures and thus it is enough to test equality at a dense subset of $C_b(\mathbb{R}^d)$. □

Using the above lemmas, we obtain that curve-velocity pairs induced by couplings have the following favorable properties.

Theorem 4.6 *Let $\mu_0, \mu_1 \in \mathcal{P}_2(\mathbb{R}^d)$ and $\alpha \in \Gamma(\mu_0, \mu_1)$. Let (μ_t, v_t) be a curve-velocity pair induced by α, i.e.*

$$\mu_t = e_{t,\sharp}\alpha, \quad \text{and} \quad v_t\mu_t = e_{t,\sharp}[(y-x)\alpha] \quad \text{for a.e. } t \in [0,1]. \tag{15}$$

Then the following holds true:

(i) *(μ_t, v_t) satisfy (CE) and*

$$\|v_t\|_{L^2(\mathbb{R}^d, \mu_t)} \leq \|y-x\|_{L^2(\mathbb{R}^d \times \mathbb{R}^d, \alpha)} \quad \text{for a.e. } t \in [0,1].$$

In particular, we have $\|v_t\|_{L^2(\mathbb{R}^d, \mu_t)} \in L^1(I)$ and μ_t is an absolutely continuous curve.
(ii) *If $\alpha \in \Gamma_o(\mu_0, \mu_1)$, then $W_2(\mu_t, \mu_{t+h}) = hW_2(\mu_0, \mu_1)$ and*

$$|\mu_t'| = W_2(\mu_0, \mu_1) = \|v_t\|_{L^2(\mathbb{R}^d, \mu_t)} \quad \text{for a.e. } t \in [0,1]. \tag{16}$$

Proof Assertion (i) follows as in [2, Theorem 17.2, Lemma 17.3], see also [7, proof of Proposition 6].

For Assertion (ii), let α be an optimal plan. Since μ_t is a geodesic, we obtain $W_2(\mu_t, \mu_{t+h}) = hW_2(\mu_0, \mu_1)$, which immediately implies the first equality (16). By i) and since α is optimal, we know that $\|v_t\|_{L^2(\mathbb{R}^d, \mu_t)} \leq W_2(\mu_0, \mu_1)$ for a.e. $t \in [0,1]$. Finally, by Theorem 3.3, we have $\|v_t\|_{L^2(\mathbb{R}^d, \mu_t)} \geq |\mu_t'| = W_2(\mu_0, \mu_1)$ for a.e. $t \in [0,1]$. □

By the above theorem, we immediately see that the Wasserstein distance between two measures can be described by the velocity field of any geodesic connecting them (curve energy).

Corollary 4.7 (Benamou-Brenier) *The Wasserstein distance is given by*

$$W_2(\mu, \nu) = \min_{(\mu_t, v_t)} \left(\int_0^1 \|v_t\|^2_{L^2(\mathbb{R}^d, \mu_t)} \, dt\right)^{\frac{1}{2}}$$

where the minimum is taken over all pairs (μ_t, v_t), where v_t is a measurable vector field, μ_t a narrowly continuous curve with $\mu_0 = \mu$, $\mu_1 = \nu$ and (μ_t, v_t) satisfy (CE).

By the following corollary, curve-velocity pairs induced by optimal plans have the favorable property of minimal vector fields. This ensures "short" curves in the related ODE.

Corollary 4.8 *Let $\mu_0, \mu_1 \in \mathcal{P}_2(\mathbb{R}^d)$ and $\alpha \in \Gamma_o(\mu_0, \mu_1)$. Let (μ_t, v_t) be curve-velocity pair induced by α. Then, $v_t \in \mathcal{T}_{\mu_t}$ for a.e. $t \in [0, 1]$.*

Proof By Theorem 4.6(ii) we know that $\|v_t\|_{L^2(\mathbb{R}^d, \mu_t)} = |\mu_t'|$ for a.e. $t \in [0, 1]$. Thus [1, Proposition 8.4.5] implies $v_t \in \mathcal{T}_{\mu_t}$ for a.e. $t \in [0, 1]$. □

If α is not only optimal, but also induced by a Monge map $T : \mathbb{R}^d \to \mathbb{R}^d$, it is often the case that the trajectories are straight, in the sense that the solution of $\partial_t \phi_t(x) = v_t(\phi_t(x))$, $\phi_0(x) = x$, is T_t. This is in particular the case, if μ_0, μ_1 are empirical measures with the same number of points. For a proof and further cases, e.g. if both measures admit densities with a certain regularity, see [7, Proposition 16].

The above minimality property, i.e. laying in the tangent space, is quite special for optimal plans. Indeed, by the following example, this is not the case for arbitrary induced velocity fields, see also [29, Example 3.5] (Fig. 5).

Example 4.9 We consider a radial density $p_0(x) := f(\|x\|^2)$ for some $f \in C^\infty(\mathbb{R})$ with corresponding measure $\mu_0 := p_0 \, dx$, and the linear maps $T_t : \mathbb{R}^2 \to \mathbb{R}^2, t \in [0, 1]$ defined by

$$T_t(x) := \begin{pmatrix} 1 & -t \\ t & 1 \end{pmatrix} x = x + t w(x), \quad w(x) := \begin{pmatrix} 0 & -1 \\ 1 & 0 \end{pmatrix} x,$$

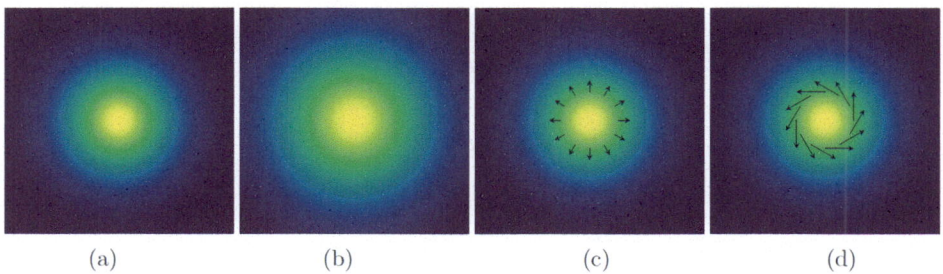

(a)   (b)   (c)   (d)

Fig. 5 Illustration to Example 4.9. Both vector fields generate the same curves μ_t, but in (**d**) mass is rotated unnecessarily, which makes the trajectories of single particles longer than for the vector field in (**c**). (**a**) μ_0 (**b**) μ_1. (**c**) $v_0 \in \mathcal{T}_{\mu_0}$. (**d**) $v_0 \notin \mathcal{T}_{\mu_0}$

which fulfill

$$T_t^{-1}(x) = \frac{1}{1+t^2}\begin{pmatrix} 1 & t \\ -t & 1 \end{pmatrix} x = \frac{1}{1+t^2}(x - tw(x)), \quad \det(\nabla T_t(x)) = \frac{1}{1+t^2}.$$

We are interested in the plan induced by $T = T_1$, i.e.,

$$\alpha := (\mathrm{Id}, T)_\sharp \mu_0 \in \Gamma(\mu_0, T_\sharp \mu_0).$$

We will show that the induced vector field $v_t \mu_t = e_{t,\sharp}((y-x)\alpha)$ is not minimal. By the change of variable formula (2), the induced curve

$$\mu_t := e_{t,\sharp}\alpha = T_{t,\sharp}\mu_0,$$

has the density

$$p_t(x) = \frac{(\rho \circ T_t^{-1})(x)}{\det(\nabla T_t(x))} = (1+t^2) f\left(\tfrac{1}{1+t^2}\|x\|^2\right)$$

which is radial again. In particular, the gradient $\nabla p_t(x)$ is a multiple of x and therefore we get by definition of w that $\langle \nabla p_t(x), w(x)\rangle = 0$. Further, we have $\nabla \cdot w = 0$ and consequently

$$\nabla \cdot (p_t w) = \langle \nabla p_t, w\rangle + p_t \nabla \cdot w = 0$$

Hence, by (8), we know that every $u_t \in \mathcal{T}_{\mu_t}$ must satisfy $\int_{\mathbb{R}^d} \langle w, u_t\rangle \, d\mu_t = 0$ for a.e. $t \in [0,1]$. Unfortunately, the induced velocity field fulfills $\int_{\mathbb{R}^d} \langle w, v_t\rangle \, d\mu_t > 0$ for an non zero subset of $[0,1]$ by the following reasons:

$$\int_{\mathbb{R}^d} \langle w, v_t\rangle d\mu_t = \int_{\mathbb{R}^d \times \mathbb{R}^d} \langle w(e_t(x,y)), y - x\rangle \, d\alpha$$

$$= \int_{\mathbb{R}^d \times \mathbb{R}^d} \langle w(e_t(x,y)), y - x\rangle \, d[(\mathrm{Id}, T)_\sharp \mu_0]$$

$$= \int_{\mathbb{R}^d} \langle w((1-t)x + tT(x)), T(x) - x\rangle \, d\mu_0$$

$$= \int_{\mathbb{R}^d} \langle w(x + tw(x)), w(x)\rangle \, d\mu_0.$$

The last expression is continuous in t and equal to $\|w\|^2_{L^2(\mathbb{R}^2,\rho)}$ for $t = 0$. Since $\|w\|^2_{L^2(\mathbb{R}^2,\rho)} > 0$, there exists an open non empty interval, such that $\int_{\mathbb{R}^d} \langle w, v_t\rangle \, d\mu_t > 0$. Thus, we have $v_t \notin \mathcal{T}_{\mu_t}$. ◇

Example 4.10 Let $\mu_0 = \frac{1}{2}\delta_{x_0} + \frac{1}{2}\delta_{x_1}$, $\mu_1 = \frac{1}{3}\delta_{y_0} + \frac{2}{3}\delta_{y_1}$ and $\alpha = \mu_0 \times \mu_1$. Then

$$\mu_t = \frac{1}{6}\delta_{e_t(x_0,y_0)} + \frac{1}{3}\delta_{e_t(x_0,y_1)} + \frac{1}{6}\delta_{e_t(x_1,y_0)} + \frac{1}{3}\delta_{e_t(x_1,y_1)}$$

$$\alpha_t = \frac{1}{6}\delta_{e_t(x_0,y_0),y_0} + \frac{1}{3}\delta_{e_t(x_0,y_1),y_1} + \frac{1}{6}\delta_{e_t(x_1,y_0),y_0} + \frac{1}{3}\delta_{e_t(x_1,y_1),y_1}.$$

Assume furthermore that for some $s \in (0, 1)$ we have that $e_s(x_0, y_1) = e_s(x_1, y_0) =: \hat{x}_s$, but $e_s(x_0, y_0) \neq \hat{x}_s \neq e_s(x_1, y_1)$, see Fig. 6. We then have that

$$\mu_s = \frac{1}{6}\delta_{e_s(x_0,y_0)} + \frac{1}{2}\delta_{\hat{x}_s} + \frac{1}{3}\delta_{e_s(x_1,y_1)}$$

$$\alpha_s = \frac{1}{6}\delta_{e_s(x_0,y_0),y_0} + \frac{1}{3}\delta_{\hat{x}_s,y_1} + \frac{1}{6}\delta_{\hat{x}_s,y_0} + \frac{1}{3}\delta_{e_s(x_1,y_1),y_1},$$

which implies $\alpha_s^{\hat{x}_s} = \frac{1}{3}\delta_{y_0} + \frac{2}{3}\delta_{y_1}$. Hence using Proposition 5.7 we obtain

$$v_s(\hat{x}_s) = \frac{1}{3}\frac{y_0 - \hat{x}_s}{1 - s} + \frac{2}{3}\frac{y_1 - \hat{x}_s}{1 - s}.$$

Note that since at μ_s the mass in $\hat{x}_s$ has to be split, μ_t cannot be described by a pushforward of a solution of an ODE and thus the assumptions of Theorem 3.5 cannot be fulfilled. ◇

We have already seen that vector fields v_t associated to optimal plans are minimal ones, meaning that $v_t \in \mathcal{T}_{\mu_t}$. This is in general not true for an independent coupling $\alpha = \mu_0 \times \mu_1$ with an arbitrary μ_0, see Fig. 6. Fortunately, the independent coupling with a Gaussian marginal μ_0 has this minimality property as the following example shows, see also [29, Section 5.1].

Proposition 4.11 *Let $\mu_0 \sim \mathcal{N}(0, I_d)$, $\mu_1 \in \mathcal{P}_2(\mathbb{R}^d)$ and $\alpha = \mu_0 \times \mu_1$. Let $\mu_t := e_{t,\sharp}\alpha$ be the induced curve and v_t the induced velocity field v_t in (13). Then μ_t admits the strictly*

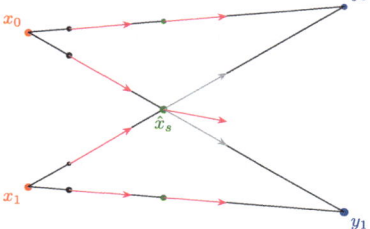

Fig. 6 Curve and vector field associated to $\alpha = \mu_0 \times \mu_1$, where $\mu_0 = \frac{1}{2}\delta_{x_0} + \frac{1}{2}\delta_{x_1}$ and $\mu_1 = \frac{1}{3}\delta_{y_0} + \frac{2}{3}\delta_{y_1}$. Vectors are scaled by 0.2 for better visibility

positive density

$$p_t(x) := (1-t)^{-d}(2\pi)^{-\frac{d}{2}} \int_{\mathbb{R}^d} e^{-\frac{\|x-ty\|^2}{2(1-t)^2}} \, d\mu_1(y), \quad t \in [0,1), \qquad (17)$$

and

$$v_t = \nabla f_t \quad \text{with} \quad f_t := \frac{1-t}{t} \log p_t + \frac{1}{2t} \|\cdot\|^2, \quad t \in (0,1), \qquad (18)$$

Further, it holds $v_t \in \mathcal{T}_{\mu_t}$.

The expression $\nabla \log p_t$ will reappear as so-called "score" in Sect. 11.

Proof

1. Relation (17) follows by directly computing for $g \in C_c^\infty(\mathbb{R}^d)$,

$$\int_{\mathbb{R}^d} g \, d\mu_t = \int_{\mathbb{R}^d \times \mathbb{R}^d} g(e_t(x,y)) \, d\alpha(x,y)$$

$$= \frac{1}{(2\pi)^{\frac{d}{2}}} \int_{\mathbb{R}^d \times \mathbb{R}^d} g((1-t)x + ty) \, e^{-\frac{\|x\|^2}{2}} \, dx d\mu_1(y)$$

and substituting x by $\frac{z-ty}{1-t}$ then

$$\int_{\mathbb{R}^d} g \, d\mu_t = \frac{1}{(1-t)^d (2\pi)^{\frac{d}{2}}} \int_{\mathbb{R}^d} g(z) e^{-\frac{\|z-ty\|^2}{2(1-t)^2}} \, d\mu_1(y) dz.$$

2. Concerning relation (18), we verify

$$\int_{\mathbb{R}^d} g \nabla f_t \, d\mu_t = \frac{1-t}{t} \int_{\mathbb{R}^d} g(x) \frac{\nabla p_t(x)}{p_t(x)} p_t(x) \, dx + \frac{1}{t} \int_{\mathbb{R}^d} g(x) \, x \, p_t(x) \, dx$$

$$= \frac{1}{t(1-t)^{d+1}(2\pi)^{\frac{d}{2}}} \int_{\mathbb{R}^d \times \mathbb{R}^d} g(x)(ty-x) e^{-\frac{\|x-ty\|^2}{2(1-t)^2}} \, d\mu_1(y) \, dx$$

$$+ \frac{1}{t(1-t)^d (2\pi)^{\frac{d}{2}}} \int_{\mathbb{R}^d \times \mathbb{R}^d} g(x) x e^{-\frac{\|x-ty\|^2}{2(1-t)^2}} \, d\mu_1(y) dx$$

$$= \frac{1}{(1-t)^{d+1}(2\pi)^{\frac{d}{2}}} \int_{\mathbb{R}^d \times \mathbb{R}^d} g(x)(y-x) e^{-\frac{\|x-ty\|^2}{2(1-t)^2}} \, d\mu_1(y) \, dx.$$

On the other hand, we have for the induced velocity field that

$$\int_{\mathbb{R}^d} g\, v_t\, d\mu_t = \int_{\mathbb{R}^d} g\, de_{t,\sharp}(y-x)\alpha = \frac{1}{(2\pi)^{\frac{d}{2}}} \int_{\mathbb{R}^d \times \mathbb{R}^d} g(e_t(x,y))(y-x) e^{-\frac{\|x\|^2}{2}} d\mu_1(y) dx$$

$$= \frac{1}{(1-t)^{d+1}(2\pi)^{\frac{d}{2}}} \int_{\mathbb{R}^d \times \mathbb{R}^d} g(z)(y-z) e^{-\frac{\|z-ty\|^2}{2(1-t)^2}} d\mu_1(y) dz,$$

which yields the assertion. Note that, in particular $\nabla f_t \in L^2(\mathbb{R}^d, \mu_t)$.

3. It remains to prove the minimality property of v_t. Since $\mu_t \in \mathcal{P}_2(\mathbb{R}^d)$, we have $\nabla \|x\|^2 \in \mathcal{T}_{\mu_t}$. Thus, we are left to show that $\nabla \log p_t \in \mathcal{T}_{\mu_t}$.

First, we have $p_t \in C^\infty(\mathbb{R}^d)$, $t \in [0,1)$ and since $p_t > 0$ also $\log p_t \in C^\infty(\mathbb{R}^d)$. Since $\nabla \|x\|^2, \nabla f_t \in L^2(\mathbb{R}^d, \mu_t)$, we conclude that also $\nabla \log p_t \in L^2(\mathbb{R}^d, \mu_t)$. Furthermore, it is easy to see that $\lim_{r \to \infty} \max\{\log p_t(x) : \|x\| \geq r\} \to -\infty$. Consequently, we obtain for functions $h_n \in C_c^\infty(\mathbb{R})$ that $\phi_n := h_n \circ \log p_t \in C_c^\infty(\mathbb{R}^d)$. Now consider especially functions with $h_n(s) = s$ for $-n \leq s \leq \|\log p_t\|_\infty$ and $|h_n'| \leq 1$. Then we obtain with $C_n := \{x \in \mathbb{R}^d : \log p_t(x) \geq -n\}$ that $\phi_n(x) = \log p_t(x)$ for $x \in C_n$ and thus

$$\int_{\mathbb{R}^d} \|\nabla \phi_n - \nabla \log p_t\|^2 d\mu_t = \int_{\mathbb{R}^d \setminus C_n} \|\nabla \phi_n - \nabla \log p_t\|^2 d\mu_t$$

$$\leq 4 \int_{\mathbb{R}^d \setminus C_n} \|\nabla \log p_t\|^2 d\mu_t.$$

Since $\nabla \log p_t \in L^2(\mathbb{R}^d, \mu_t)$, the latter converges to zero. Hence, by (7), this yields $\nabla \log p_t \in \mathcal{T}_{\mu_t}$. □

Example 4.12 (Exploding Velocity Field) We consider $\mu_0 \sim \mathcal{N}(0,1)$, $\mu_1 := \delta_0$ and $\alpha = \mu_0 \times \mu_1$. Then, by (17), the curve $\mu_t := e_{t,\sharp}\alpha$ admits the strictly positive density

$$p_t(x) = (1-t)^{-1}(2\pi)^{-\frac{1}{2}} e^{-\frac{x^2}{2(1-t)^2}}, \quad t \in [0,1),\ x \in \mathbb{R}.$$

Following (18), we calculate

$$f_t(x) := \frac{1-t}{t} \log p_t(x) + \frac{1}{2t} x^2$$

$$= \frac{1-t}{t} \Big(\log\big((1-t)^{-1}(2\pi)^{-\frac{1}{2}}\big) - \frac{x^2}{2(1-t)^2} \Big) + \frac{1}{2t} x^2$$

$$= \frac{1-t}{t} \log\big((1-t)^{-1}(2\pi)^{-\frac{1}{2}}\big) - \frac{x^2}{2t(1-t)} + \frac{1}{2t} x^2$$

and the induced velocity field

$$v_t(x) = \nabla f_t = \frac{1}{t}x - \frac{x}{t(1-t)}$$
$$= \frac{x}{t-1}.$$

Hence, for any $x \in \mathbb{R} \setminus \{0\}$, the absolute value $|v_t(x)|$ explodes for $t \to 1$. Note that despite the exploding velocity field, the velocity of a trajectory remains constant over time. More precisely, the map $T_t(x) = (1-t)x$ is the solution of the flow ODE $\partial_t \phi(t, x) = v_t(\phi(t, x))$, $\phi(0, x) = x$ and thus the speed of a trajectory starting at x is constant in time and equal to $v_t(T_t(x)) = -x$.

5 Curves via Markov Kernels

Unfortunately, the approach of Lipman et al. [28], see Example 5.4, is not covered by the previous section. Indeed, we need a more general viewpoint via Markov kernels which is given in the following Sect. 5.1. We will see in Sect. 5.2 that curve-velocity pairs induced by couplings fit into this general setting.

5.1 General Construction

The main observation is the following theorem, which describes how to construct a curve velocity pair (v_t, v_t) when we are given a family of Markov kernels $\mathcal{K}_t(y, \cdot)$ and a measure μ_1. Note that in this construction it is not necessarily the case that ν_0 is tractable and $\nu_1 = \mu_1$. But we will see later that for suitable chosen $\mathcal{K}_t(y, \cdot)$, we can achieve $\nu_0 = \mathcal{N}(0, 1)$ and $\nu_1 \cong \mu_1$.

Theorem 5.1 *Let $(\mathcal{K}_t)_{t \in [0,1]}$ be a family of Markov kernels $\mathcal{K}_t : \mathbb{R}^d \times \mathcal{B}(\mathbb{R}^d) \to \mathbb{R}$. For a given measure $\mu_1 \in \mathcal{P}_2(\mathbb{R}^d)$, we consider*

$$\alpha_t := \mathcal{K}_t(y, \cdot) \times_y \mu_1$$

and introduce the measure

$$\nu_t := \pi^1_\# \alpha_t \quad \text{so that} \quad \alpha_t = \alpha_t^x \times_x \nu_t.$$

Assume that $\mathcal{K}_t(y, \cdot)$ is an absolutely continuous curve in $\mathcal{P}_2(\mathbb{R}^d)$ for μ_1-a.e. $y \in \mathbb{R}^d$ with a vector field v_t^y such that $(\mathcal{K}_t(y, \cdot), v_t^y)$ fulfills (CE) for a.e. y. Further, let $(t, x, y) \mapsto v_t^y$ be measurable and $\|v_t^y\|_{L^2(\mathbb{R}^d \times \mathbb{R}^d, \alpha_t)} \in L^1([0, 1])$. Then α_t and ν_t are narrowly continuous. The curve v_t is absolutely continuous and (ν_t, v_t) satisfies (CE), where

$$v_t(x) := \int_{\mathbb{R}^d} v_t^y \, d\alpha_t^x(y). \tag{19}$$

Proof

1. For $f \in C_b(\mathbb{R}^d \times \mathbb{R}^d)$ with $\|f\|_\infty \leq C$ we have $\left|\int_{\mathbb{R}^d} f \, d\mathcal{K}_t(y, \cdot)\right| \leq C$. Thus, using the dominated convergence theorem and the fact that $\mathcal{K}_t(y, \cdot)$ is absolutely continuous for μ_1-a.e. $y \in \mathbb{R}^d$, we conclude

$$\lim_{t \to t'} \int_{\mathbb{R}^d \times \mathbb{R}^d} f \, d\alpha_t = \int_{\mathbb{R}^d} \lim_{t \to t'} \int_{\mathbb{R}^d} f \, d\mathcal{K}_t(y, \cdot) d\mu_1(y) = \int_{\mathbb{R}^d} \int_{\mathbb{R}^d} f \, d\mathcal{K}_{t'}(y, \cdot) \, d\mu_1(y)$$

$$= \int_{\mathbb{R}^d \times \mathbb{R}^d} f \, d\alpha_{t'} \quad \text{for all} \quad f \in C_b(\mathbb{R}^d \times \mathbb{R}^d),$$

so that α_t is narrowly continuous and the same holds true for $\nu_t = \pi_\sharp^1 \alpha_t$.

2. For any $\varphi \in C_c^\infty([0, 1] \times \mathbb{R}^d)$, we obtain

$$\int_0^1 \int_{\mathbb{R}^d} \left\langle \nabla_x \varphi(t, x), \int_{\mathbb{R}^d} v_t^y(x) \, d\alpha_t^x(y) \right\rangle d\nu_t(x) dt$$

$$= \int_0^1 \int_{\mathbb{R}^d \times \mathbb{R}^d} \langle \nabla_x \varphi(t, x), v_t^y(x) \rangle \, d\alpha_t(x, y) \, dt$$

$$= \int_0^1 \int_{\mathbb{R}^d} \int_{\mathbb{R}^d} \langle \nabla_x \varphi(t, x), v_t^y(x) \rangle \, d\mathcal{K}_t(y, \cdot)(x) \, d\mu_1(y) \, dt$$

$$= \int_{\mathbb{R}^d} \int_0^1 \int_{\mathbb{R}^d} \langle \nabla_x \varphi(t, x), v_t^y(x) \rangle \, d\mathcal{K}_t(y, \cdot)(x) \, dt \, d\mu_1(y)$$

$$= -\int_{\mathbb{R}^d} \int_0^1 \int_{\mathbb{R}^d} \partial_t \varphi(t, x) \, d\mathcal{K}_t(y, \cdot)(x) \, dt \, d\mu_1(y)$$

$$= -\int_0^1 \int_{\mathbb{R}^d \times \mathbb{R}^d} \partial_t \varphi(t, x) \, d\alpha_t \, dt = \int_0^1 \int_{\mathbb{R}^d \times \mathbb{R}^d} \partial_t \varphi(t, x) \, d(\pi_\sharp^1 \alpha_t) dt$$

$$= -\int_0^1 \int_{\mathbb{R}^d} \partial_t \varphi(t, x) \, d\nu_t(x) dt,$$

so that (ν_t, v_t) fulfills (CE).

3. Finally, we conclude

$$\int_0^1 \|v_t\|_{L^2(\mathbb{R}^d, \nu_t)} \, dt = \int_0^1 \left(\int_{\mathbb{R}^d} \|v_t\|^2 \, d\nu_t \right)^{\frac{1}{2}} dt$$

$$\leq \int_0^1 \left(\int_{\mathbb{R}^d} \int_{\mathbb{R}^d} \|v_t^y(x)\|^2 \, d\alpha_t^x(y) d\nu_t(x) \right)^{\frac{1}{2}} dt$$

$$= \int_0^1 \left(\int_{\mathbb{R}^d \times \mathbb{R}^d} \|v_t^y(x)\|^2 \, d\alpha_t \right)^{\frac{1}{2}} dt < \infty,$$

which finishes the proof, where the measurability of v is left to the reader. □

Remark 5.2 In general, every curve $\mu_t \in \mathcal{P}_2(\mathbb{R}^d)$ can be written as in Theorem 5.1. Namely, $\mathcal{K}_t(y, \cdot) := \mu_t(\cdot)$ is a Markov kernel and for $\alpha_t := \mathcal{K}_t(y, \cdot) \times_y \mu_1$ it holds that $\nu_t := \pi_\sharp^1 \alpha_t = \mu_t$. Of course, we don't win anything using this construction since $v_t^y = v_t$ and thus we do not obtain a more tractable vector field. ◇

Remark 5.3 (Relation to Lipman et al. [28]) The authors in [28] argue only with densities. Then the notation of the so-called "conditional probability path" translates as

$$\mathcal{K}_t(y, \cdot) \longleftrightarrow p_t(x|y).$$

Further, we have

$$\alpha_t = \mathcal{K}_t(y, \cdot) \times_y \mu_1 \longleftrightarrow p_t(x, y), \quad \alpha_t^x \longleftrightarrow p_t(y|x), \quad \pi_\sharp^2 \alpha_t = \mu_1 \longleftrightarrow q(y)$$

and

$$\int_{\mathbb{R}^d} \varphi \, d\nu_t(x) = \int_{\mathbb{R}^d \times \mathbb{R}^d} \varphi(x) \, d\mathcal{K}_t(y, \cdot) d\mu_1(y) \longleftrightarrow p_t(x) = \int_{\mathbb{R}^d} p_t(y|x) q(y) \, dy$$

so that $\nu_t \longleftrightarrow p_t$. By the Bayesian law, we know that $p_t(y|x) = \frac{p_t(x|y)}{p_t(x)} q(y)$, so that we finally obtain the correspondence

$$v_t(x) = \int_{\mathbb{R}^d} v_t^y(x) \, d\alpha_t^x(y) \longleftrightarrow v_t = \int_{\mathbb{R}^d} v_t(x|y) \frac{p_t(x|y)}{p_t(x)} q(y) \, dy. \quad ◇$$

We recap the example in [28, Example II] with our approach.

Example 5.4 For fixed $r \in (0, 1)$, we consider the family of Markov kernels

$$\mathcal{K}_t(y, \cdot) := \mathcal{N}(ty, (1 - rt)^2 I_d), \quad t \in [0, 1].$$

By (5), there is an optimal transport map $T : \mathbb{R}^d \to \mathbb{R}^d$ between the Gaussians $\mathcal{K}_0(y, \cdot) = \mathcal{N}(0, I_d) =: \mu_0$ and $\mathcal{K}_1(y, \cdot) = \mathcal{N}(y, (1-r)^2 I_d)$ given by

$$T(x) = y + (1-r)x$$

with corresponding plan $\gamma = (\text{Id}, T)_\sharp \mu_0$ and

$$T_t(x) = e_{t,\sharp}(x, T(x)) = (1-tr)x + ty, \quad T_t^{-1}(z) = \frac{z - ty}{1 - tr}.$$

Let $\mu_1 \in \mathcal{P}_2(\mathbb{R}^d)$. Further, it can be easily verified that

$$\mathcal{K}_t(y, \cdot) = e_{t,\sharp}\big((\text{Id}, T)_\sharp \mu_0\big) = (T_t)_\sharp \mu_0.$$

By Corollary 5.10, we obtain that $(\mathcal{K}_t(y, \cdot), v_t^y)$ with the velocity field

$$v_t^y(x) = T(T_t^{-1}(x)) - T_t^{-1}(x) = \frac{y - rx}{1 - tr},$$

fulfills (CE). Now, let $\mu_1 \in \mathcal{P}_2(\mathbb{R}^d)$ and consider the family of plans

$$\alpha_t := \mathcal{K}_t(y, \cdot) \times_y \mu_1 \quad \text{and} \quad v_t := \pi_\sharp^1 \alpha_t.$$

It is easy to check that $v_0 = \mathcal{N}(0, I_d)$. Then we know by Proposition 5.1 that the curve v_t is absolutely continuous and fulfills (CE) with the velocity field v_t in (19). However, if we want to sample from the target density μ_1, this can only be done approximately by following the path of v_t, since

$$v_1 = \pi_\sharp^1 \Big[\mathcal{N}(y, (1-r)^2 I_d) \times_y \mu_1(y)\Big] = \mathcal{N}(0, (1-r)^2 I_d) * \mu_1 \neq \mu_1.$$

By [1, Lemma 7.1.10] we have that $W_2(v_1, \mu_1) \leq (1-r)\sqrt{\int \|x\|^2 d\mu_1}$ and thus v_1 converges to μ_1 in $\mathcal{P}_2(\mathbb{R}^d)$ as $r \to 1$. The authors of [28] then learn the velocity field v_t of v_t as explained in Example 7.2. ◇

A curve that can be described by Markov kernels, but is not induced by any coupling is given in the following example.

Example 5.5 Let μ_t be a curve induced by $\alpha \in \Gamma(\mu_0, \mu_1)$. Then, by Proposition 4.6 and Corollary 4.7 of Benamou-Brenier, we obtain

$$W_2(\mu_t, \mu_{t+h})^2 \leq h^2 \|x - y\|_{L^2(\mathbb{R}^d \times \mathbb{R}^d, \alpha)}^2 \leq h^2 \big(\|x\|_{L^2(\mathbb{R}^d, \mu_0)} + \|x\|_{L^2(\mathbb{R}^d, \mu_1)}\big)^2,$$

which is a bound independent of the plan. Thus, if μ_t is not constant, we can find a monotone function $f \in C^\infty([0, 1])$ with $f(0) = 0, f(1) = 1$, such that $\gamma_t := \mu_{f(t)}$ cannot be induced by a plan by simply speeding up μ_t such that the above inequality for $W_2(\gamma_t, \gamma_{t+h})^2$ is not satisfied. However, we can construct it via a Markov kernel. Consider $\bar{\alpha}_t = \alpha_{f(t)}$, where as usual $\alpha_t = (e_t, \pi^2)_\sharp \alpha$ and $\alpha_t^y \times_y \mu_1 = \alpha_t$. Then $\bar{\alpha}_t^y := \alpha_{f(t)}^y$ is absolutely continuous. Furthermore, for $\bar{v}_t^y := f'(t) v_{f(t)}^y$ we obtain that $(\bar{\alpha}_t^y, \bar{v}_t^y)$ fulfills the continuity equation. The curve $\bar{\mu}_t$ associated to the Markov kernel $\bar{\alpha}_t^y$ and μ_1 is γ_t, which cannot be induced by a plan. ◊

5.2 Curves Induced by Couplings via Disintegration

Next, let us see how curve-velocity pairs induced by couplings $\alpha \in \Gamma(\mu_0, \mu_1)$ fit into the general setting of the previous subsection. Let $(e_t, \pi^2) : \mathbb{R}^d \times \mathbb{R}^d \to \mathbb{R}^d \times \mathbb{R}^d$ be given by $(x, y) \mapsto (e_t(x, y), y)$ and define a family of couplings

$$\alpha_t := (e_t, \pi^2)_\sharp \alpha, \quad t \in [0, 1]. \tag{20}$$

Then α_t has the marginals

$$\pi_\sharp^1 \alpha_t = e_{t,\sharp} \alpha = \mu_t \quad \text{and} \quad \pi_\sharp^2 \alpha_t = \mu_1,$$

and corresponding disintegrations

$$\alpha_t = \alpha_t^x \times_x \mu_t \quad \text{and} \quad \alpha_t = \alpha_t^y \times_y \mu_1 = \mathcal{K}_t(y, \cdot) \times_y \mu_1. \tag{21}$$

By the following proposition, α_t^y is a curve induced by the independent coupling of α^y and δ_y.

Proposition 5.6 *The disintegration α_t^y in (21) fulfills*

$$\alpha_t^y = e_{t,\sharp}(\alpha^y \times \delta_y) \quad \text{for } \mu_1\text{-a.e. } y \tag{22}$$

and

$$v_t^y(x) := \frac{y - x}{1 - t}, \quad t \in (0, 1) \tag{23}$$

is an induced velocity field of $\alpha^y \times \delta_y$. In particular, α_t^y is absolutely continuous for μ_1-a.e. $y \in \mathbb{R}^d$, $v_t^y \in \mathcal{T}_{\alpha_t^y}$ and (α_t^y, v_t^y) fulfills (CE).

Proof For any measurable, bounded function $f: \mathbb{R}^d \to \mathbb{R}^d$, we obtain

$$\int_{\mathbb{R}^d \times \mathbb{R}^d} f(x, y) \, d\alpha_t = \int_{\mathbb{R}^d \times \mathbb{R}^d} f(e_t(x, y), y) \, d\alpha = \int_{\mathbb{R}^d \times \mathbb{R}^d} f(e_t(x, y), y) d\alpha^y(x) d\mu_1(y)$$

$$= \int_{\mathbb{R}^d \times \mathbb{R}^d} f(e_t(x, z), y) \, d(\alpha^y \times \delta_y)(x, z) d\mu_1(y)$$

$$= \int_{\mathbb{R}^d \times \mathbb{R}^d} f(x, y) de_{t,\sharp}[\alpha^y \times \delta_y](x) \, d\mu_1(y),$$

which implies (22). Furthermore, we get for v_t^y in (23) that

$$\int_{\mathbb{R}^d} f(x) v_t^y(x) \, d\alpha_t^y = \int_{\mathbb{R}^d} f(x) \frac{y-x}{1-t} \, de_{t,\sharp}(\alpha^y \times_y \delta_y) = \int_{\mathbb{R}^d} f(e_t(x, y))(y-x) \, d\alpha^y(x)$$

$$= \int_{\mathbb{R}^d \times \mathbb{R}^d} f(e_t(x, z))(z-x) \, d(\alpha^y \times \delta_y)$$

$$= \int_{\mathbb{R}^d \times \mathbb{R}^d} f \, de_{t,\sharp}[(z-x)(\alpha^y \times \delta_y)].$$

Hence, by (13), the velocity field v_t^y is associated to $\alpha^y \times \delta_y$. Since $\Gamma(\alpha^y, \delta_y) = \Gamma_0(\alpha^y, \delta_y) = \{\alpha^y \times \delta_y\}$, we obtain by Theorem 4.6 that α_t^y is absolutely continuous. By Corollary 4.8, we know that $v_t^y \in \mathcal{T}_{\alpha_t^y}$. □

By the next proposition, we can rewrite the α-induced velocity field in (13) using the disintegration α_t^x.

Proposition 5.7 *For $\mu_0, \mu_1 \in \mathcal{P}_2(\mathbb{R}^d)$, let $\alpha \in \Gamma(\mu_0, \mu_1)$ and $\mu_t := e_{t,\sharp}\alpha$. Then $v_t : \mathbb{R}^d \to \mathbb{R}^d$ defined by*

$$v_t(x) := \int_{\mathbb{R}^d} \frac{y-x}{1-t} \, d\alpha_t^x(y), \quad t \in (0, 1). \tag{24}$$

is the induced velocity field by α.

Proof The assertion follows by straightforward computation:

$$\int_{\mathbb{R}^d} f(x) v_t(x) \, d\mu_t = \int_{\mathbb{R}^d} \int_{\mathbb{R}^d} f(x) \frac{y-x}{1-t} \, d\alpha_t^x(y) d\mu_t(x)$$

$$= \int_{\mathbb{R}^d \times \mathbb{R}^d} f(x) \frac{y-x}{1-t} \, d\alpha_t(x, y)$$

$$= \int_{\mathbb{R}^d} f((1-t)x + ty)(y-x) \, d\alpha(x,y)$$

$$= \int_{\mathbb{R}^d \times \mathbb{R}^d} f \, de_{t,\sharp}\, [(y-x)\alpha].$$

$\square$

Note that it is not a priori clear, that there are representatives of v_t such that v_t seen as a function $[0,1] \times \mathbb{R}^d \to \mathbb{R}^d$ is measurable. We show in Proposition B.1 why this is the case.

We summarize the last two propositions in the following theorem.

Theorem 5.8 *Let* $\mu_0, \mu_1 \in \mathcal{P}_2(\mathbb{R}^d)$ *and* $\alpha \in \Gamma(\mu_0, \mu_1)$. *Then the pair*

$$\mathcal{K}_t(y, \cdot) = e_{t,\sharp}(\alpha^y \times_y \delta_y), \quad v_t^y = \frac{y-x}{1-t}$$

fulfills the conditions of Theorem 5.1 and in this case

$$\mu_t := e_{t,\sharp}\alpha = \nu_t := \pi_{\sharp}^1(\mathcal{K}_t(y,\cdot) \times_y \mu_1), \quad v_t(x) = \int_{\mathbb{R}^d} v_t^y \, d\alpha_t^x(y).$$

is a curve-velocity pair induced by α.

Here is the relation to random variables.

Remark 5.9 (Random Variables) Consider two random variables X_0, X_1 with law $\mu_0, \mu_1 \in \mathcal{P}_2(\mathbb{R}^d)$, $X_t = (1-t)X_0 + tX_1$ and $\alpha = P_{X_0,X_1}$. Then $\alpha_t = P_{X_t,X_1}$ and $\alpha_t^x = P_{X_1|X_t=x}$, where $P_{X_1|X_t=x}$ is defined via $P_{X_t,X_1} = P_{X_1|X_t=x} \times_x P_{X_t}$. Thus (24) reads as

$$v_t(x) = \int_{\mathbb{R}^d} \frac{y-x}{1-t} \, dP_{X_1|X_t=x}(y).$$

$\diamond$

If the coupling $\alpha \in \Gamma(\mu_0, \mu_1)$ comes from a map T, we can characterize the induced velocity field using the map as in the following corollary.

Corollary 5.10 *Let* $\mu_0, \mu_1 \in \mathcal{P}_2(\mathbb{R}^d)$ *and let* $T : \mathbb{R}^d \to \mathbb{R}^d$ *be an invertible measurable map such that* $T_\sharp \mu_0 = \mu_1$. *Consider the coupling* $\alpha := (\mathrm{Id}, T)_\sharp \mu_0$ *and the curve* (11) *induced by α. Assume that* $T_t(x) := (1-t)x + tT(x)$ *is invertible for* $t \in [0,1]$ *Then it* holds that $\alpha_t = (T_t, T)_\sharp \mu_0$ with disintegrations

$$\alpha_t^x = \delta_{T(T_t^{-1}(x))} \quad \text{and} \quad \alpha_t^y = \delta_{T_t(T^{-1}(y))}$$

and the α-velocity field in (24) reads as $v_t(x) = T(T_t^{-1}(x)) - T_t^{-1}(x)$. In other words, we have a linear velocity field $v_t(T_t(x)) = T(x) - x$.

Proof Then we see that

$$\int_{\mathbb{R}^d \times \mathbb{R}^d} f(x, y) \, \mathrm{d}\alpha_t = \int_{\mathbb{R}^d \times \mathbb{R}^d} f(e_t(x, y), y) \, \mathrm{d}\alpha$$

$$= \int_{\mathbb{R}^d} f(e_t((x, T(x)), T(x)) \, \mathrm{d}\mu_0$$

$$= \int_{\mathbb{R}^d} f(T_t(x), T(x)) \, \mathrm{d}\mu_0, \qquad (25)$$

which yields $\alpha_t = (T_t, T)_\sharp \mu_0$ and by (21) also $\mu_t = (T_t)_\sharp \mu_0$. Further, this implies

$$\int_{\mathbb{R}^d \times \mathbb{R}^d} f(x, y) \, \mathrm{d}\alpha_t = \int_{\mathbb{R}^d} f\left(T_t(x), T\left(T_t^{-1} T_t(x)\right)\right) \mathrm{d}\mu_0$$

$$= \int_{\mathbb{R}^d \times \mathbb{R}^d} f\left(e_t(x, y), T(T_t^{-1}(e_t(x, y)))\right) \mathrm{d}\alpha$$

$$= \int_{\mathbb{R}^d \times \mathbb{R}^d} f\left(x, T(T_t^{-1}(x))\right) \mathrm{d}\alpha_t$$

$$= \int_{\mathbb{R}^d} f\left(x, T(T_t^{-1}(x))\right) \mathrm{d}\pi_\sharp^1 \alpha_t,$$

so that $\alpha_t^x = \delta_{T(T_t^{-1}(x))}$. On the other hand, we obtain by (25) that

$$\int_{\mathbb{R}^d \times \mathbb{R}^d} f(x, y) \, \mathrm{d}\alpha_t = \int_{\mathbb{R}^d} f((1-t)y + tT(y), T(y)) \, \mathrm{d}\mu_0$$

$$= \int_{\mathbb{R}^d} f((1-t)T^{-1}(T(y)) + tT(y), T(y)) \, \mathrm{d}\mu_0$$

$$= \int_{\mathbb{R}^d \times \mathbb{R}^d} f((1-t)T^{-1}(y) + ty, y) \, \mathrm{d}\alpha$$

$$= \int_{\mathbb{R}^d \times \mathbb{R}^d} f((1-t)T^{-1}(y) + ty, y) \, \mathrm{d}\alpha_t$$

$$= \int_{\mathbb{R}^d \times \mathbb{R}^d} f((1-t)T^{-1}(y) + ty, y) \, \mathrm{d}\pi_\sharp^2 \alpha_t,$$

meaning that $\alpha_t^y = \delta_{T_t(T^{-1}(y))}$. Finally, we get by (24) that

$$v_t(x) = \int_{\mathbb{R}^d} \frac{y-x}{1-t} \, d\delta_{T(T_t^{-1}(x))} = \frac{T(T_t^{-1}(x)) - x}{1-t} = T(T_t^{-1}(x)) - T_t^{-1}(x).$$

□

Example 5.11 In Example 5.4, we have seen that the curve v_t is unfortunately not induced by the coupling $\mathcal{N}(0, I_d) \times \mu_1$. This would be only the case for $r = 0$. But then $\mathcal{K}_1(y, \cdot) = \delta_y$ has no longer a density. However, also for $r \in (0, 1)$, the curve v_t is induced by another plan, namely

$$\tilde{\alpha} := \tilde{e}_\sharp(\mathcal{N}(0, I_d) \times \mu_1) \quad \text{with} \quad \tilde{e}(x, y) := (x, (1-r)x + y),$$

since

$$\int_{\mathbb{R}^d} f(x) \, dv_t(x) = \int_{\mathbb{R}^d} f(x) \, d\pi_\sharp^1(\mathcal{K}_t(y, \cdot) \times_y \mu_1(y))$$

$$= \int_{\mathbb{R}^d \times \mathbb{R}^d} f(x) \, d\mathcal{N}(ty, (1-rt)^2) d\mu_1(y)$$

$$= \int_{\mathbb{R}^d \times \mathbb{R}^d} f((1-rt)x + ty) \, d\mathcal{N}(0, 1)(y) d\mu_1(y)$$

$$= \int_{\mathbb{R}^d} f(x) \, d(e_t \circ \tilde{e})_\sharp(\mathcal{N}(0, 1) \times \mu_1)(x),$$

i.e., $v_t = e_{t,\sharp}(\tilde{\alpha})$. By Example 7.2 the vector field induced by $\tilde{\alpha}$ coincides with the vector field constructed in (19).

However, considering the family $\tilde{\alpha}_t = (e_t, \pi^2)_\sharp \tilde{\alpha}$, we have that $\pi_\sharp^2 \tilde{\alpha}_t = \mathcal{N}(0, (1-r)^2 \text{Id}) * \mu_1$, which is not the construction of $\pi_\sharp^2 \alpha_t = \mu_1$ from Example 5.4. In particular, the family of Markov kernels inducing a curve is not unique. ◇

6 Curves via Stochastic Processes

This section resembles the results from the papers of Liu et al. [29, 30] with our notation. As already mentioned in the introduction, these authors use a stochastic process $(X_t)_t$ to determine an absolutely continuous curve μ_t and an associated vector field v_t based on the conditional expectation

$$\mu_t := P_{X_t} \quad \text{and} \quad v_t(x) := \mathbb{E}[\partial_t X_t | X_t = x],$$

see Theorem 6.3. For the special process $X_t := (1-t)X_0 + tX_1$, this will again result in a curve-velocity pair induced by a coupling, namely $\alpha = P_{X_0,X_1}$, see Corollary 6.4. We start by recalling the notation of conditional expectation in Sect. 6.1 and use this in Sect. 6.2 to deduce appropriate curve-velocity pairs.

6.1 Conditional Expectation

Throughout this section, let $(\Omega, \Sigma, \mathbb{P})$ be a probability space and $X, Y : \Omega \to \mathbb{R}^d$ be square integrable random variables, i.e., $\int_\Omega \|X\|^2 \, d\mathbb{P} < \infty$. Furthermore, let

$$\sigma(Y) := \{Y^{-1}(A) : A \in B(\mathbb{R}^d)\}$$

be the Σ-algebra generated by Y. The *conditional expectation* $\mathbb{E}[X|Y] : \Omega \to \mathbb{R}^d$ is a square integrable random variable on $(\Omega, \sigma(Y), \mathbb{P})$ characterized by

$$\mathbb{E}_\mathbb{P}[f \, \mathbb{E}[X|Y]] = \mathbb{E}_\mathbb{P}[fX]$$

for all $\sigma(Y)$-measurable, bounded functions $f : \Omega \to \mathbb{R}$.

Conditional expectations $\mathbb{E}[X|Y]$ of random vectors can also be interpreted as Borel measurable functions $\psi : \mathbb{R}^d \to \mathbb{R}^d$ via the Doob–Dynkin Lemma, see [23, Lemma 1.14].

Proposition 6.1 (Doob–Dynkin Lemma) *Let $(\Omega, \Sigma, \mathbb{P})$ be a measure space and let C and D be metric spaces endowed with the Borel Σ-algebra. Let $f : \Omega \to C$, $g : \Omega \to D$ be measurable functions. Then the following are equivalent:*

(i) *f is $\sigma(g)$-measurable.*
(ii) *There exists Borel measurable function $\psi : D \to C$ such that $f = \psi \circ g$.*

If $C = \mathbb{R}^d$, $D = \mathbb{R}^m$ and $f \in L^2(\Omega, \mathbb{R}^d, \mathbb{P})$, then we have $\psi \in L^2(\mathbb{R}^m, \mathbb{R}^d, g_\sharp \mathbb{P})$.

By the Doob-Dynkin Lemma, there exists a function $\psi \in L^2(\mathbb{R}^d, Y_\sharp \mathbb{P})$ such that

$$\mathbb{E}[X|Y] = \psi \circ Y.$$

We also write $\mathbb{E}[X|Y=y]$ for $\psi(y)$. Strictly speaking, this notation only makes sense for $y \in Y(\Omega)$, but every measurable subset of $Y(\Omega)^C$ is a $Y_\sharp \mathbb{P}$ zero set. By definition it holds that $\mathbb{E}[X|Y=\cdot] \circ Y = \mathbb{E}[X|Y]$.

We can express $\mathbb{E}[X|Y=y]$ also by disintegration of measures.

Proposition 6.2 *Consider the disintegration*

$$\alpha := P_{X,Y} = \alpha^y \times_y P_Y.$$

Then it holds $\mathbb{E}[X|Y] = \psi \circ Y$ *with*

$$\psi(y) := \int_{\mathbb{R}^d} x \, d\alpha^y(x). \tag{26}$$

Proof We check that $\mathbb{E}_\mathbb{P}[f \psi \circ Y] = \mathbb{E}_\mathbb{P}[fX]$ for every $\sigma(Y)$-measurable bounded function $f : \Omega \to \mathbb{R}$. By the Doob-Dynkin Lemma there exists a Borel measurable function $g : \mathbb{R}^d \to \mathbb{R}^d$ such that $f = g \circ Y$. Thus we can compute, changing the integral order by Fubini,

$$\begin{aligned}
\mathbb{E}_\mathbb{P}[f\psi \circ Y] &= \int_\Omega (g \circ Y)(\omega) \int_{\mathbb{R}^d} x \, d\alpha^{Y(\omega)}(x) d\mathbb{P}(\omega) \\
&= \int_{\mathbb{R}^d \times \mathbb{R}^d} g(y) x \, d\alpha^y(x) d(Y_\sharp \mathbb{P})(y) \\
&= \int_{\mathbb{R}^d \times \mathbb{R}^d} g(y) x \, d\alpha^y(x) dP_Y(y) = \int_{\mathbb{R}^d \times \mathbb{R}^d} g(y) x \, dP_{X,Y}(x,y) \\
&= \int_\Omega f(\omega) X(\omega) \, d\mathbb{P}(\omega) = \mathbb{E}_\mathbb{P}[fX].
\end{aligned}$$

□

6.2 Curve via Conditional Expectation

A family of random variables $X_t : \Omega \to \mathbb{R}^d$, $t \in [0, 1]$, is called a *stochastic process*. We say that X_t is *continuously differentiable in* $t_0 \in [0, 1]$, if, for a.e. $\omega \in \Omega$, the map $t \mapsto X_t(\omega)$ is continuously differentiable in t_0, where we mean the continuous extension to the boundary if $t_0 \in \{0, 1\}$. We denote the derivative with respect to t by $\partial_t X_t$.

In [30, Theorem 3.3] it was proven that for well-behaved stochastic processes $(X_t)_t$ the conditional expectation can be used to obtain a vector field v_t such that $(X_{t,\sharp}\mathbb{P}, v_t)$ fulfills the continuity equation. For convenience, we include the proof.

Theorem 6.3 *Let* $X_t \in L^2(\Omega, \mathbb{R}^d, \mathbb{P})$ *be continuously differentiable in every* $t \in [0, 1]$. *Let*

$$\mu_t := X_{t,\sharp}\mathbb{P} = P_{X_t}, \quad \text{and} \quad v_t := \mathbb{E}[\partial_t X_t | X_t = \cdot]. \tag{27}$$

Assume that μ_t is a narrowly continuous curve, $v_t \in L^2(\mathbb{R}^d, \mu_t)$ for $t \in [0, 1]$ and $\|v_t\|_{L^2(\mathbb{R}^d, \mu_t)} \in L^1([0, 1])$. Then μ_t is an absolutely continuous curve in $\mathcal{P}_2(\mathbb{R}^d)$ and (μ_t, v_t) fulfills the continuity equation.

Proof For $X_t \in L^2(\Omega, \mathbb{R}^d, \mathbb{P})$, we know that $\mu_t = X_{t,\sharp}\mathbb{P} \in \mathcal{P}_2(\mathbb{R}^d)$. Since μ_t is narrowly continuous and $\|v_t\|_{L^2(\mu_t, \mathbb{R}^d)} \in L^1([0, 1])$, it suffices by Theorem 3.3 to show that (μ_t, v_t) satisfy the continuity equation (CE). For $\varphi \in C_c^\infty((0, 1) \times \mathbb{R}^d)$, we have

$$\frac{d}{dt}(\varphi_t \circ X_t) = (\partial_t \varphi_t) \circ X_t + \langle \nabla_x \varphi_t \circ X_t, \partial_t X_t \rangle,$$

so that

$$0 = \int_\Omega (\varphi_1 \circ X_1 - \varphi_0 \circ X_0) \, d\mathbb{P} = \int_\Omega \int_0^1 \frac{d}{dt}(\varphi_t \circ X_t) \, dt \, d\mathbb{P}$$

$$= \int_0^1 \int_\Omega (\partial_t \varphi_t) \circ X_t + \langle \nabla_x \varphi_t \circ X_t, \partial_t X_t \rangle \, d\mathbb{P} \, dt$$

$$= \int_0^1 \int_{\mathbb{R}^d} \partial_t \varphi_t \, d\mu_t \, dt + \int_0^1 \mathbb{E}[\langle \nabla_x \varphi_t \circ X_t, \partial_t X_t \rangle] \, dt$$

$$= \int_0^1 \int_{\mathbb{R}^d} \partial_t \varphi_t \, d\mu_t \, dt + \int_0^1 \mathbb{E}[\langle \nabla_x \varphi_t \circ X_t, \mathbb{E}[\partial_t X_t | X_t] \rangle] \, dt$$

$$= \int_0^1 \int_{\mathbb{R}^d} \partial_t \varphi_t \, d\mu_t \, dt + \int_0^1 \mathbb{E}[\langle \nabla_x \varphi_t \circ X_t, \mathbb{E}[\partial_t X_t | X_t = x] \circ X_t \rangle] \, dt$$

$$= \int_0^1 \int_{\mathbb{R}^n} \partial_t \varphi_t \, d\mu_t \, dt + \int_0^1 \int_{\mathbb{R}^d} \langle \nabla_x \varphi_t, \mathbb{E}[\partial_t X_t | X_t = x] \rangle \, d\mu_t \, dt.$$

It remains to verify the measurability of v which can be done similarly as in the proof of Proposition 4.5. Set $X : [0, 1] \times \Omega \to [0, 1] \times \mathbb{R}^d$ with $(t, \omega) \mapsto (t, X_t(\omega))$. Then X is measurable, since it can be written as limit of measurable functions by approximating it in t by step functions. Similarly, we have that $d_t X : [0, 1] \times \Omega \to \mathbb{R}^d$ given by $(t, \omega) \mapsto \partial_t X_t(\omega)$ is measurable. Thus, we can define a measurable function $v : [0, 1] \times \mathbb{R}^d \to \mathbb{R}^d$ by $(t, x) \mapsto \mathbb{E}[d_t X | X = (t, x)]$. Then it holds $v(t, \cdot) = \mathbb{E}[\partial_t X_t | X_t = \cdot]$ for a.e. $t \in [0, 1]$ by the following reason: let $\{g_n\}_{n \in \mathbb{N}} \subset C_b(\mathbb{R}^d)$ be a dense subset and $h \in C_b([0, 1])$. Then we have

$$\mathbb{E}_{(t,\omega) \sim \mathcal{L}_{[0,1]} \times \mathbb{P}}[((hg_n) \circ X)(v \circ X)] = \int_0^1 h(t) \mathbb{E}[(g_n \circ X_t)(v(t, \cdot) \circ X_t)] \, dt$$

and

$$\mathbb{E}_{(t,\omega)\sim\mathcal{L}_{[0,1]}\times\mathbb{P}}[((hg_n)\circ X)\,d_tX] = \int_0^1 h(t)\mathbb{E}[(g_n\circ X_t)\,\partial_tX_t]\,dt$$

and the left hand sides are equal by definition of v. This means that

$$\mathbb{E}[(g_n\circ X_t)\,(v(t,\cdot)\circ X_t)] = \mathbb{E}[(g_n\circ X_t)\,\partial_tX_t]$$

for a.e. $t \in [0, 1]$ and all $n \in \mathbb{N}$. Since $\{g_n\}_{n\in\mathbb{N}}$ is dense in $C_b(\mathbb{R}^d)$ and $C_b(\mathbb{R}^d)$ is dense in $L^2(\mathbb{R}^d, X_{t,\sharp}\mathbb{P})$, which contains the bounded measurable functions, we can conclude that $v(t,\cdot) = \mathbb{E}[\partial_tX_t|X_t = \cdot]$ for a.e. $t \in [0, 1]$. □

The following special case of Theorem 4.6 gives a relation to curve-velocity pairs induced by couplings.

Corollary 6.4 *For the special stochastic process $X_t = (1-t)X_0 + tX_1$, $t \in [0, 1]$ and the plan $\alpha = P_{X_0,X_1}$, the curve-velocity field (27) coincides with those induced by α in (15).*

Proof Let (μ_t, v_t) be given by (27). By (26), we can rewrite

$$v_t(x) = \mathbb{E}[\partial_tX_t|X_t = x] = \int_{\mathbb{R}^d} z\,dP_{\partial_tX_t|X_t=x}.$$

Since $X_t = (1-t)X_0 + tX_1$, $t \in [0, 1]$, we obtain $\partial_tX_t = X_1 - X_0$ and further, for any $f \in C_b(\mathbb{R}^d)$, that

$$\int_{\mathbb{R}^d} fv_t\,d\mu_t = \int_{\mathbb{R}^d} f(x)\int_{\mathbb{R}^d} z\,dP_{\partial_tX_t|X_t=x}dP_{X_t} = \int_{\mathbb{R}^d\times\mathbb{R}^d} f(x)z\,dP_{\partial_tX_t,X_t}$$

$$= \int_{\mathbb{R}^d\times\mathbb{R}^d} f(x)z\,dP_{X_1-X_0,X_t} = \int_{\mathbb{R}^d\times\mathbb{R}^d} f(e_t(x,y))(y-x)\,dP_{X_0,X_1}$$

$$= \int_{\mathbb{R}^d} f\,de_{t,\sharp}((y-x)\alpha).$$

□

Remark 6.5 Let $X_t = (1-t)X_0 + tX_1$ for independent random variables X_0, X_1, where $X_0 \sim \mathcal{N}(0, I_d)$. Then in [45, Appendix A1] it is shown that the formula for the vector field associated with μ_t,

$$v_t = \frac{1-t}{t}\nabla_x\log p_t(x) + \frac{x}{t}, \quad t \in (0, 1)$$

from Proposition 4.11 can also be derived via the velocity field from Theorem 6.3 and Tweedie's formula. Recall that Tweedie's formula [11, Lemma 3.2] for X_t reads as

$$\nabla_x \log p_t(x) = \frac{t \mathbb{E}[X_1 | X_t = x] - x}{(1-t)^2},$$

and thus $\mathbb{E}[X_1 | X_t = x] = \frac{(1-t)^2}{t} \nabla_x \log p_t(x) + \frac{x}{t}$. Using $X_0 = \frac{X_t - t X_1}{1-t}$ for $t \in (0, 1)$ we obtain

$$\begin{aligned} v_t(x) &= \mathbb{E}[\partial_t X_t | X_t = x] = \mathbb{E}[X_1 - X_0 | X_t = x] \\ &= \mathbb{E}\left[\frac{X_1}{1-t} - \frac{X_t}{1-t} \bigg| X_t = x\right] \\ &= \frac{1-t}{t} \nabla_x \log p_t(x) + \frac{x}{t(1-t)} - \frac{x}{1-t} \\ &= \frac{1-t}{t} \nabla_x \log p_t(x) + \frac{x}{t}, \end{aligned}$$

where we used Tweedie's formula and the fact that $\mathbb{E}[X_t | X_t] = X_t$. ◇

7 Flow Matching

In this section, we will see how we can approximate velocity fields of certain absolutely continuous curves $\mu_t : [0, 1] \to \mathcal{P}(\mathbb{R}^d)$ by neural networks. If the velocity field is known and the starting measure μ_0 is such that we can easily sample from, like the standard Gaussian one, then we can use the ODE (9) to produce samples from μ_t.

We consider the velocity fields v_t associated to the absolutely continuous curves v_t from Proposition 5.1. By the next proposition, these velocity fields can be obtained as minimizers of a certain loss function.

Proposition 7.1 *Let $\mu_1 \in \mathcal{P}_2(\mathbb{R}^d)$, and let $\mathcal{K}_t(y, \cdot) : [0, 1] \to \mathcal{P}_2(\mathbb{R}^d)$ be absolutely continuous curves with associated velocity fields v_t^y for μ_1-a.e. $y \in \mathbb{R}^d$. Further, assume that $(t, x, y) \mapsto v_t^y$ is measurable and $\|v_t^y\|_{L^2(\mathbb{R}^d \times \mathbb{R}^d, \alpha_t)} \in L^2([0, 1])$. For $\alpha_t := \mathcal{K}_t(y, \cdot) \times_y \mu_1$, we consider the absolutely continuous curve $v_t := \pi_\sharp^1 \alpha_t$. Then its associated velocity field $v : [0, 1] \times \mathbb{R}^d \to \mathbb{R}^d$ defined by (19) fulfills*

$$v = \underset{u \in L^2(\mathbb{R}^d, v_t \times \mathcal{L}_{(0,1)})}{\arg\min} \mathbb{E}_{(t,x,y) \sim \alpha_t \times \mathcal{L}_{(0,1)}} \left[\|u_t(x) - v_t^y(x)\|^2 \right].$$

Proof Starting with

$$\|u_t(x) - v_t^y(x)\|^2 = \|u_t(x)\|^2 - 2\langle u_t(x), v_t^y(x)\rangle + \|v_t^y(x)\|^2,$$

we deal with each part individually. It holds

$$\mathbb{E}_{(t,x,y)\sim\alpha_t\times_t\mathcal{L}_{(0,1)}}\left[\|u_t(x)\|^2\right] = \mathbb{E}_{t\sim\mathcal{L}_{(0,1)}}\mathbb{E}_{x\sim\nu_t}\left[\mathbb{E}_{y\sim\alpha_t^x}\left[\|u_t(x)\|^2\right]\right]$$
$$= \mathbb{E}_{(t,x)\sim\nu_t\times_t\mathcal{L}_{(0,1)}}\left[\|u_t(x)\|^2\right].$$

For the second term, we obtain with Fubini's theorem

$$\mathbb{E}_{(t,x,y)\sim\alpha_t\times_t\mathcal{L}_{(0,1)}}\left[\langle u_t(x), v_t^y(x)\rangle\right] = \mathbb{E}_{(t,x)\sim\nu_t\times_t\mathcal{L}_{(0,1)}}\left[\int_{\mathbb{R}^d}\langle u_t(x), v_t^y(x)\rangle\,\mathrm{d}\alpha_t^x(y)\right]$$
$$= \mathbb{E}_{(t,x)\sim\nu_t\times_t\mathcal{L}_{(0,1)}}\left[\langle u_t(x), \int_{\mathbb{R}^d}v_t^y(x)\,\mathrm{d}\alpha_t^x(y)\rangle\right]$$
$$= \mathbb{E}_{(t,x)\sim\nu_t\times_t\mathcal{L}_{(0,1)}}\left[\langle u_t(x), v_t^y(x)\rangle\right].$$

Putting everything together, we get

$$\mathbb{E}_{(t,x,y)\sim\alpha_t\times_t\mathcal{L}_{(0,1)}}\left[\|u_t(x) - v_t^y(x)\|^2\right] = \mathbb{E}_{(t,x)\sim\nu_t\times_t\mathcal{L}_{(0,1)}}\left[\|u_t(x) - v_t(x)\|^2\right]$$
$$- \mathbb{E}_{(t,x)\sim\nu_t\times_t\mathcal{L}_{(0,1)}}\left[\|v_t(x)\|^2\right] + \mathbb{E}_{(t,x,y)\sim\alpha_t\times_t\mathcal{L}_{(0,1)}}\left[\|v_t^y(x)\|^2\right].$$

Since the last two terms do not depend on u and the minimizer of the first term is v, we obtain the assertion. □

In order to learn v, the proposition suggests to use the loss

$$\mathrm{FM}_{\alpha_t}(\theta) := \mathbb{E}_{(t,x,y)\sim\alpha_t\times_t\mathcal{L}_{(0,1)}}[\|u_t^\theta(x) - v_t^y(x)\|^2]. \tag{28}$$

The name "FM" comes from "flow matching". In the original paper [28], the notation "conditional flow matching" was used which refers to a "conditional flow path", see Remark 5.3. In practice, we work with the empirical expectation. To this end, we will need to be able to sample from $\alpha_t = \mathcal{K}_t(y,\cdot)\times_y\mu_1$ and to compute $v_t^y(x)$. A typical scenario is that samples y_i, $i = 1,\ldots,n$ from the data distribution μ_1 are available and that it is easy to sample from the Markov kernel $\mathcal{K}_t(y_i,\cdot)$, $i = 1,\ldots,n$.

Example 7.2 A setting, where FM_{α_t} is tractable is Example 5.4. Here $\mu_0 = \mathcal{N}(0, I_d)$ and

$$\mathcal{K}_t(y, \cdot) = \mathcal{N}(ty, (1-rt)^2 I_d) \quad \text{and} \quad v_t^y = \frac{y - rx}{1 - tr},$$

and we obtain

$$\begin{aligned}
\mathrm{FM}_{\alpha_t}(\theta) &= \mathbb{E}_{(t,x,y) \sim \alpha_t \times \mathcal{L}_{(0,1)}} \left[\| u_t^\theta(x) - v_t^y(x) \|^2 \right] \\
&= \mathbb{E}_{t \sim [0,1]} \mathbb{E}_{y \sim \mu_1} \mathbb{E}_{x \sim \mathcal{N}(ty, (1-rt)^2 I_d)} \left[\| u_t^\theta(x) - \frac{y - rx}{1 - tr} \|^2 \right] \\
&= \mathbb{E}_{t \sim [0,1]} \mathbb{E}_{y \sim \mu_1} \mathbb{E}_{x \sim \mathcal{N}(0, I_d)} \left[\| u_t^\theta((1-tr)x + ty) - (y - rx) \|^2 \right] \\
&= \mathbb{E}_{t \sim [0,1], (x,y) \sim \mu_0 \times \mu_1} \left[\| u_t^\theta((1-tr)x + ty) - (y - rx) \|^2 \right].
\end{aligned}$$

For α_t coming from a coupling $\alpha \in \Gamma(\mu_0, \mu_1)$ as in (20) with disintegration kernel $\mathcal{K}_t(y, \cdot)$, the associated velocity field v_t^y has the simple form (23). In this important case, the loss function (28) can be simplified.

Proposition 7.3 *Let $\alpha \in \Gamma(\mu_0, \mu_1)$ and $\mu_t = e_{t,\sharp}\alpha$. Then the velocity field $v : [0, 1] \times \mathbb{R}^d \to \mathbb{R}^d$ induced by α fulfills*

$$v = \operatorname*{argmin}_{u \in L^2(\mathbb{R}^d, \mu_t \times \mathcal{L}_{(0,1)})} \mathbb{E}_{(t,x,y) \sim \mathcal{L}_{(0,1)} \times \alpha} \left[\| u_t(e_t(x, y)) - (y - x) \|^2 \right].$$

Proof By Proposition 5.6, we know that $\mathcal{K}_t(y, \cdot)$ given via

$$\alpha_t := (e_t, \pi^2)_{\sharp} \alpha = \mathcal{K}_t(y, \cdot) \times_y \mu_1$$

is an absolutely continuous curve for μ_1-a.e. $y \in \mathbb{R}^d$ with associated simple velocity field $v_t^y(x) := \frac{y-x}{1-t}$, $t \in [0, 1)$. Further, $\mu_t = \nu_t := \pi^1_{\sharp} \alpha_t$ and the velocity field induced by α in (24) coincides with those in (19). Thus, by Proposition 7.1, we can calculate

$$\begin{aligned}
v &= \operatorname*{argmin}_{u \in L^2(\mathbb{R}^d, \mu_t \times \mathcal{L}_{(0,1)})} \mathbb{E}_{(t,x,y) \sim \alpha_t \times \mathcal{L}_{(0,1)}} \left[\| u_t(x) - v_t^y(x) \|^2 \right] \\
&= \operatorname*{argmin}_{u \in L^2(\mathbb{R}^d, \mu_t \times \mathcal{L}_{(0,1)})} \mathbb{E}_{(t,x,y) \sim \alpha_t \times \mathcal{L}_{(0,1)}} \left[\left\| u_t(x) - \frac{y-x}{1-t} \right\|^2 \right] \\
&= \operatorname*{argmin}_{u \in L^2(\mathbb{R}^d, \mu_t \times \mathcal{L}_{(0,1)})} \mathbb{E}_{(t,x,y) \sim \alpha \times \mathcal{L}_{(0,1)}} \left[\| u_t(e_t(x, y)) - (y - x) \|^2 \right].
\end{aligned}$$

$\square$

Thus, for curves and velocity fields induced by plans $\alpha \in \Gamma(\mu_0, \mu_1)$ we can use the simplified flow matching objective

$$\mathrm{FM}_\alpha(\theta) = \mathbb{E}_{t \sim [0,1], (x,y) \sim \alpha}[\|u_t^\theta(e_t(x, y)) - (y - x)\|^2]. \tag{29}$$

Computing the empirical expectation requires only samples from the coupling $\alpha \in \Gamma(\mu_0, \mu_1)$. Two canonical choices of couplings, which we can usually sample from, are

- the independent coupling $\alpha = \mu_0 \times \mu_1$, which requires only samples of both μ_0 and μ_1, and
- an optimal coupling $\alpha \in \Gamma_o(\mu_0, \mu_1)$, which is usually approximated by applying an optimal transport solver on empirical measures approximating μ_i, $i = 0, 1$.

Finally, for a curve arising from a stochastic process also the description of the velocity field by conditional expectation in (27),

$$v_t := \mathbb{E}[\partial_t X_t | X_t = \cdot]$$

can be interpreted as a minimization problem. Namely, it is well-known, see, e.g., [6, 10.1.5 (5)], that

$$\mathbb{E}[X | Y = \cdot] = \underset{f \in L^2(\mathbb{R}^d, Y_\sharp \mathbb{P})}{\mathrm{argmin}} \ \mathbb{E}_\mathbb{P}[\|X - f(Y)\|^2].$$

Thus, we obtain

$$v = \underset{u \in L^2(\mathbb{R}^d, \mu_t \times_t \mathcal{L}_{(0,1)})}{\mathrm{argmin}} \ \mathbb{E}_{\mathcal{L}_{(0,1)} \times \mathbb{P}}[\|\partial_t X_t - u_t(X_t)\|^2].$$

For the special stochastic process $X_t = (1-t)X_0 + tX_1$ in Corollary 6.4, leading to curves induced by plans, this becomes

$$v = \underset{u \in L^2(\mathbb{R}^d, \mu_t \times_t \mathcal{L}_{(0,1)})}{\mathrm{argmin}} \ \mathbb{E}_{\mathcal{L}_{(0,1)} \times \mathbb{P}}[\|X_1 - X_0 - u_t(X_t)\|^2]$$

$$= \underset{u \in L^2(\mathbb{R}^d, \mu_t \times_t \mathcal{L}_{(0,1)})}{\mathrm{argmin}} \ \mathbb{E}_{(t,x,y) \sim \mathcal{L}_{(0,1)} \times P_{X_0, X_1}}[\|y - x - u_t(e_t(x, y))\|^2]$$

and we are back at Proposition 7.3.

8 Numerical Examples for Flow Matching

8.1 Flow Matching Algorithms

Sampling from a product plan $\alpha = \mu_0 \times \mu_1$ can be reduced to sampling from the individual measures $z_i \sim \mu_0$, $x_i \sim \mu_1$ and simply building the product $(z_i, x_i) \sim \alpha$. This is the reason that training a flow matching model from a product plan is particularly easy, as can be seen in Algorithm 1.

Given two measures $\mu_0, \mu_1 \in \mathcal{P}_2(\mathbb{R}^d)$, sampling from an optimal transport plan $\alpha \in \Gamma_o(\mu_0, \mu_1)$ is usually not possible. Often both measures are not known analytically but only an approximation by samples, so that an optimal plan cannot be obtained. But even the plan for the approximation is often not computable since usually the number of samples is so high that solving the OT problem is computationally too expensive. In this case a popular method is minibatch OT flow matching [42]. The idea is to partition the samples $x_1, \ldots, x_{N_t} \sim \mu_1$ into minibatches (smaller subsets), on which solving the OT problem is computational feasible. Details are described in Algorithm 2.

Once a flow vector field u^θ is learned, sampling is then done by sampling $z_i \sim \mu_0$ from the source distribution and using an ODE solver for $f'(t) = u_t^\theta(f(t))$, $f(0) = z_i$ to compute $f(1)$.

Algorithm 1 Learning u^θ from the plan $\alpha = \mu_0 \times \mu_1 \in \Gamma(\mu_0, \mu_1)$

N_b: batch size, N_e: number of epochs, N_t number of samples from the target distribution
$x_1, \ldots, x_{N_t}$ samples from the target distribution μ_1
Network $u^\theta : [0, 1] \times \mathbb{R}^d \to \mathbb{R}^d$ depending on a parameter $\theta \in \mathbb{R}^m$.
for $e = 1, \ldots, N_e$ **do**
    for $s = 1, \ldots, \frac{N_t}{N_b}$ **do**
        Draw $(z_i) \sim \mu_0$ for $1 \leq i \leq N_b$
        Draw $t_i \sim \mathcal{L}_{(0,1)}$ for $1 \leq i \leq N_b$
        Compute $x_{t_i}^i = (1 - t_i)z_i + t_i x_{i+(s-1)N_b}$
        Compute $\mathcal{L}(\theta) = \frac{1}{N_b} \sum_{i=1}^{N_b} \left\| u^\theta(t, x_{t_i}^i) - (x_{i+(s-1)N_b} - z_i) \right\|^2$ and $\nabla_\theta \mathcal{L}(\theta)$
        Update θ using $\nabla_\theta \mathcal{L}(\theta)$
    end for
end for
Return: θ

Algorithm 2 Learning u^θ from minibatch OT flow matching for μ_0 the source distribution and μ_1 the target distribution

N_b: batch size, N_{bOT} OT batch size, N_e: number of epochs, N_t number of samples from the target distribution
$x_1, \ldots, x_{N_t}$ samples from the target distribution μ_1
Network $u^\theta : [0, 1] \times \mathbb{R}^d \to \mathbb{R}^d$ depending on a parameter $\theta \in \mathbb{R}^m$.
for $e = 1, \ldots, N_e$ do
    for $o = 1, \ldots, \frac{N_t}{N_{bOT}}$ do
        Draw $z_i \sim \mu_0$ for $1 \leq i \leq N_{bOT}$
        Compute optimal transport map for

$$W_2 \left(\frac{1}{N_{bOT}} \sum_{i=1}^{N_{bOT}} \delta_{z_i}, \frac{1}{N_{bOT}} \sum_{i=1}^{N_{bOT}} \delta_{x_{i+(o-1)N_{bOT}}} \right)$$

        described by a function $T_o : \{1, \ldots, N_{bOT}\} \to \{1 + (o-1)N_{bOT}, \ldots, oN_{bOT}\}$
        for $s = 1, \ldots, \frac{N_{bOT}}{N_b}$ do
            Draw $t_i \sim \mathcal{L}_{(0,1)}$ for $1 \leq i \leq N_b$
            Compute $x_{t_i}^i = (1-t_i)z_i + t_i x_{T_o(i)}$
            Compute $\mathcal{L}(\theta) = \frac{1}{N_b} \sum_{i=1}^{N_b} \left\| u^\theta(t_i, x_{t_i}^i) - (x_{T_o(i)} - z_i) \right\|^2$ and $\nabla_\theta \mathcal{L}(\theta)$
            Update θ using $\nabla_\theta \mathcal{L}(\theta)$
        end for
    end for
end for
Return: θ

8.2 Numerical Examples

For our first example we chose standard $2d$ Gaussian distribution as source distribution and a Gaussian mixture model with eight equality weighted modes as target distribution, see Fig. 7. In the middle, we can see the trajectories created by a vector field learned with minibatch OT flow matching, using a simple Euler method for solving the flow ODE. Here the trajectories are mostly straight, making sampling easier since the sampling quality does not depend as much on time discretization or the ODE solving method. On the left, the trajectories for a vector field learned via flow matching for a product plan is displayed, where again an Euler method was used to solve the flow ODE. Here the trajectories are much less straight and sample quality depends very much on the ODE solving quality. On the right we see the result for Neural ODEs, described in Sect. 10, which has trajectories differing from both of the flow matching approaches.

Finally, we show a high dimensional application of flow matching in Fig. 8. Here we used the flow vector field from [31], which was already trained via minibatch OT flow matching on images of cat faces. For a comparison of the performance for the flow matching with different plans see [42, Table 2].

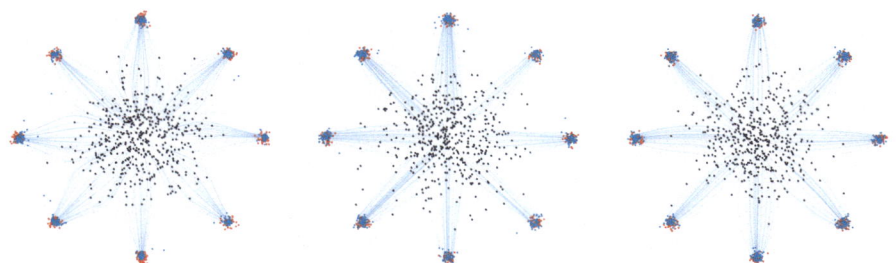

Fig. 7 Trajectories of points from a vector field u^θ obtained via flow matching from a coupling $\alpha \in \Gamma(\mu_0, \mu_1)$. We chose $\mu_0 = \mathcal{N}(0, 1)$ and as target distribution μ_1 a Gaussian mixture model with 8 modes. Black: points $\{z_i\}_{i=1}^{300}$ drawn from the source distribution $\mu_0 = \mathcal{N}(0, 1)$. Blue: points sampled via the vector field from $\{z_i\}_{i=1}^{300}$. Red: points sampled from the target distribution. Blue lines: trajectories of the vector field. The left image shows the results for u^θ trained using the independent coupling and the middle image shows the result when trained with minibatch OT flow matching. The right image shows the result from training a continuous normalizing flow as in Sect. 10

Fig. 8 Single trajectory for a flow matching model trained on cat images. For the trained model, see [31]. Note that for the standard Gaussian distribution we can sample in each direction separately, so that the left starting sample contains the one-dimensional coordinates in each of the $d = 256^2$ directions for the red-green-blue channels

9 Flow Matching in Bayesian Inverse Problems

In this section, we are interested in Bayesian inverse problems

$$W = F(X) + \Xi$$

for random variables $W : \Omega \to \mathbb{R}^m$, $X : \Omega \to \mathbb{R}^d$, a forward operator $F : \mathbb{R}^d \to \mathbb{R}^m$ and a random variable $\Xi : \Omega \to \mathbb{R}^m$ which is responsible for noisy outputs. The aim is to approximate the posterior distribution $P_{X|W=w}$. We propose to learn a map

$$g : \mathbb{R}^m \times \mathbb{R}^d \to \mathbb{R}^m \times \mathbb{R}^d, \quad (w, x) \mapsto (w, g_w(x))$$

such that

$$g_\sharp(P_W \times \mathcal{N}(0, 1)) = P_{W,X}.$$

This implies $g_{w,\sharp}\mathcal{N}(0,1) = P_{X|W=w}$, since we have for any $f \in C_b(\mathbb{R}^m \times \mathbb{R}^d)$ that

$$\int_{\mathbb{R}^m \times \mathbb{R}^d} f \, dP_{W,X} = \int_{\mathbb{R}^m \times \mathbb{R}^d} f \, dg_\sharp(P_W \times \mathcal{N}(0,1)) = \int_{\mathbb{R}^m} \int_{\mathbb{R}^d} f(w,x) \, d(g_{w,\sharp}\mathcal{N}(0,1)) dP_W.$$

In particular, we can sample from $P_{X|W=w}$ by sampling from $\mathcal{N}(0,1)$ and applying g_w. Again, we want to describe the generator g via a curve in a space of probability measures. However, now we need a curve, where the W-marginal does not change, i.e., a curve in the space

$$\mathcal{P}_{P_W} := \{\mu \in \mathcal{P}_2(\mathbb{R}^m \times \mathbb{R}^d) : \pi^1_\sharp \mu = P_W\}.$$

In the next subsection, we will equip the space $\mathcal{P}_{P_W}$ with a metric turning it into a complete metric space and determine couplings such that the corresponding curve stays in $\mathcal{P}_{P_W}$. Then we will use these curves for flow matching in order to learn conditional generating networks that approximate the posterior distribution $P_{X|W=w}$ for w sampled from P_W. This section is based on our paper [7], see also [24].

9.1 Conditional Wasserstein Distance

Conditional Wasserstein distances appear already in [26, Section 2.4], [32] as well as in the PhD thesis of Gigli [13, Chapter 4], who studied a metric on the so-called "geometric tangent space" of the Wasserstein-2 space. Recently, conditional Wasserstein distances were studied in the realm of machine learning in [4] as well as in functional optimal transport in [21]. We will focus on its application in Bayesian flow matching as presented in our paper [7], where also the proofs of the following theorems and propositions are provided.

To fix the notation, let $\mathfrak{u} = (w,x) \in \mathbb{R}^m \times \mathbb{R}^d$ and denote the projections by

$$\pi^1(\mathfrak{u}) := w, \quad \pi^2(\mathfrak{u}) := x.$$

For an arbitrary, fixed reference measure $\eta \in \mathcal{P}_2(\mathbb{R}^m)$, we consider the space

$$\mathcal{P}_\eta(\mathbb{R}^m \times \mathbb{R}^d) := \{\mu \in \mathcal{P}_2(\mathbb{R}^m \times \mathbb{R}^d) : \pi^1_\sharp \mu = \eta\}.$$

In order to define a conditional Wasserstein distance on $\mathcal{P}_\eta$, we introduce, for $\mu, \nu \in \mathcal{P}_\eta(\mathbb{R}^m \times \mathbb{R}^d)$, the set of 4-plans

$$\Gamma_\eta = \Gamma_\eta(\mu,\nu) := \left\{\alpha \in \Gamma(\mu,\nu) : \pi^{1,1}_\sharp \alpha = \Delta_\sharp \eta\right\},$$

where

$$\pi^{1,1}(u_1, u_2) := (\pi^1(u_1), \pi^1(u_2)) = \left(\pi^1((w_1, x_1)), \pi^1((w_2, x_2))\right) = (w_1, w_2)$$

and $\Delta : \mathbb{R}^m \to \mathbb{R}^m \times \mathbb{R}^m, w \mapsto (w, w)$ is the diagonal map. Note that

$$\Delta^{-1}(w_1, w_2) = \emptyset \text{ if } w_1 \neq w_2 \quad \text{and} \quad \Delta^{-1}(w_1, w_2) = w \text{ if } w_1 = w_2 = w.$$

Now we define the *conditional Wasserstein(-2) distance* by

$$W_{2,\eta}(\mu, \nu) := \left(\min_{\alpha \in \Gamma_\eta(\mu,\nu)} \int_{(\mathbb{R}^m \times \mathbb{R}^d)^2} \|u_1 - u_2\|^2 \, d\alpha\right)^{\frac{1}{2}}.$$

Indeed, the above minimum is obtained and the set of optimal couplings is denoted by $\Gamma_{\eta,o}$. The following theorem collects important properties of the conditional Wasserstein distance.

Theorem 9.1 *The conditional Wasserstein space $\left(\mathcal{P}_\eta(\mathbb{R}^m \times \mathbb{R}^d), W_{2,\eta}\right)$ is a complete metric space. The conditional Wasserstein distance has the following properties:*

(i) *For $\mu, \nu \in \mathcal{P}_\eta(\mathbb{R}^m \times \mathbb{R}^d)$, it holds*

$$W_{2,\eta}^2(\mu, \nu) = \mathbb{E}_{w \sim \eta}\left[W_2^2(\mu^w, \nu^w)\right],$$

where μ^w, ν^w denote the disintegrations of μ, ν with respect to η, respectively.

(ii) *There exist optimal plans $\alpha_w \in \Gamma_o(\mu^w, \nu^w)$, $w \in \mathbb{R}^m$ such that*

$$\alpha := \underbrace{(\delta_w \times \alpha_w)}_{\in \mathcal{P}_2(\mathbb{R}^m \times \mathbb{R}^d \times \mathbb{R}^d)} \times_w \eta \in \Gamma_{\eta,o}(\mu, \nu).$$

Note that (ii) means, for any $f \in C_b(\mathbb{R}^d)$, that

$$\int_{(\mathbb{R}^m \times \mathbb{R}^d)^2} f \, d\alpha = \int_{\mathbb{R}^m} \int_{\mathbb{R}^m \times \mathbb{R}^d \times \mathbb{R}^d} f(w_1, x_1, w_2, x_2) \, d\delta_{w_1}(w_2) d\alpha_{w_1}(x_1, x_2) d\eta(w_1)$$

$$= \int_{\mathbb{R}^m} \int_{\mathbb{R}^d \times \mathbb{R}^d} f(w, x_1, w, x_2) \, d\alpha_w(x_1, x_2) d\eta(w).$$

Fig. 9 Consider $\eta = \frac{1}{2}\delta_0 + \frac{1}{2}\delta_1$, $\mu_n = \frac{1}{2}\delta_{0,n} + \frac{1}{2}\delta_{1,0}$ and $\nu_n = \frac{1}{2}\delta_{0,0} + \frac{1}{2}\delta_{1,n}$. Then $W_2(\mu_n, \nu_n) = 1$ is the length of $\longrightarrow$ and $\longrightarrow$ indicates the optimal transport map. Furthermore $W_{2,\eta}(\mu_n, \nu_n) = n$ is the length of $\longrightarrow$ and $\longrightarrow$ indicates the optimal transport for $W_{2,\eta}$

Remark 9.2 Proposition 9.1(*ii*) indicates another use of the conditional Wasserstein distance. Namely if we have two measures $\mu_0, \mu_1 \in \mathcal{P}_\eta$ and we have that $W_2(\mu_0, \mu_1)$ is small that does not necessarily mean that the posteriors are close. We can see in Fig. 9 that there exist $\eta \in \mathcal{P}_2(\mathbb{R})$, $\mu_n, \mu_n \in \mathcal{P}_\eta(\mathbb{R} \times \mathbb{R})$ such that $W_2(\mu_n, \nu_n) = 1$ and $\mathbb{E}_{w \sim \eta}[W_2^2(\mu_{0,w}, \mu_{1,w})] = W_{2,\eta}(\mu_n, \nu_n) = n$. This means that W_2 is not a good measure in order to access if a generator g is approximating the posteriors well. In contrast if we use $W_{2,\eta}$ we know that the posteriors are, at least in expectation, well approximated.

As shown in [24, Theorem 3,4] and [4], there is a counterpart of Theorem 3.3 for the conditional Wasserstein distance.

Theorem 9.3 *Let $\mu_t : I \to \mathcal{P}_\eta(\mathbb{R}^m \times \mathbb{R}^d)$ be a narrowly continuous curve. Then μ_t is absolutely continuous in with respect to $W_{2,\eta}$ in the sense of (6), if and only if there exists a measurable vector field $v : I \times \mathbb{R}^m \times \mathbb{R}^d \to \mathbb{R}^m \times \mathbb{R}^d$ such that the following three conditions are fulfilled:*

(i) $(v_t)_j = 0$ *as element in* $L^2(\mathbb{R}, \mu_t)$ *for a.e. $t \in [0, 1]$ and for all $j \leq m$,*
(ii) $\|v_t\|_{L^2(\mathbb{R}^m \times \mathbb{R}^d, \mu_t)} \in L^1(I)$,
(iii) (μ_t, v_t) *fulfills* (CE).

In this case, the metric derivative $|\mu_t'|$ exists and we have $|\mu_t'| \leq \|v_t\|_{L^2(\mu_t, \mathbb{R}^m \times \mathbb{R}^d)}$ for a.e. $t \in I$.

The following theorem ensures that the curve μ_t associated to a plan $\alpha \in \Gamma_\eta(\mu_0, \mu_1)$ fulfills $\mu_t \in \mathcal{P}_\eta(\mathbb{R}^m \times \mathbb{R}^d)$ for all $t \in [0, 1]$ and the corresponding vector field does not move mass in the w component.

Theorem 9.4 Let $\mu_0, \mu_1 \in \mathcal{P}_\eta(\mathbb{R}^m \times \mathbb{R}^d)$ and let $\alpha \in \Gamma_\eta(\mu_0, \mu_1)$. Let (μ_t, v_t) be the curve and measurable velocity field induced by α, i.e.

$$\mu_t = e_{t,\sharp}\alpha, \quad \text{and} \quad v_t \mu_t = e_{t,\sharp}[(\mathfrak{u}_2 - \mathfrak{u}_1)\alpha] \quad \text{for a.e. } t \in [0, 1].$$

Then the following holds true:

(i) $\mu_t \in \mathcal{P}_\eta(\mathbb{R}^m \times \mathbb{R}^d)$ for all $t \in [0, 1]$,
(ii) $(v_t)_j = 0$ as element in $L^2(\mathbb{R}, \mu_t)$ for a.e. $t \in [0, 1]$ and for all $j \leq m$,
(iii) (μ_t, v_t) satisfy the continuity equation (CE) and

$$\|v_t\|_{L^2(\mathbb{R}^m \times \mathbb{R}^d, \mu_t)} \leq \|y - x\|_{L^2((\mathbb{R}^m \times \mathbb{R}^d)^2, \alpha)} \quad \text{for a.e. } t \in [0, 1].$$

In particular, we have $\|v_t\|_{L^2(\mathbb{R}^m \times \mathbb{R}^d, \mu_t)} \in L^1(I)$ and μ_t is an absolutely continuous curve with respect to $W_{2,\eta}$.

Proof To proof (i) and (ii) follow from [7, Lemma 5, Proposition 6]. The first statements of (iii) follow from Proposition 4.6 and finally the absolutely continuity with respect to $W_{2,\eta}$ is implied by Theorem 9.3. □

Remark 9.5 Note that for an absolutely curve $\mu_t \in \mathcal{P}_\eta$ and a vector field $v_t \in L^2(\mu_t, \mathbb{R}^m \times \mathbb{R}^d)$ it is not necessarily true that $v_j = 0$ for every $j \leq m$ since these components could encode η preserving maps. However, Proposition 9.4 ensures that for a curve associated to a plan $\alpha \in \Gamma_\eta(\mu_0, \mu_1)$ this is the case and in particular for a map ϕ which is a solution of the ODE $\partial_t \phi_t = v_t(\phi_t); \phi_0 = \text{Id}$ we can conclude that $\pi^1(\phi_t(\mathfrak{u})) = \pi^1(\mathfrak{u})$.

The counterpart of Proposition 4.4 reads as follows, see [7, Lemma 5].

Proposition 9.6 Let $\mu_0, \mu_1 \in \mathcal{P}_\eta(\mathbb{R}^m \times \mathbb{R}^d)$ and let $\alpha \in \Gamma_{\eta,o}(\mu_0, \mu_1)$. Then the curve $\mu_t := (e_t)_\sharp \alpha$ is a geodesic in $(\mathcal{P}_\eta(\mathbb{R}^m \times \mathbb{R}^d), W_{2,\eta})$. In particular, $(\mathcal{P}_\eta(\mathbb{R}^m \times \mathbb{R}^d), W_{2,\eta})$ is a geodesic space.

An illustration of the difference between geodesics with respect to W_2 and geodesics with respect to $W_{2,\eta}$ is given in Fig. 10.

9.2 Almost Conditional Couplings

One drawback of the space $\mathcal{P}_\eta(\mathbb{R}^m \times \mathbb{R}^d)$ is that we can in general not approximate $\mu \in \mathcal{P}_\eta(\mathbb{R}^m \times \mathbb{R}^d)$ by an empirical measure, if η is not empirical. In other words, an empirical

Fig. 10 Consider
$\eta = \frac{1}{2}\delta_0 + \frac{1}{2}\delta_1$,
$\mu_0 = \frac{1}{2}\delta_{0,5} + \frac{1}{2}\delta_{1,0}$ and
$\mu_1 = \frac{1}{2}\delta_{0,0} + \frac{1}{2}\delta_{1,5}$. (**a**)
Geodesic with respect to W_2,
green: $\mu_{\frac{1}{2}}$. (**b**) Geodesic with
respect to $W_{2,\eta}$, green: $\mu_{\frac{1}{2}}$

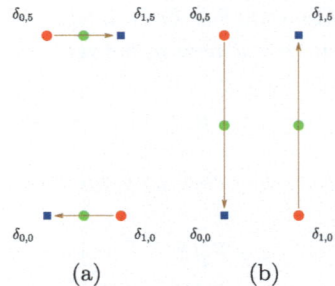

(a)   (b)

approximation $\mu_n = \frac{1}{n}\sum_{i=1}^n \delta_{(w_i,x_i)}$ of $\mu \in \mathcal{P}_\eta(\mathbb{R}^m \times \mathbb{R}^d)$ with respect to W_2 is usually not an element of $\mathcal{P}_\eta(\mathbb{R}^m \times \mathbb{R}^d)$. Even if we approximate two measures $\mu, \nu \in \mathcal{P}_\eta(\mathbb{R}^m \times \mathbb{R}^d)$ with respect to W_2 by empirical measures μ_n, ν_n with $\pi^1_\sharp \mu_n = \pi^1_\sharp \nu_n$, it is not necessarily true that there is a sequence $\alpha_n \in \Gamma_{\pi^1_\sharp \mu_n, o}(\mu_n, \nu_n)$ such that α_n converges to an $\alpha \in \Gamma_{\eta,o}(\mu, \nu)$ as [7, Example 9] shows.

One way to overcome the above drawback is to relax the hard constraint $\pi^{1,1}\alpha = \Delta_\sharp \eta$ in the definition of the conditional Wasserstein distance. To this end, we define

$$d_\beta(\mathfrak{u}_1, \mathfrak{u}_2) := \|x_1 - x_2\|^2 + \beta\|w_1 - w_2\|^2, \quad \beta > 0.$$

Then

$$W_\beta(\mu, \nu) := \min_{\alpha \in \Gamma(\mu,\nu)} \left(\int_{(\mathbb{R}^m \times \mathbb{R}^d)^2} d_\beta(\mathfrak{u}_1, \mathfrak{u}_2) \, d\alpha \right)^{\frac{1}{2}}$$

defines a metric on $\mathcal{P}_2(\mathbb{R}^m \times \mathbb{R}^d)$. It turns out that for $0 \ll \beta$ the optimal couplings α in this metric are "close" to fulfilling $\pi^{1,1}\alpha = \Delta \eta$ in the following sense, see [7, Proposition 10].

Proposition 9.7 *Let $\mu_0, \mu_1 \in \mathcal{P}_\eta(\mathbb{R}^m \times \mathbb{R}^d)$ and let $(\alpha_\beta)_\beta$ be a sequence of optimal transport plans from μ_0 to μ_1 with respect to W_β. Then we have*

$$\int_{\mathbb{R}^m \times \mathbb{R}^m} \|w_1 - w_2\|^2 \, d\pi^{1,1}_\sharp \alpha_\beta = \int_{(\mathbb{R}^m \times \mathbb{R}^d)^2} \|w_1 - w_2\|^2 \, d\alpha_\beta \to 0 \quad as \quad \beta \to \infty.$$

The distance W_β successfully addresses the issue of approximating measures by empirical measures as the following proposition from [7, Proposition 12] shows.

Proposition 9.8 *Let $\mu, \nu \in \mathcal{P}_\eta(\mathbb{R}^m \times \mathbb{R}^d)$ and let μ_n, ν_n be empirical measures which converge narrowly to μ, ν. Then, for a sequence $\beta_k \to \infty$, there exists an increasing* subsequence n_k and optimal plans $\alpha_{n_k} \in \Gamma(\mu_{n_k}, \nu_{n_k})$ for $W_{\beta_k}(\mu_{n_k}, \nu_{n_k})$ such that α_{n_k} converges narrowly to an optimal plan $\alpha \in \Gamma_{\eta,o}(\mu, \nu)$.

Remark 9.9 For $\beta > 0$ let $a_\beta : \mathbb{R}^m \times \mathbb{R}^d \to \mathbb{R}^m \times \mathbb{R}^d$ be defined by $a_\beta(w, x) = (\sqrt{\beta}w, x)$. Then an easy calculation shows that $W_\beta(\mu, \nu) = W_2(a_{\beta,\sharp}\mu, a_{\beta,\sharp}\nu)$ and there is a one-to-one correspondence between optimal plans for $W_\beta(\mu, \nu)$ and $W_2(a_{\beta,\sharp}\mu, a_{\beta,\sharp}\nu)$ given by $\alpha \mapsto (a_\beta, a_\beta)_\sharp \alpha$. Thus W_β and optimal plans can be computed by standard W_2 solvers. ◇

9.3 Bayesian Flow Matching

Since $\Gamma_\eta(\mu_0, \mu_1) \subseteq \Gamma(\mu_0, \mu_1)$ we can learn the vector field v_t corresponding to $\alpha \in \Gamma_\eta(\mu_0, \mu_1)$ with the extended flow matching objective (29), i.e.,

$$\text{CFM}(\theta) := \mathbb{E}_{t \sim \mathcal{L}[0,1], (w, x_0, w, x_1) \sim \alpha} \left[\| v^\theta(t, e_t((w, x_0, w, x_1))) - (0, x_1 - x_0) \|^2 \right].$$

Since $\pi_\sharp^{1,1} \alpha = \Delta_\sharp \eta$, we only have to consider samples (w, x_0, w, x_1) from α. Further, since $v_j = 0$ for $j \leq m$, we can parametrize v_t^θ as function with values in $\mathbb{R}^d$ instead of $\mathbb{R}^m \times \mathbb{R}^d$. In this way, we have for a solution of

$$\partial_t \phi_t(x) = v_t(\phi_t(x)), \quad \phi_0(x) = x$$

automatically that $\pi_\sharp^1(\phi_{t,\sharp}\mu_0) = \eta$.

Remark 9.10 In order to get absolutely continuous curves from plans, we have used (i) direct product of measures or (ii) optimal transport plans in the previous sections. Let us see how similar settings look in the Bayesian case.

(i) Unfortunately, for $\mu_0, \mu_1 \in \mathcal{P}_\eta(\mathbb{R}^m \times \mathbb{R}^d)$ it does not hold that $\mu_0 \times \mu_1 \in \Gamma_\eta(\mu_0, \mu_1)$. A remedy is to use $\mu_0 = \eta \times \nu \in \mathcal{P}_2(\mathbb{R}^m \times \mathbb{R}^d)$ with some $\nu \in \mathcal{P}_d(\mathbb{R}^d)$ and the plan

$$\alpha := \tilde{\Delta}_\sharp(\nu \times \mu_1) \in \Gamma_\eta(\mu_0, \mu_1),$$

where

$$\tilde{\Delta} : \mathbb{R}^d \times \mathbb{R}^m \times \mathbb{R}^d \to (\mathbb{R}^m \times \mathbb{R}^d)^2, \quad \tilde{\Delta}(x_0, w_1, x_1) := (w_1, x_0, w_1, x_1).$$

In order to sample from α we only need to be able to sample from ν and μ_1.

(ii) We can use Proposition 9.1 (ii) and solve the optimal transport problem for fixed samples w from η, to obtain an optimal coupling $\alpha \in \Gamma_{\eta,o}(\mu_0, \mu_1)$. In order to ensure that the $\alpha_w \in \Gamma_o(\mu_0^w, \mu_1^w)$ can be used to build $\alpha \in \Gamma_{\eta,o}(\mu_0, \mu_1)$, some measurability conditions must be ensured. For examples where these conditions are fulfilled, see e.g. [14, Corollary 1.2] and [7, Proposition 8].

Of course we can also choose $\beta \gg 0$ and compute an optimal transport plan α for $W_\beta(\mu_0, \mu_1)$. We then use the loss

$$\mathrm{CFM}(\theta) := \mathbb{E}_{t \sim \mathcal{L}[0,1], (w_0, x_0, w_1, x_1) \sim \alpha} \left[\| v^\theta(t, e_t((w_0, x_0, w_1, x_1)) - (0, x_1 - x_0) \|^2 \right].$$

Although this is not exactly the loss for the plan α and we only have approximately $w_0 \cong w_1$, in practice this leads to good results, see also next section. Furthermore, we use minibatch OT described in Algorithm 2 to approximately minimize CFM. We will call this strategy minibatch OT Bayesian flow matching with respect to W_β.

9.4 Numerical Examples of Bayesian Flow Matching

In order to show the feasibility of Bayesian flow matching, we provide two examples.

Conditional Image Generation We used the dataset Cifar10 [27], which consists of 10 classes of color images of size $32 \times 32 \times 3$. Here $\eta = \mathcal{P}_2(\mathbb{R}^{10})$ is defined by $\eta = \frac{1}{10} \sum_{i=1}^{10} \delta_{e_i}$, where e_i is the standard basis of $\mathbb{R}^{1}0$. Then samples from μ_1 are of the form $(e_{l(x_i)}, x_i)$, where $l(x_i)$ is the class of x_i. The source measure is $\mu_0 := \eta \times \mathcal{N}(0, 1)$. For conditional image generation, we used trained flow matching models from [7]. For training with respect to $W_{2,\beta}$ the minibatch OT training algorithm 2 was used for Fig. 11a,b. For the plan $\alpha \in \Gamma_\eta(\mu_0, \mu_1)$ in Fig. 11c, we applied the coupling from Remark 9.10 and for training Algorithm 1.

Figure 11 shows generated images, where in row i the samples are generated from (e_i, z) where 10 different samples $z \sim \mathcal{N}(0, 1)$ are drawn. For $\beta = 1$, Fig. 11a illustrates that the generation is not class conditional, i.e. in every row there are images of multiple classes. For $\beta = 100$ and for the conditional plan from Remark 9.10, we can see in Fig. 11

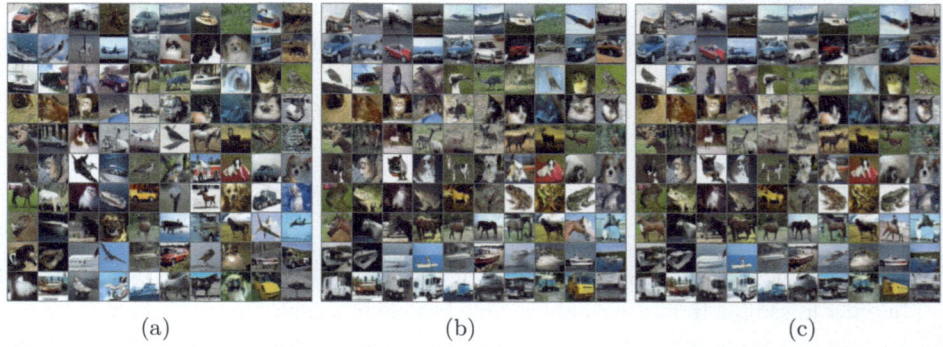

(a)   (b)   (c)

Fig. 11 (Bayesian) Flow matching on Cifar10. (**a**) α for $W_{2,1}(\mu_0, \mu_1)$. (**b**) α for $W_{2,100}(\mu_0, \mu_1)$. (**c**) $\alpha \in \Gamma_\eta(\mu_0, \mu_1)$

that the generation is class conditional, meaning that every row only contains samples from one class.

Simple Bayesian Inverse Problem The following example can be found in [7, Section 8.2]. Let $X \in \mathbb{R}^5$ be a random variable for which the law P_X is a Gaussian mixture model with 10 modes with means uniformly chosen in $[-1, 1]^5$ and standard derivations 0.1. We consider the Bayesian inverse problem

$$Y = AX + \Xi \quad \text{with} \quad A := \Big(\frac{0.1}{i}\delta_{i,j}\Big)_{i,j=1}^5.$$

The noise variable Ξ, which is independent from X, is distributed as $\mathcal{N}(0, 0.1)$. Then by [18, Lemma 11] the posterior distribution $P_{X|Y=y}$ is a Gaussian mixture model which can be analytically computed. This allows us to sample from $P_{X|Y=y}$.

Training Bayesian Flow Matching with α from Remark 9.10 Here we sample from $\alpha \in \Gamma(P_Y \times \mathcal{N}(0, I_5), P_{Y,X})$ by sampling $x_i \sim P_X$, computing $y_i := f(x_i) + \xi_i \sim P_Y$ for $\xi_i \sim \mathcal{N}(0, 0.1 \, \text{Id})$ and $z_i \sim \mathcal{N}(0, \text{Id})$. We then use (y_i, z_i, y_i, x_i) in order to approximate α and to compute $\text{CFM}(\theta)$. Batching in i and t is done as in Algorithm 1.

Training Minibatch OT Bayesian Flow Matching with Respect to $W_{2,100}$ Here we use Algorithm 2 for $\mu_0 = P_Y \times \mathcal{N}(0, \text{Id})$ and $\mu_1 = P_{Y,X}$. In order to be able to sample from μ_0 for a minibatch $(y_i, x_i)_{i=1}^{N_{bOT}} \sim P_{Y,X}$, where N_{bOT} is the minibatch size, we sample $z_i \sim \mathcal{N}(0, \text{Id})$ and use $(y_i, z_i)_{i=1}^{N_{bOT}} \sim \mu_0$.

The results are depicted in Fig. 12. To create this figure, we draw $x \sim P_X$ and computed an observation $y = Ax + \xi \sim P_Y$. Then we draw samples $x^{y,n}$, $1 \leq n \leq 1000$ from $P_{X|Y=y}$. The orange colored histograms on the diagonal are the one dimensional

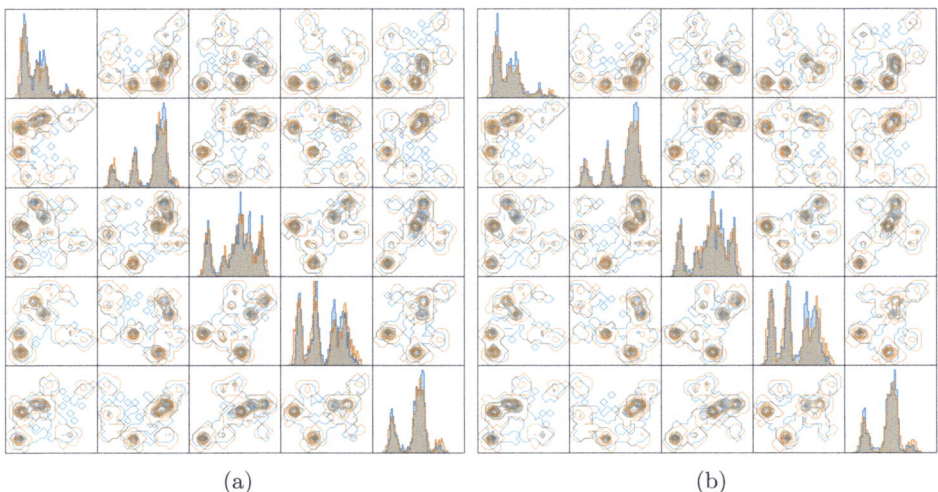

Fig. 12 Bayesian flow matching for an inverse problem. (**a**) minibatch OT flow matching for $W_{2,100}$. (**b**) α from Remark 9.10

histograms of the projection onto the $i - th$ component $\langle e_i, x^{y,n} \rangle$. For $i \neq j$ the orange colored two dimensional histograms of $(\langle e_i, x^{y,n} \rangle, \langle e_j, x^{y,n} \rangle)$ are shown in position (i, j). The blue histograms are the output obtained by the flow matching procedures, i.e. instead of $x^{y,n}$, we use $\phi_1(y, z^n)$ for $z^n \sim \mathcal{N}(0, I_5)$, $1 \leq n \leq 1000$, where ϕ_t solves $\partial_t \phi_t = u_t^\theta(\phi_t)$, $\phi_0(y, z) = (y, z)$.

10 Continuous Normalizing Flows

Flow matching is related to continuous normalizing flows, which historically were introduced earlier, but appears to be more complicated when learned with a likelihood loss. To provide a more circumvent view, we add this section. A remark on (noncontinuous) normalizing flows is given in Appendix C.

Continuous normalizing flows aim to find a vector field v_t such that the solution $\psi : [0, 1] \times \mathbb{R}^d \to \mathbb{R}^d$ of

$$\partial_t \psi(t, x) = v_t(\psi(t, x)), \quad \psi(0, x) = x, \tag{30}$$

satisfies $\psi(1, \cdot)_\sharp \mu_0 = \mu_1$. More precisely, for every $t \in [0, 1]$, the solution $\psi(t, \cdot) : \mathbb{R}^d \to \mathbb{R}^d$ has to be a diffeomorphism. In contrast to flow matching, continuous normalizing flow techniques work without previously determining the curve via plans, Markov kernels or a stochastic process, but approximate the vector field by a neural network using an appropriate loss function, namely the log-likelihood one.

Therefore, we first derive general properties of vector fields and associated curves to make the process invertible in the next Sect. 10.1. In other words, we will take care about the time reverse ODE. This will facilitate the finding of an appropriate loss function in Sect. 10.2. In Sect. 10.3, we deal with the minimization of the loss by computing gradients with the adjoint method.

In this section, we will exclusively work with absolutely continuous measures and switch from the notation of measures to those of densities. In particular, we will write $p_t := \psi(t, \cdot)_\sharp \rho_0$ for the density p_t of $\psi(t, \cdot)_\sharp \mu_0$, if it exists.

10.1 Curves of Probability Measures from ODEs

We start with classical results from the theory of ODEs, for existence and regularity see, e.g., [41, Corollary 2.6, Theorem 2.10].

Theorem 10.1 *Assume that $v_t \in C^l([0, 1] \times \mathbb{R}^d, \mathbb{R}^d)$, $l \geq 1$ fulfills a global Lipschitz condition in the second variable, i.e., there exists $K > 0$ such that*

$$\|v_t(x) - v_t(y)\| \leq K \|x - y\| \quad \text{for all } x, y \in \mathbb{R}^d, t \in [0, 1].$$

Then, for any fixed $t_0 \in [0, 1]$ and $x_0 \in \mathbb{R}^d$, the initial value problem $f(t_0) = x_0$ and

$$f'(t) = v_t(f(t)), \qquad (31)$$

admits a unique global solution $f : [0, 1] \to \mathbb{R}^d$. Furthermore, if we define $\psi(t, s, x) := f(t)$ for the solution f of (31) with initial condition $f(s) = x$, then it holds $\psi \in C^l([0, 1] \times [0, 1] \times \mathbb{R}^d, \mathbb{R}^d)$.

Concerning the reverse ODE, we obtain the following corollary.

Corollary 10.2 *In the setting of Theorem 10.1, we have that $\psi(t, s, \psi(s, t, x)) = x$. In particular, for $(t, s) \in [0, 1] \times [0, 1]$, the function $\psi^{t,s} : \mathbb{R}^d \to \mathbb{R}^d$ defined by $\psi^{t,s}(x) := \psi(t, s, x)$ is a $C^l(\mathbb{R}^d, \mathbb{R}^d)$ diffeomorphism with inverse $\psi^{s,t}$.*

Proof Let f and g be a solution of (31) with initializations $f(t) = x$, and $g(s) = f(s) = \psi(s, t, x)$, respectively. By Theorem 10.1, we know that $f = g$ on $[0, 1]$ and therefore

$$g(t) = \psi(t, s, \psi(s, t, x)) = f(t) = x.$$

$\square$

The next proposition shows that under the above smoothness assumptions $\psi(t, \cdot) : \mathbb{R}^d \to \mathbb{R}^d$ is indeed a C^2 diffeomorphism.

Proposition 10.3 *Assume that $v_t \in C^l([0, 1] \times \mathbb{R}^d, \mathbb{R}^d)$, $l \geq 2$ globally Lipschitz in the second variable. Then (30) admits a unique global solution and $\psi \in C^l([0, 1] \times \mathbb{R}^d, \mathbb{R}^d)$. If $\mu_0 = p_0 \mathcal{L}$ with $p_0 \in C^{l-1}(\mathbb{R}^d)$, then the curve $\mu_t := \psi(t, \cdot)_\sharp \mu_0$ admits a density $p_t \in C^{l-1}([0, 1] \times \mathbb{R}^d)$. Furthermore, if there exists $t_0 \in [0, 1]$ such that $p_{t_0}(x) > 0$ for all $x \in \mathbb{R}^d$, then $p_t(x) > 0$ for all $x \in \mathbb{R}^d$ and all $t \in [0, 1]$.*

Proof The existence and uniqueness follow directly from Theorem 10.1. Since $\psi(t, \cdot)$ is a $C^l(\mathbb{R}^d, \mathbb{R}^d)$ diffeomorphism by Corollary 10.2 we know that $\det(\nabla_x \psi(t, \cdot)(x)) \neq 0$. Furthermore using (2) we conclude that μ_t is absolutely continuous with density

$$p_t(x) = \frac{p_0 \circ \psi(t, \cdot)^{-1}(x)}{\det(\nabla_x \psi(t, \cdot)(x))}.$$

The latter equation implies the remaining claims. $\square$

Corollary 10.4 *Let $v_t \in C^2([0, 1] \times \mathbb{R}^d, \mathbb{R}^d)$ be globally Lipschitz and $\mu_0 = p_0 \mathcal{L}$ with $p_0 \in C^1(\mathbb{R}^d)$. Let $\psi \in C^l([0, 1] \times \mathbb{R}^d, \mathbb{R}^d)$ be the solution of (30) and $p_t := \psi(t, \cdot)_\sharp p_0$.*

Then (p_t, v_t) fulfills the continuity equation in a strong sense

$$\partial_t p_t + \mathrm{div}(p_t\, v_t) = 0, \quad x \in \mathbb{R}^d, t \in [0, 1].$$

Proof By Remark 3.6, we know that (30) implies for all $\varphi \in C_c^\infty((0, 1) \times \mathbb{R}^d)$ that

$$0 = \int_0^1 \int_{\mathbb{R}^d} \langle \nabla_x \varphi, v_t \rangle + \partial_t \varphi \, \mathrm{d}[\psi(t, \cdot)_\sharp \mu_0] \mathrm{d}t = \int_0^1 \int_{\mathbb{R}^d} \left(\mathrm{div}(v_t p_t) + \partial_t p_t \right) \varphi \, \mathrm{d}x \mathrm{d}t.$$

By Theorem 10.3, we know that $p_t \in C^1(\mathbb{R}^d)$, so that the function in the inner brackets is continuous, which yields the assertion. □

10.2 Likelihood Loss for Continuous Normalizing Flow

Opposite to flow matching, we assume that the initial density of the ODE is the target density, and the final one approximates the latent density, i.e.,

$$p_0 \approx p_{\text{data}}, \quad p_1 = p_{\text{latent}}.$$

We assume that the vector field and thus, also the corresponding ODE solution depend on a parameter vector $\theta \in \mathbb{R}^n$, i.e.,

$$\partial_t \psi^\theta(t, x) = v_t^\theta(\psi(t, x)), \quad \psi^\theta(0, x) = x.$$

By the previous subsection, we know for $v_t^\theta \in C^2([0, 1] \times \mathbb{R}^d, \mathbb{R}^d)$ with Lipschitz continuous second component for all $t \in [0, 1]$, that $S^\theta := \psi^\theta(1, \cdot)$ is a C^2 diffeomorphism and $T^\theta := \left(S^\theta\right)^{-1}$ fulfills $p_0 := T_\sharp^\theta p_1$. Moreover, we have for $p_t := \psi^\theta(t, \cdot)_\sharp p_0$ that

$$p_1 = S_\sharp^\theta p_0 = S_\sharp^\theta T_\sharp^\theta p_Z = p_Z \quad \text{and} \quad T_\sharp^\theta p_1 = T_\sharp^\theta S_\sharp^\theta p_0 = p_0.$$

As *loss function*, the log-likelihood function appears to be a reasonable choice

$$\mathcal{L}(\theta) = \mathbb{E}_{x \sim p_X}[-\log T_\sharp^\theta p_1] = \mathbb{E}_{x \sim p_X}[-\log p_0].$$

Alternatively, this may be reformulated using the Kullback-Leibler divergence between p_X and $T_\sharp^\theta p_1$, see Remark 2.3. In order to minimize the loss function using backpropagation, we must compute the gradient of $\log(T_\sharp^\theta p_1) = \log p_0(x)$ with respect to θ. To this end, we introduce the function $\ell : [0, 1] \times \mathbb{R}^d \to \mathbb{R}^d$ by

$$\ell(t, x) := \log\left(p_t(\psi(t, x))\right) - \log\left(p_0(x)\right), \tag{32}$$

so that the loss function can be rewritten setting $t = 1$ as

$$\mathbb{E}_{x \sim p_X}[-\log p_0(x)] = \mathbb{E}_{x \sim p_X}[\ell(1, x) - \log p_1(\psi(1, x))]. \tag{33}$$

Interestingly, ℓ is the solution of an ODE which includes the velocity field v_t.

Proposition 10.5 *Let $v \in C^2([0, 1] \times \mathbb{R}^d, \mathbb{R}^d)$ be Lipschitz continuous in the second component for all $t \in [0, 1]$ and let $p_0 \in C^1(\mathbb{R}^d)$ be a strictly positive density. For the solution $\psi \in C^l([0, 1] \times \mathbb{R}^d, \mathbb{R}^d)$ of (30), let $p_t := \psi(t, \cdot)_\sharp p_0$. Then $\ell : [0, 1] \times \mathbb{R}^d \to \mathbb{R}^d$ in (32) is the solution of the ODE*

$$\partial_t \ell(t, x) = -(\nabla \cdot v_t)(\psi(t, x)), \quad \ell(x, 0) = 0$$

Proof By the chain rule, we obtain

$$\partial_t \ell(t, x) = \frac{d}{dt} \log\left(p_t(\psi(t, x))\right) = \frac{1}{p_t(\psi(t, x))} \frac{d}{dt} p_t(\psi(t, x))$$

$$= \frac{1}{p_t(\psi(t, x))} \Big(\langle \nabla_x p_t(\psi(t, x)), \partial_t \psi(t, x) \rangle + \partial_t p_t(\psi(t, x))\Big).$$

and further by applying the definition of ψ for the first term and the continuity equation from Corollary 10.4 for the second one,

$$\partial_t \ell(t, x) = \frac{1}{p_t(\psi(t, x))} \Big(\langle \nabla_x p_t(\psi(t, x)), v_t(\psi(t, x)) \rangle - \text{div}\big(p_t(\psi(t, x))v_t(\psi(t, x))\big)\Big)$$

$$= \frac{1}{p_t(\psi(t, x))} \Big(\langle \nabla_x p_t(\psi(t, x)), v_t(\psi(t, x)) \rangle$$

$$- \langle \nabla_x p_t(\psi(t, x)), v_t(\psi(t, x)) \rangle - p_t(\psi(t, x))(\text{div } v_t)(\psi(t, x))\Big)$$

$$= -\text{div}(v_t(\psi(t, x))).$$

$\square$

Now we can combine the ODEs for ψ and ℓ into the ODE system

$$\begin{pmatrix} \partial_t \Psi_1(t, x, y, \theta) \\ \partial_t \Psi_2(t, x, y, \theta) \\ \partial_t \Psi_3(t, x, y, \theta) \end{pmatrix} = \begin{pmatrix} v_t\big(\Psi_1(t, x, y, \theta), \theta\big) \\ -(\text{div}_x v_t)\big(\Psi_1(t, x, y, \theta), \theta\big) \\ 0 \end{pmatrix}, \quad \begin{pmatrix} \Psi_1(0, x, y, \theta) \\ \Psi_2(0, x, y, \theta) \\ \Psi_3(0, x, y, \theta) \end{pmatrix} = \begin{pmatrix} x \\ 0 \\ \theta \end{pmatrix}$$

where $x \in \mathbb{R}^d$, $y \in \mathbb{R}$. Note that Ψ_2 corresponds to ℓ from (32) and Ψ_3 remains θ. Making the velocity field dependent on a parameter $\theta \in \mathbb{R}^n$, we can rewrite the loss function (33) as

$$\mathcal{L}(\theta) = \mathbb{E}_{x \sim p_X}\big[\Psi_2(1, x, 0, \theta) - \log p_1\big(\Psi_1(1, x, 0, \theta)\big)\big].$$

Now the main challenge is to find $\nabla_\theta \mathcal{L}(\theta)$. To this end, consider

$$F(x, y, \theta) := y - \log p_1(x).$$

Then

$$F \circ (\Psi(1, \cdot, \cdot, \cdot))(x, y, \theta) = \Psi_2(1, x, 0, \theta) - \log p_1\big(\Psi_1(1, x, 0, \theta)\big)$$

and we can compute by the Leibniz rule for measure spaces

$$\nabla_\theta \mathcal{L}(\theta) = \mathbb{E}_{x \sim p_X}[\nabla_\theta (F \circ (\Psi(1, \cdot, \cdot, \cdot))(x, y, \theta)]. \tag{34}$$

In the next subsection, we show how to compute $\nabla_\theta (F \circ \Psi(1, \cdot, \cdot, \cdot))(x, y, \theta)$ by solving a system of ODEs.

10.3 Computing Gradients with the Adjoint Method

This section is based on the original paper [9] as well as the blog [37].

To start from an arbitrary time $s \in [0, 1]$, we use the following notation. For $v_t \in C^2([0, 1] \times \mathbb{R}^d, \mathbb{R}^d)$, let $\boldsymbol{\psi} : [0, 1] \times [0, 1] \times \mathbb{R}^d \to \mathbb{R}^d$ be the solution of

$$\partial_t \boldsymbol{\psi}(t, s, x) = v_t(\boldsymbol{\psi}(t, s, x)), \quad \boldsymbol{\psi}(s, s, x) = x. \tag{35}$$

For fixed $s = 0$, we have $\psi(t, x) = \boldsymbol{\psi}(t, 0, x)$. For an arbitrary $F \in C^2(\mathbb{R}^d)$, we define a function $a : [0, 1] \times \mathbb{R}^d \to \mathbb{R}^d$ by

$$a_t(x) := \nabla_x\big(F \circ \boldsymbol{\psi}(1, t, \cdot)\big)(\psi(t, x)).$$

Note that $a_t(x)$ is a row vector here.

Proposition 10.6 *Let $v_t \in C^2([0, 1] \times \mathbb{R}^d, \mathbb{R}^d)$ be Lipschitz continuous in the second variable, and let $\psi : [0, 1] \times \mathbb{R}^d \to \mathbb{R}^d$ be the solution of (30). Then it holds*

$$\partial_t a_t(x) = -a_t(x)\,(\nabla_x v_t)(\psi(t, x)). \tag{36}$$

Proof Since $v_t \in C^2([0, 1] \times \mathbb{R}^d, \mathbb{R}^d)$, we have by Theorem 10.1 that $\boldsymbol{\psi} \in C^2([0, 1] \times [0, 1] \times \mathbb{R}^d, \mathbb{R}^d)$. Noting that

$$\boldsymbol{\psi}(1, 0, x) = \boldsymbol{\psi}(1, t, \cdot) \circ \boldsymbol{\psi}(t, 0, x) = \boldsymbol{\psi}(1, t, \cdot)(\psi(t, x)),$$

we obtain

$$\begin{aligned} a_0(x) &= \nabla_x \big(F \circ \boldsymbol{\psi}(1, 0, \cdot)\big)(\psi(0, x)) \\ &= \nabla_x \big(F \circ (\boldsymbol{\psi}(1, t, \cdot) \circ \boldsymbol{\psi}(t, 0, \cdot))\big)(\psi(0, x)) \\ &= \nabla_x \big((F \circ \boldsymbol{\psi}(1, t, \cdot)) \circ \psi(t, \cdot)\big)(x) = a_t(x) \, \nabla_x \psi(t, x). \end{aligned}$$

Since the left hand side does not depend on t, we obtain after differentiation with respect to t that

$$0 = \partial_t (a_t(x)) \, \nabla_x \psi(t, x) + a_t(x) \, \partial_t (\nabla_x \psi(t, x)). \tag{37}$$

Next, we compute

$$\partial_t (\nabla_x \psi(t, x)) = \nabla_x (\partial_t \psi(t, x)) = \nabla_x (v_t(\psi(t, x))) = (\nabla_x v_t)(\psi(t, x)) \nabla_x \psi(t, x).$$

Then we get in (37) that

$$0 = \partial_t (a_t(x)) \, \nabla_x \psi(t, x) + a_t(x) \, (\nabla_x v_t)(\psi(t, x)) \nabla_x \psi(t, x).$$

By Corollary 10.2, we know that $\psi(t, x)$ invertible with differentiable inverse, so that the matrix $\nabla_x \psi(t, x)$ is invertible and we obtain the assertion (36). □

To compute gradients of v_t^θ with respect $\theta \in \mathbb{R}^n$, we extend the ODE (35) for $\boldsymbol{\psi}$: $[0, 1] \times [0, 1] \times \mathbb{R}^d \times \mathbb{R}^n \to \mathbb{R}^d \times \mathbb{R}^n$ and $v : \times [0, 1] \times \mathbb{R}^d \times \mathbb{R}^n \to \mathbb{R}^d$ as

$$\partial_t \boldsymbol{\psi}(t, s, x, \theta) = \begin{pmatrix} v_t(\boldsymbol{\psi}(t, s, x, \theta)) \\ 0 \end{pmatrix}, \quad \boldsymbol{\psi}(s, s, x, \theta) = (x, \theta)$$

where we use the same symbols $\boldsymbol{\psi}$ and v_t for convenience. Again, we write ψ for $\boldsymbol{\psi}(\cdot, 0, \cdot, \cdot)$. Now let $F \in C^2(\mathbb{R}^d \times \mathbb{R}^n)$ be a function *which does not depend on the second component* and define $a : [0, 1] \times \mathbb{R}^d \times \mathbb{R}^n \to \mathbb{R}^d \times \mathbb{R}^n$ by

$$a_t(x, \theta) := \nabla_{x,\theta}(F \circ \boldsymbol{\psi}(1, t, \cdot, \cdot))(\psi(t, x, \theta)) = \big(a_t^x, a_t^\theta\big).$$

Note that in the loss function (34) we need

$$a_0^\theta = \nabla_\theta (F \circ \psi(1,\cdot,\cdot))(x,\theta). \tag{38}$$

Then (36) modifies to

$$\partial_t a_t(x) = -a_t(x) \nabla_{x,\theta} \begin{pmatrix} v_t \\ 0 \end{pmatrix} (\psi(t,x,\theta)).$$

Since

$$\nabla_{x,\theta} \begin{pmatrix} v_t \\ 0 \end{pmatrix} = \begin{pmatrix} \nabla_x v_t & \nabla_\theta v_t \\ 0 & 0 \end{pmatrix},$$

we obtain finally

$$\begin{pmatrix} \partial_t \psi(t,x,\theta) \\ \partial_t a_t^x(x,\theta) \\ \partial_t a_t^\theta(x,\theta) \end{pmatrix} = \begin{pmatrix} v_t(\psi(t,x,\theta),\theta) \\ a_t^x(x,\theta)(\nabla_x v_t)(\psi(t,x,\theta)) \\ a_t^x(x,\theta)(\nabla_\theta v_t)(\psi(t,x,\theta)) \end{pmatrix} \tag{39}$$

As noted above, we need to compute a_0^θ to get $\nabla_\theta \mathcal{L}(\theta)$ in (34). In (39) the initial conditions at $t = 0$ are implicitly already encoded, since we used for our calculations that $\psi(0,x,\theta) = (x,\theta)$. However, to get a_0^θ, we need to find the appropriate condition for $t = 1$ such that the solution of (39) matches the initial condition at $t = 0$. Since F does not depend on θ and we know that $\boldsymbol{\psi}(1,1,x,\theta) = (x,\theta)$, we can conclude

$$a_1^\theta = \nabla_\theta (F \circ \boldsymbol{\psi}(1,1,\cdot,\cdot)) = \nabla_\theta F = 0.$$

Furthermore, with $x^1 := \psi(1,x,\theta)$ we have that

$$a_1^x(x,\theta) = \nabla_x (F \circ \boldsymbol{\psi}(1,1,\cdot))(x^1) = (\nabla_x F)(x^1).$$

Hence we can obtain $a_0^\theta = \nabla_\theta (F \circ \psi(1,\cdot,\cdot))(x,\theta)$ by solving (39) with initial conditions

$$\psi(1,x,\theta) = (x^1,\theta); \quad a_1^x(x,\theta) = (\nabla_x F)(x^1); \quad a_1^\theta(x,\theta) = 0.$$

For the special loss (34) the variable x consists of two parts, namely (x,y) and ψ corresponds to Ψ and thus we can evaluate the gradient of the loss function.

10.4 Numerical Examples of Continuous Normalizing Flows

By (34) and (38), training a continuous normalizing flow needs the computation of

$$\nabla_\theta \mathcal{L}(\theta) = \mathbb{E}_{x \sim p_X}[a_0^\theta(x)],$$

where we approximate the expectation by an empirical expectation. In order to compute a_0^θ we need to solve (39). The only ingredients for solving this equation are the vector field $v_t(x, y, \theta)$, the gradients thereof, and the initial conditions. The velocity field v_t and its gradients are available since we parametrize it by a neural network. For the first initial condition we need $\psi(1, x, 0, \theta)$ which we can obtain by solving $\partial_t f(t) = v_t(f(t), \theta)$, $f(0) = (x, 0)$. For the second initial condition we need $\nabla_{x,y} F$ for $F(x, y) = y - \log p_1(x)$. Thus we need to be given a density where the gradient $\nabla_x \log p_1$ is tractable, which is the case, e.g., for the standard Gaussian distribution. In total we need samples from p_X and a density p_1 for which $\nabla_x \log p_1$ is tractable. The computation of $a_0^\theta(x)$ can then be done by solving (39), for which we used the library [8]. Figure 13 shows the trajectories for a vector field obtained via continuous normalizing flow training with three different target distributions. For Fig. 13a we chose a GMM with 8 equally weighted modes. The trajectories differ from the ones obtained by flow matching in Fig. 7. For Fig. 13b resp. Figure 13c we chose the moons resp. spirals data set from [35].

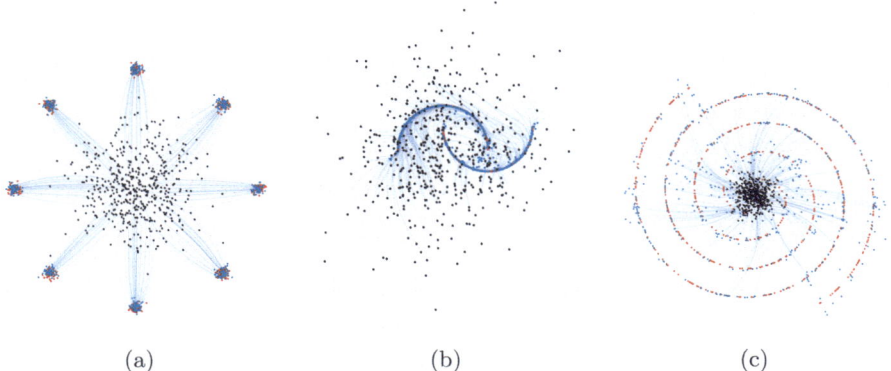

Fig. 13 Trajectories of points from a vector field v_t^θ obtained via a vector field obtained by training a CNF. We chose $\mu_0 = \mathcal{N}(0, 1)$. Black: points $\{z_i\}_{i=1}^{300}$ drawn from the source distribution $\mu_0 = \mathcal{N}(0, 1)$. Blue: points sampled via the vector field from $\{z_i\}_{i=1}^{300}$. Red: points sampled from the target distribution. Blue lines: trajectories of the vector field. (**a**) GMM. (**b**) Moons. (**c**) Spirals

11 A Glimpse at Score-Based Diffusion

Diffusion models go back to [38] and [39]. They can be transformed into CNFs, thus allowing tractable likelihood computation with numerical ODE solvers [40]. However, diffusion models are based on stochastic differential equations (SDEs). Roughly speaking, the forward SDE computes $(X_t)_{t \in [0,T]}$ starting with the target distribution by

$$dX_t = f(t, X_t) \, dt + g(t) \, dW_t, \quad X_0 \sim P_{\text{data}}, \tag{40}$$

where W_t is a standard Brownian motion. A usual choice is

$$f(t, X_t) := -\frac{1}{2} \beta_t X_t \quad \text{and} \quad g(t) := \sqrt{\beta_t}$$

with a positive, increasing so-called "time schedule" β, e.g., $\beta_t = \beta_{\min} + t(\beta_{\max} - \beta_{\min})$ for some $0 < \beta_{\min} \ll \beta_{\max}$. Then the SDE becomes a linear one

$$dX_t = -\frac{1}{2} \beta_t X_t \, dt + \sqrt{\beta_t} \, dW_t, \quad X_0 \sim P_{\text{data}},$$

which has the closed form solution

$$X_t = \sqrt{1 - e^{-h(t)}} Z + e^{\frac{-h(t)}{2}} X_0, \quad h(t) := \int_0^t \beta_s \, ds, \tag{41}$$

where $Z \sim \mathcal{N}(0, I_d)$. Hence, X_t in (41) reaches Z for $t \to \infty$. Then $X_t \sim P_{X_t}, t > 0$ has a density p_{X_t}, and under certain assumptions, a reverse process becomes

$$dY_t = \left(-f(T-t, Y_t) + g(T-t)^2 \nabla \log p_{X_{T-t}}(Y_t) \right) dt + g(T-t) \, dW_t, \quad Y_0 \sim X_T. \tag{42}$$

However, the reverse SDE depends on the so-called *score* $\nabla \log p_{X_t}$. Score-based models intend to approximate the score by a neural network s^θ by minimizing, for $T > 0$ large enough

$$\mathcal{L}(\theta) := \min_\theta \mathbb{E}_{t \sim \mathcal{L}_{[0,T]}} \mathbb{E}_{x \sim P_{X_t}} \left[\| s_t^\theta(x) - \nabla \log p_{X_t}(x) \|^2 \right].$$

Usually, we do not have access to $\nabla \log p_t$. Fortunately, by [43, Appendix], the loss function can be rewritten, up to a constant, as

$$\mathcal{L}(\theta) = \mathbb{E}_{x_0 \sim P_{\text{data}}, t \sim \mathcal{L}_{[0,T]}} \mathbb{E}_{x \sim P_{X_t | X_0 = x_0}} \left[\| s_t^\theta(x) - \nabla \log p_{X_t | X_0 = x_0}(x) \|^2 \right],$$

which does not involve the score itself, but instead the conditional distribution. For (41), this is simply a Gaussian

$$P_{X_t|X_0=x_0} = \mathcal{N}\left(b_t\, x_0, (1-b_t^2)\, I_d\right) \quad \text{with} \quad b_t = e^{-\frac{h(t)}{2}}$$

since X_0 and Z are independent. This implies

$$\nabla \log p_{X_t|X_0=x_0}(x) = \nabla \left(-\frac{1}{2(1-b_t^2)} \|x - b_t x_0\|^2\right) = -\frac{x - b_t x_0}{1 - b_t^2}.$$

Plugging this into the loss function, we get

$$\mathcal{L}(\theta) = \mathbb{E}_{t \sim \mathcal{L}_{[0,T]}} \mathbb{E}_{x_0 \sim P_{\text{data}}} \mathbb{E}_{x \sim \mathbb{P}_{X_t|X_0=x_0}} \left[\left\| s_t^\theta(x) + \frac{x - b_t x_0}{1 - b_t^2} \right\|^2 \right].$$

Once the score is computed, we can use it in the reverse SDE (42), starting with the $Z \sim \mathcal{N}(0, I_d)$ which approximates X_T for T large enough.

For the forward PDE (40), the corresponding densities $p_t = p_{X_t}$ fulfill the *Fokker-Planck equation*

$$\begin{aligned}\partial_t p_t(x) &= -\nabla \cdot (f(t,x) p_t(x)) + g(t)^2 \Delta p_t(x) \\ &= -\nabla \cdot \big(\underbrace{\left(f(t,x) - g(t)^2 \nabla_x \log p_t(x)\right)}_{v_t} p_t(x) \big)\end{aligned} \qquad (43)$$

which is the continuity equation for P_{X_t}. This gives the relation to the previous sections. In particular, given the score, we can also use the flow ODE (9) for sampling. Moreover, we could compute the log density via Proposition 10.5.

For $f(t,x) = -\frac{1}{2}\beta_t x$ as above, we obtain $v_t(x) = -\frac{1}{2}\beta_t x - g(t)^2 \nabla \log p_t(x)$. This is closely related to the vector field associated to the independent coupling which was computed to be $v_t(x) = \frac{x}{t} + \frac{1-t}{t} \nabla \log p_t(x)$ in Proposition 4.11 and Remark 6.5.

Finally, we like to mention that the above (Fokker-Planck) continuity equation (43) arises from a so-called Wasserstein gradient flow of the KL divergence $KL(p, p_{\text{data}})$, i.e., the velocity field v_t is in the subdifferential of the above KL function at p_t, see e.g. [1, Chapter 10.4].

Acknowledgments Many thanks to R. Beinert and R. Duong for reading the manuscript and to R. Duong for pointing out Example 4.12.

Appendix A: Proof of Theorem 3.1

Recall that a family $\mathcal{A}$ of subsets of a set X is called *monotone class*, if

- $\bigcup_{i=1}^{\infty} A_n \in \mathcal{A}$ for every increasing sequence $A_i \in \mathcal{A}$, and
- $\bigcap_{i=1}^{\infty} A_i \in \mathcal{A}$ for every decreasing sequence $A_i \in \mathcal{A}$.

For a family of subsets B of X, we call the smallest monotone class containing B the *monotone class generated by* B. The following theorem can be found, e.g., in [6, Theorem 1.9.3].

Theorem A.1 (Monotone Class Theorem) *Let $\mathcal{A}$ be an algebra of sets, i.e., a collection of sets such that $A \in \mathcal{A}$ implies $X \setminus A \in \mathcal{A}$, and $A, B \in \mathcal{A}$ implies $A \cup B \in \mathcal{A}$. Then the σ-algebra generated by $\mathcal{A}$ coincides with the monotone class generated by $\mathcal{A}$. In particular, any monotone class containing $\mathcal{A}$ also contains the σ-algebra generated by $\mathcal{A}$.*

Theorem 3.1 Let $\mu_t : I \to \mathcal{P}_2(\mathbb{R}^d)$ be a narrowly continuous curve. Then, for every Borel set $B \subseteq \mathbb{R}^d$, we have that $t \mapsto \mu_t(B)$ is measurable, i.e., $\mu_t : I \times \mathcal{B}(\mathbb{R}^d) \to \mathbb{R}$ is a Markov kernel.

Proof Step 1: Let $\mathcal{A}$ be the set of all measurable sets $B \subseteq \mathbb{R}^d$ such that there exists a series $f_n \in C_b(\mathbb{R}^d)$, $\|f\|_\infty \leq 1$ such that $\|f_n - 1_B\|_{L^1(\mu_t)} \to 0$ for all $t \in I$. Note that for such sequences with $\|f_n - 1_A\|_{L^1(\mu_t)} \to 0$, $\|g_n - 1_B\|_{L^1(\mu_t)} \to 0$, we have that $\|\min\{f_n, g_n\} - 1_{A \cap B}\|_{L^1(\mu_t)} \to 0$ and $\|(1 - f_n) - 1_{X \setminus A}\|_{L^1(\mu_t)} \to 0$. The former follows from $\min\{a, b\} = \frac{a+b}{2} - \frac{|a-b|}{2}$ and the fact that the L^1-limit is linear and interchanges with the absolute value by the reverse triangle inequality. The latter uses $1 \in L^1(\mu_t)$, i.e. the finiteness of the measure μ_t. Since trivially, $\emptyset \in \mathcal{A}$, it follows that $\mathcal{A}$ is an algebra of sets and we checked the first requirement of the monotone class theorem.

Step 2: We will show that for $A \in \mathcal{A}$, the map $t \mapsto \mu_t(A)$ is measurable. Let $f_n \in C_b(\mathbb{R}^d)$, $\|f_n\|_\infty \leq 1$ be such that $\|f_n - 1_A\|_{L^1(\mu_t)} \to 0$ for all t. For all t we have by dominated convergence and $\|f_n - 1_A\|_{L^1(\mu_t)} \to 0$ that

$$t \to \mu_t(A) = \int 1_A \mathrm{d}\mu_t = \int \lim_{n \to \infty} f_n \mathrm{d}\mu_t = \lim_{n \to \infty} \int f_n \mathrm{d}\mu_t.$$

Consequently, since $t \to \int f_n \mathrm{d}\mu_t$ is continuous, we can conclude that $t \mapsto \mu_t(A)$ is measurable as a pointwise limit of continuous functions.

Furthermore it is easy to show that any open measurable set is in $\mathcal{A}$ since for an open set A we can approximate 1_A pointwise from below by a series of continuous bounded functions. Thus the σ-algebra generated by $\mathcal{A}$ contains $\mathcal{B}(\mathbb{R}^d)$.

Step 3: We show that $\mathcal{C}$, which contains all $C \in \mathcal{B}(\mathbb{R}^d)$ such that $t \mapsto \mu_t(C)$ is measurable, is a monotone class. Thus we have to show that for an increasing sequence of measurable sets $C_i \in \mathcal{C}$ and $C = \cup_{i \in \mathcal{N}} C_i$, we have that $C \in \mathcal{C}$, i.e. $t \mapsto \mu_t(A)$ is measurable and the same for intersections. But this is true since $\mu_t(C) = \lim_{i \to \infty} \mu_t(C_i)$ and pointwise limits of measurable functions are measurable.

Finally, since by Step 2 it holds $\mathcal{A} \subseteq \mathcal{C}$, we can conclude that $\mathcal{B}(\mathbb{R}^d) \subseteq \mathcal{C}$ by the monotone class theorem and thus the claim. □

Appendix B: Measurability of v_t in Lemma 5.7

Proposition B.1 *For $\mu_0, \mu_1 \in \mathcal{P}_2(\mathbb{R}^d)$, let $\alpha \in \Gamma(\mu_0, \mu_1)$ and $\mu_t := e_{t,\sharp}\alpha$. Let $\bar{\alpha} = a_\sharp(\mathcal{L}_{(0,1)} \times \alpha)$, where $a(t, x, y) = (t, e_t(x, y), y)$ and denote the disintegration with respect to $\pi^{t,1}(t, x, y) = (t, x)$ by $\bar{\alpha}^{t,x}$. Finally, define*

$$v(t, x) := \int_{\mathbb{R}^d} \frac{y - x}{1 - t} d\bar{\alpha}^{t,x}(y)$$

for $(t, x) \in (0, 1) \times \mathbb{R}^n$. Then v is the velocity field induced by α. For $\alpha_t = (e_t, \pi^2)_\sharp \alpha$, $\alpha_t = \alpha_t^x \times_x \mu_t$ and $v_t(x) = \int_{\mathbb{R}^d} \frac{y-x}{1-t} d\alpha_t^x(y)$ we have that

$$v(t, \cdot) = v_t$$

for a.e. $t \in (0, 1)$.

Proof Note that $\pi^{t,1}_\sharp \bar{\alpha} = (t, e_t)_\sharp (\mathcal{L}_{(0,1)} \times \alpha) = \mu_t \times_t \mathcal{L}_{(0,1)}$. Hence

$$\int f(t, x) \int \frac{y - x}{1 - t} d\bar{\alpha}^{t,x}(y) d(\mu_t \times_t \mathcal{L}_{(0,1)}) = \int f(t, x) \frac{y - x}{1 - t} d\bar{\alpha}(t, x, y)$$

$$= \int f(t, e_t(x, y))(y - x) d(\alpha \times \mathcal{L}_{(0,1)})$$

$$= \int f(t, x) de_\sharp((y - x)\alpha \times \mathcal{L}_{(0,1)}).$$

Thus, v is the vector field induced by α since we verified (12). As in the proof of Lemma 4.5 it follows $v(t, \cdot) = v_t$ for almost all $t \in (0, 1)$. More precisely, by carefully checking the definition of disintegration we will show that $\bar{\alpha}^{t,x} = \alpha_t^x$ for a.e. $t \in [0, 1]$. Let $\{f_n\}_{n \in \mathbb{N}} \subset C_b(\mathbb{R}^d \times \mathbb{R}^d)$ be a dense subset. Then for every $h \in C_b([0, 1])$ we have that

$$\int_0^1 h(t) \int_{\mathbb{R}^d} \int_{\mathbb{R}^d} f_n(x, y) d\bar{\alpha}^{t,x}(y) d\mu_t(x) dt = \int_0^1 h(t) \int_{\mathbb{R}^d \times \mathbb{R}^d} f_n(e_t(x, y), y) d\alpha dt$$

$$= \int_0^1 h(t) \int_{\mathbb{R}^d \times \mathbb{R}^d} f_n(x, y) d\alpha_t dt$$

and therefore

$$\int_{\mathbb{R}^d} \int_{\mathbb{R}^d} f_n(x,y) \mathrm{d}\bar{\alpha}^{t,x}(y) \mathrm{d}\mu_t(x) = \int_{\mathbb{R}^d \times \mathbb{R}^d} f_n(x,y) \mathrm{d}\alpha_t \quad \text{for a.e. } t \in [0,1].$$

Thus there exists a zero set N such that

$$\int_{\mathbb{R}^d} \int_{\mathbb{R}^d} f_n(x,y) \mathrm{d}\bar{\alpha}^{t,x}(y) \mathrm{d}\mu_t(x) = \int_{\mathbb{R}^d \times \mathbb{R}^d} f_n(x,y) \mathrm{d}\alpha_t$$

for all $t \in [0,1] \setminus N$ and all $n \in \mathbb{N}$. Using that $\{f_n\}_{n \in \mathbb{N}} \subset C_b(\mathbb{R}^d \times \mathbb{R}^d)$ is dense we can conclude that $\bar{\alpha}^{t,x} = \alpha_t^x$ as Markov kernels for a.e. $t \in [0,1]$. Hence, $v(t,\cdot) = v_t$ for a.e. $t \in (0,1)$. $\square$

Appendix C: Remark on Normalizing Flows

Let us have a quick look at normalizing flows which are invertible neural networks. Other invertible neural networks are residual networks [10, 19], which we will not address in this paper.

A *normalizing flow* between two distributions $\mu_i \in \mathcal{P}(\mathbb{R}^d)$, $i = 0, 1$ is a C^1 diffeomorphism $T : \mathbb{R}^d \to \mathbb{R}^d$ such that

$$\mu_1 = T_\sharp \mu_0.$$

Normalizing flows were approximated by neural networks T^θ which have a special architecture in order to be invertible [3, 12]. To this end, a loss function is minimized, e.g. the KL divergence between the push-forward of the latent distribution $T_\sharp^\theta \mu_0$ and the data distribution μ_1. Due to the special network architecture, normalizing flows are not suited for in high dimensional problems. Moreover, they suffer from a limited expressiveness, e.g., when trying to map a unimodal (Gaussian) distribution to a multimodal one, their Lipschitz constants explodes [16]. Stochastic normalizing flows [44] circumvent this problem by introducing stochastic layers. For the definition of stochastic normalizing flows via Markov kernels and an overview, we refer to [17, 18].

In contrast to normalizing flows, recent generative models like flow matching and continuous normalizing flows as well as score based diffusion models use, instead of learning a map $T : \mathbb{R}^d \to \mathbb{R}^d$, curves in the space of probability measures, which slowly transform the latent distribution into the data distribution. In this way, at every time step only a small change of probability measures has to be learned as opposed to learning everything in one step.

References

1. L. Ambrosio, N. Gigli, G. Savaré, *Gradient Flows: In Metric Spaces and in the Space of Probability Measures* (Springer Science & Business Media, New York, 2005)
2. L. Ambrosio, E. Brué, D. Semola, *Lectures on Optimal Transport*, UNITEXT (Springer International Publishing, Cham, 2021)
3. L. Ardizzone, J. Kruse, C. Rother, U. Köthe, Analyzing inverse problems with invertible neural networks, in *7th International Conference on Learning Representations, ICLR 2019, New Orleans, LA, USA, May 6–9, 2019* (2019)
4. R. Barboni, G. Peyré, F.-X. Vialard, Understanding the training of infinitely deep and wide resnets with conditional optimal transport. Preprint. arXiv:2403.12887 (2024)
5. Q. Bertrand, R. Emonet, A. Gagneux, S. Martin, M. Massias, A visual dive into conditional flow matching. https://dl.heeere.com/conditional-flow-matching/blog/conditional-flow-matching/
6. V.I. Bogachev, M.A.S. Ruas, *Measure Theory*, vol. 1 (Springer, Berlin, 2007)
7. J. Chemseddine, P. Hagemann, C. Wald, G. Steidl, Conditional Wasserstein distances with applications in Bayesian OT flow matching. Preprint. arXiv:2403.18705 (2024)
8. R.T.Q. Chen, torchdiffeq (2018)
9. R. Chen, Y. Rubanova, J. Bettencourt, D. Duvenaud, Neural ordinary differential equations, in *Advances in Neural Information Processing Systems*, vol. 31 (2018)
10. R. Chen, J. Behrmann, D.K. Duvenaud, J.-H. Jacobsen, Residual flows for invertible generative modeling, in *Advances in Neural Information Processing Systems*, vol. 32 (Curran Associates, New York, 2019)
11. G. Daras, A.G. Dimakis, C. Daskalakis, Consistent diffusion meets tweedie: Training exact ambient diffusion models with noisy data. Preprint. arXiv:2404.10177 (2024)
12. L. Dinh, J. Sohl-Dickstein, S. Bengio, Density estimation using real NVP, in *5th International Conference on Learning Representations, ICLR 2017, Toulon, France, April 24–26, 2017, Conference Track Proceedings* (2017)
13. N. Gigli, On the geometry of the space of probability measures endowed with the quadratic Optimal Transport distance. PhD Thesis, 2008. cvgmt preprint
14. A. González-Sanz, S. Sheng, Linearization of Monge-Ampère equations and data science applications. Preprint. arXiv:2408.06534 (2024)
15. I.J. Goodfellow, J. Pouget-Abadie, M. Mirza, B. Xu, D. Warde-Farley, S. Ozair, A. Courville, Y. Bengio, Generative adversarial nets, in *Advances in Neural Information Processing Systems* (2014), pp. 2672–2680
16. P. Hagemann, S. Neumayer, Stabilizing invertible neural networks using mixture models. Inverse Probl. **37**(8), 085002 (2021)
17. P. Hagemann, J. Hertrich, G. Steidl, Generalized normalizing flows via Markov chains, in *Non-local Data Interactions: Foundations and Applications* (Cambridge University Press, Cambridge, 2022)
18. P. Hagemann, J. Hertrich, G. Steidl, Stochastic normalizing flows for inverse problems: A Markov chains viewpoint. SIAM/ASA J. Uncertainty Quantif. **10**(3), 1162–1190 (2022)
19. K. He, X. Zhang, S. Ren, J. Sun, Deep residual learning for image recognition, in *Proceedings of the IEEE Conference on Computer Vision and Pattern Recognition* (2016), pp. 770–778
20. P. Holderrieth, M. Havasi, J. Yim, N. Shaul, I. Gat, T. Jaakkola, B. Karrer, R. T. Q. Chen, Y. Lipman, Generator matching: Generative modeling with arbitrary Markov processes. ICLR (2025)
21. B. Hosseini, A.W. Hsu, A. Taghvaei, Conditional optimal transport on function spaces. Preprint. arXiv:2311.05672 (2024)

22. T. Jahn, J. Chemseddine, P. Hagemann, C. Wald, G. Steidl, Trajectory generator matching for time series. Preprint. arXiv:2505.23215 (2025)
23. O. Kallenberg, O. Kallenberg, *Foundations of Modern Probability*, vol. 2 (Springer, 1997)
24. G. Kerrigan, G. Migliorini, P. Smyth, Dynamic conditional optimal transport through simulation-free flows. Preprint. arXiv:2404.04240 (2024)
25. D.P. Kingma, M. Welling, Auto-encoding variational bayes. *Preprint. arXiv:1312.6114* (2013)
26. B.R. Kloeckner, Extensions with shrinking fibers. Ergodic Theory Dyn. Syst. **41**(6), 1795–1834 (2021)
27. A. Krizhevsky, G. Hinton et al., Learning multiple layers of features from tiny images (2009)
28. Y. Lipman, R.T.Q. Chen, H. Ben-Hamu, M. Nickel, M. Le, Flow matching for generative modeling, in *The Eleventh International Conference on Learning Representations* (2023)
29. Q. Liu, Rectified flow: A marginal preserving approach to optimal transport. Preprint. arXiv:2209.14577 (2022)
30. X. Liu, C. Gong, Q. Liu, Flow straight and fast: Learning to generate and transfer data with rectified flow, in *The Eleventh International Conference on Learning Representations* (2023)
31. S. Martin, A. Gagneux, P. Hagemann, G. Steidl, PnP-flow: Plug-and-play image restoration with flow matching. ICLR *Preprint arXiv:2410.02423* (2025)
32. J. Peszek, D. Poyato, Heterogeneous gradient flows in the topology of fibered optimal transport. Calculus of Variations and Partial Differential Equations **62**(9), 258 (2023)
33. G. Peyré, M. Cuturi et al., Computational optimal transport: With applications to data science. Found. Trends® Mach. Learn. **11**(5–6), 355–607 (2019)
34. G. Plonka, D. Potts, G. Steidl, M. Tasche, *Numerical Fourier Analysis*. Applied and Numerical Harmonic Analysis, 2nd edn. (Birkhäuser, Boston, 2023)
35. M. Poli, S. Massaroli, A. Yamashita, H. Asama, J. Park, S. Ermon, Torchdyn: Implicit models and neural numerical methods in pytorch
36. F. Santambrogio, *Optimal Transport for Applied Mathematicians* (Birkäuser, Boston, 2015)
37. I. Schurov, Adjoint state method, backpropagation and neural odes. https://ilya.schurov.com/post/adjoint-method/
38. J. Sohl-Dickstein, E. Weiss, N. Maheswaranathan, S. Ganguli, Deep unsupervised learning using nonequilibrium thermodynamics, in *Proceedings of the 32nd International Conference on Machine Learning*, ed. by F. Bach, D. Blei, volume 37 of Proceedings of Machine Learning Research, Lille, France, 07–09 Jul 2015. PMLR (2015), pp. 2256–2265
39. Y. Song, S. Ermon, Generative modeling by estimating gradients of the data distribution. ArXiv 1907.05600 (2019)
40. Y. Song, C. Durkan, I. Murray, S. Ermon, Maximum likelihood training of score-based diffusion models, in *Advances in Neural Information Processing Systems*, ed. by A. Beygelzimer, Y. Dauphin, P. Liang, J.W. Vaughan (2021)
41. G. Teschl, *Ordinary Differential Equations and Dynamical Systems*, vol. 140 (American Mathematical Society, Providence, 2024)
42. A. Tong, N. Malkin, G. Huguet, Y. Zhang, J. Rector-Brooks, K. Fatras, G. Wolf, Y. Bengio, Improving and generalizing flow-based generative models with minibatch optimal transport, in *ICML Workshop on New Frontiers in Learning, Control, and Dynamical Systems* (2023)
43. P. Vincent, A connection between score matching and denoising autoencoders. Neural Comput. **23**(7), 1661–1674 (2011)
44. H. Wu, J. Köhler, F. Noé, Stochastic normalizing flows, in *Advances in Neural Information Processing Systems 2020*, ed. by H. Larochelle, M. A. Ranzato, R. Hadsell, M. Balcan, H. Lin (2020)
45. Y. Zhang, P. Yu, Y. Zhu, Y. Chang, F. Gao, Y.N. Wu, O. Leong, Flow priors for linear inverse problems via iterative corrupted trajectory matching. Preprint. arXiv:2405.18816 (2024)

The manufacturer's authorised representative in the EU is Springer Nature Customer Service Centre GmbH, Europaplatz 3, 69115 Heidelberg, Germany. If you have any concerns regarding our products, please contact ProductSafety@springernature.com

Printed and bound by CPI Group (UK) Ltd, Croydon, CR0 4YY

26/03/2026

02078943-0017